Epipolar Geometry in Stereo, Motion and Object Recognition

Computational Imaging and Vision

Volume 6

Epipolar Geometry in Stereo, Motion and Object Recognition

A Unified Approach

by

Gang Xu

Department of Computer Science,
Ritsumeikan University,
Kusatsu, Japan

and

Zhengyou Zhang

INRIA Sophia-Antipolis,
Sophia-Antipolis, France

KLUWER ACADEMIC PUBLISHERS

DORDRECHT / BOSTON / LONDON

A C.I.P. Catalogue record for this book is available from the Library of Congress.

ISBN 978-90-481-4743-4

Published by Kluwer Academic Publishers,
P.O. Box 17, 3300 AA Dordrecht, The Netherlands.

Kluwer Academic Publishers incorporates
the publishing programmes of
D. Reidel, Martinus Nijhoff, Dr W. Junk and MTP Press.

Sold and distributed in the U.S.A. and Canada
by Kluwer Academic Publishers,
101 Philip Drive, Norwell, MA 02061, U.S.A.

In all other countries, sold and distributed
by Kluwer Academic Publishers Group,
P.O. Box 322, 3300 AH Dordrecht, The Netherlands.

Printed on acid-free paper

For ***Akiyo*** and ***Xinglai Seira***

G.X.

For ***Mingyue*** and ***Rosaline Shuting***

Z.Z.

CONTENTS

Forward by Olivier Faugeras

The book by Xu and Zhang attacks some of the oldest problems in Computer Vision: to recover the 3-D geometric and kinematic structures of the world from two images and to recognize a 3-D object in a cluttered scene from one or several views of this object in a different setting. Not only are these problems in some sense the founding problems of the discipline now known as Computer Vision but they are also among the most difficult.

The authors propose to look at them from the point of view of geometry which naturally leads them to consider the intricate and beautiful projective geometric structure which relates two images of the same object. This structure gives its title to the book, it is the epipolar structure. From the practitioner's standpoint it allows to reduce the difficult 2-D search tasks which are encountered in the previous founding problems to much simpler 1-D search tasks and, for the theoretically inclined person, it ties the projective, affine, and Euclidean structures of the scene to those of the images.

I had great pleasure in reading the book which conveys the feeling that several different problems have been unified through the use of a unique geometric concept. Moreover, the authors have managed to achieve the "tour de force" to completely avoid using projective geometry in their exposition and to guide the reader through the various aspects of epipolar geometry, stereo vision, motion analysis and object recognition using only the standard tools of linear algebra. This makes the book usable by a much broader audience than otherwise and will, I hope, allow the quick development of many new and exciting applications of 3-D Computer Vision.

Olivier Faugeras
Research Director, INRIA
Maître de Conférence, Ecole Polytechnique
Adjunct Professor of Computer Science, MIT

Forward by Saburo Tsuji

Many years ago, artists and scientists studied the rules of perspective by which a great number of pictures we now enjoy have been drawn. Recently, a number of scientists have studied computer vision to explore how a computer can recognize a three-dimensional world by analyzing its images. The study has yielded epipolar geometry which is more than a simple extension of the rules of perspective but provides general rules between three-dimensional object shapes and their images.

Gang Xu and Zhengyou Zhang, two young scientists working with institutes in different countries, have deeply explored the theory and application of the epipolar geometry. Through discussions via the computer network, they have written this interesting book which provides a unified approach to important issues of computer vision, such as stereo, motion and object recognition. This book will appeal to graduate students and researchers in computer science, cognitive science, robotics and applied mathematics.

Saburo Tsuji
Professor and Dean
Faculty of Systems Engineering
Wakayama University

PREFACE

The Philosophies of this century will be
the common sense for the next.
— anonymous

It is impossible to recover, geometrically, the 3D information from a single image due to the loss of depth information during the image formation. However, many vision-based applications, such as object modeling, vehicle navigation and geometric inspection, require the 3D (either metric or non-metric) information. One of the solutions is to use multiple views. But then, a number of other problems arise: how to identify the same features among different views; how to fuse the highly redundant information present in different views; how to deal with multiple object motions because the environment is usually dynamic; how to efficiently represent the perceived scene; etc. The epipolar constraint, which is the only geometric constraint available from any two views, constitutes the basis in solving these problems.

This book is to give a comprehensive and hopefully complete treatment of the epipolar geometry and its use in a wide spectrum of computer vision problems. So far the epipolar constraint has not been made full use of in solving motion and object recognition problems, and their combinations. The reason is considered to lie in the difficulty in recovering the epipolar geometry from images.

This book begins with a general introduction to the epipolar geometry underlying camera and/or object motions. A relatively complete description of different camera models is provided. We then propose algorithms to recover the epipolar equations from images under both full perspective and affine projection models. Later chapters describe how to use the recovered epipolar equations to solve the binocular stereo problem, the motion correspondence and segmentation problem, and model view based 3D object recognition and segmentation problem.

Since the epipolar geometry underlies every problem that involves multiple views, and many of the vision problems do involve multiple views, this book should be of value to many researchers and practitioners in the field of computer vision and robotic vision.

Knowledge of Projective Geometry is not required at all to read this book. This arrangement, of course, sometimes sacrifices the elegance of the formulations if Projective Geometry were used; but we think that it is worth doing because most of the vision researchers and practitioners do not have easy access to Projective Geometry. We do assume, however, a basic knowledge of linear algebra, geometry and calculus. The book has been written with the aim to solve several practical vision problems. Details have been provided such that readers can easily reproduce the results to solve their own problems. For those who are interested in more theoretical aspects of multiple view problem, we highly recommend the book by Luong and Faugeras, entitled *A Projective Geometric Approach to 3D Vision: Fundamental Matrix and Self-Calibration*, also published by Kluwer Academic Publishers.

Acknowledgment

We, each of us working in a different country, would like to thank our colleagues in Osaka University, Japan, and INRIA (Institut National de Recherche en Informatique et Automatique), France, and in particular, Saburo Tsuji and Olivier Faugeras who deserve much credit for their constant help, inspirational discussion and outstanding contributions to the work described in this book.

Gang Xu's thanks also go to his students Eigo Segawa, Eiji Nishimura, Mihoko Takahashi, Takashi Ono, Takashi Naruse and Toshiaki Tomii for helping implement the algorithms, to Toshio Hayase, Hiroshi Ishiguro, Shigang Li, Yukio Osawa, Tomohiro Yamaguchi and Yoshio Iwai for maintaining a constructive research environment, to Heisuke Hironaka, Takeshi Kasai, Seiji Inokuchi, Masahiko Yachida, Minoru Asada, Hiromi T. Tanaka, Hitoshi Ogawa, Chil-Woo Lee, Seiji Yamada, Naoki Saiwaki and Qiang Xu for various help and constant encouragement over many years, to Alan Yuille, David Mumford, Mark Nitzberg and Takahiro Shiota for a number of discussions on using epipolar geometry to match uncalibrated views during his sabbatical stay in Harvard University, and to Song De Ma, Tomaso Poggio, Yoshiaki Shirai, Anil K. Jain, Takeo Kanade, Roberto Cipolla and Andrew Zisserman for their helpful comments and discussions at conferences and via email. His work reported in this book was mostly done in Osaka University, and was partially supported by grants from the Inamori Foundation, the Nissan Science Foundation and the Japanese Ministry of Education.

Zhengyou Zhang extends his thanks to Rachid Deriche, Tuan Luong, and Ph.D. students in his lab (Gabriella Csurka, Frédric Devernay, Reyes Enciso, Stéphane Laveau, Théo Papadopoulo and Cyril Zeller) for discussions and contributions, to Nicholas Ayache, Mike Brady, Song De Ma, Roger Mohr, and Hans-Hellmut Nagel for their kindness and constant encouragement, to Yiannis Aloimonos, Robert Azencott, Patrick Bouthemy, Mike Brooks, Radu Horaud, Martial Hébert, Steve Maybank, Gérard Medioni, Luc Robert, Charlie Rothwell, Thierry Viéville, Andrew Zisserman, and many others for their interest in his work, and to Jean-Luc Szpyrka for system management. His research was partially supported by grants under Eureka project IARES and Esprit project VIVA.

We thank Ray Zhang and Ron Zhang of China News Digest who helped proofread the English of Preface and Introduction of this book.

Lastly, albeit mostly, our gratitude goes to our wives and families, to whom the book is dedicated, for their love, and their understanding of many holidays and evenings spent on this book.

1

INTRODUCTION

1.1 VISION RESEARCH

Brain, including the perceptual systems, is probably the last hardest nut for scientists to crack. There are different approaches to this end. Neurophysiologists and psychophysicists have been investigating brain at different but complementary levels for many years. Since vision is the most "visible" part of the brain, it is also the most intensively studied part of the brain.

There is an immensely large amount of work in these fields, providing some answers to some problems in the case of biological vision systems. One well-known example is the elegant theory about the spatial correspondence between feature detectors and the array of cells in the retina, which later won a Nobel prize [73].

However, neither neurophysiological nor psychophysical approaches directly answer the questions of what (in the representational sense) is computed in vision and how these representations are computed. These questions have to be answered from the computational point of view.

Over the last two decades, computer vision scientists have taken a new approach to vision [96]. They build different computational models of what *should* be computed, what *can* really be computed, and *how* these computations can be realized by computer programs, and they use computers to test if their models are correct. The result is a better understanding of vision from a different point of view, and at the same time some working artificial vision systems are built that can be used in industry, medicine, etc.

The knowledge obtained in neurophysiology and psychophysics have given hints to and influenced computer vision scientists, helping find solutions to the design of specific algorithms and implementation of vision systems. On the other hand, computational vision has also given neurophysiologists and psychophysicists a mathematical framework for modeling vision processes.

The philosophy developed in computer vision is that vision can be studied at three different levels–computational theory level, algorithm level and implementation level [96]. At the computational theory level, each vision task is described in terms of functions of information processing, which are common for and independent of specific algorithm to perform the computation. At the algorithm level, there are usually many algorithms; and humans and computers do not necessarily employ the same. Algorithms can be independent of the hardware on which they are implemented, it might be the case that certain algorithms can only be implemented on a particular hardware system. At the implementation level, of course, the hardware of human vision and that of computer vision are completely different.

This philosophy brings two distinguished advantages to the understanding of vision. Firstly, it provides a mathematical framework, a mathematical language and precision with which both computer vision scientists and scientists in neurophysiology and psychophysics can think and exchange ideas. Secondly, as a result of the first advantage, simple and working systems can be built to test the computational models and to be used in industrial applications.

This book deals with vision as a computational problem. Little reference is made to neurophysiology and psycophysics or other disciplines which also deal with vision.

1.2 MULTIPLE VIEW PROBLEMS IN VISION

According to Marr, vision is a process of seeing *what* is *where* [96]. He proposes to divide the vision problem into a hierarchical system of subproblems. At the lowest level is the *Primal Sketch*, which is composed of significant image features, like edges, corners, line segments, etc. At the next level, properties of the visible surfaces like surface shape are computed from the Primal Sketch. Since the 3D shape information is described in the coordinate system defined by the camera, he calls it the $2\frac{1}{2}D$ *Sketch*. At the third level, Object-Centered representations of the objects are computed from the viewer-centered $2\frac{1}{2}$D Sketch. Object recognition and other high level cognitive tasks are then made possible by using the third level representations.

Research in the past two decades has been done mainly under the influence of his framework. The advantage of his division of vision into a system of visual modules has enabled us to study vision as independent problems (to extend Marr's paradigm, Aloimonos proposes integration of visual modules [3]). Of the work done so far, the most extensively studied modules are feature detection, shape from stereo, shape from motion, shape from contour, shape from shading, shape from texture, other shape-from-X modules, and object recognition. There has been a wide recognition that many of the visual problems are generally ill-posed, that is, solutions are not unique [117, 116, 3]. Techniques to select solutions from among the possible ones have been proposed through regularization [117, 116], Maximizing A Priori Probability [44] and Active Vision [3].

Of the vision modules, many are related to multiple views. A multiple view problem is inherently a vision problem involving more than one view of the same scene. By this definition, stereo, motion and object recognition can all be considered as multiple view problems. Binocular stereo is a problem to determine the 3D shape of visible surfaces in a static scene from images taken of the same scene by two cameras or one camera at two different positions. The basic assumption here is that the geometry of the cameras is known, thus the epipolar lines are *given*. The first step is to match the points in the two images along the *known* epipolar lines, so that one can then determine the 3D position of each point by triangulation. Motion can be thought of as a problem to determine the 3D shape of visible surfaces in a dynamic scene from images taken by a dynamic camera or a static camera. Here again, the first step is to match points in consecutive images so that one can compute the 3D coordinates of each point. Object recognition is a problem to match the input image with a particular model in the object database which contains the models of the objects that have to be recognized. Although there are still debates about whether 3D data or 2D views should be used as object models, if we employ 2D views as object models, the object recognition problem is to match the input view onto one (or one set) of the model views in the database. In this sense, all the above three vision problems share the common problem of image matching between two views, and thus also share any constraints that arise from the geometry of multiple views.

So far all these three problems have been studied as separate and different problems in computer vision. However, as discussed above, by refining the problems from a new perspective, they become the same problem. More specifically, for any two view problems, they share the common epipolar constraint, which is the only geometrical constraint available regardless of specific objects. It is shown that recovery of structure is coupled with recovery of motion, of which epipolar geometry is a subset. For full perspective projection, recovery of motion is equivalent to recovery of epipolar geometry. For linear approximations to the nonlinear full perspective projection, recovery of epipolar geometry reduces the space of motion to another smaller space. For image matching problems, by the epipolar constraint, search for matches degrades from 2-dimensionality to 1-dimensionality. Moreover, by examining which points share the same epipolar geometry, we can segment the images into regions which correspond to different rigid motions. This is especially important when there are multiple motions in the same scene.

Although the epipolar constraint was widely used in stereo matching, it has not been fully exploited in motion and object recognition until recently. The reason is thought to be the difficulty in recovering it from images, especially when there are multiple motions.

1.3 ORGANIZATION OF THIS BOOK

In the following parts of this book, we will proceed from a general and comprehensive introduction to the epipolar geometry under various projection models in Chapter 2. Derivation of epipolar equations is given, and their geometric interpretations are described.

Chapter 3 discusses how to reliably and robustly recover the epipolar geometry and structure from point matches. This is an important step in order to make effective use of epipolar geometry for other vision tasks, as determining the coefficients of epipolar equations can be very sensitive to noise in the image data. Outlier rejection and clustering techniques are used, and various criteria are computed and compared in terms of reliability and robustness.

Chapter 4 presents techniques for the recovery of epipolar geometry and structure from line segments or lines. We show that it is possible to recover the epipolar geometry from the line segments by requiring the matched line segments to overlap. Traditionally, the endpoints of line segments are completely ignored because of their high instability, and the straight support lines are then used. In this case, a minimum of 3 views are necessary. The epipolar geometry of 3 views, known as *trifocal constraint*, is also discussed.

In Chapter 5, the problems of stereo, motion and object recognition are re-examined from the perspective of epipolar geometry. With recovered epipolar equations, search for correspondence can be posed as a 1-dimensional search along epipolar lines. And we propose to represent the ambiguity along the epipolar lines in Spatial Disparity Space, and to resolve the ambiguity by maximizing the smoothness of surfaces in that space.

Chapter 6 discusses the problem of stereo. Given two uncalibrated binocular stereo images, the first step is to recover the epipolar geometry underlying the two images, and then use this constraint to determine correspondence between the two images.

In Chapter 7, we discuss the problem of motion correspondence and segmentation. Given two images of a dynamic scene involving multiple rigid motions, we propose a procedure to recover the epipolar geometry for each motion, and further propose an algorithm to determine the correspondence for each point and to determine which motion each point belongs to. By the epipolar constraint, the 2D search problem is reduced to a 1D problem, thus eliminating the *aperture problem* defined in motion.

In Chapter 8, we discuss the problem of object recognition and localization using 2D model views. The problem is formulated as one of finding the epipolar geometry which should exist between the input view and the model view(s). The cases of using a single model view and of using multiple model views are examined. In either case, the recovery of epipolar equations between the uncalibrated views is essential. With recovered epipolar geometry, edge images can be matched, thus recognizing and localizing the objects.

1.4 NOTATION

The following convention is used throughout the book. Vectors are in bold lower case letters, e.g. $\mathbf{x}$. 3D vectors corresponding to space points are also denoted by an upper case letter, but are typeset in typewriter font, e.g. $\mathtt{M}$. Bold upper case letters are used to denote matrices, e.g. $\mathbf{R}$. Tilde $\tilde{}$ is used to denote an augmented vector such that

$$\tilde{\mathbf{x}} = \lambda[\mathbf{x}^T \ 1]^T ,$$

where λ is a scalar and is usually chosen to be 1. When multiple images are concerned, objects in the second image are denoted with a prime $'$, and those in the third, with a double prime $''$.

In Table 1.1, we provide a list of symbols commonly used in this book.

Table 1.1 List of commonly used symbols

Symbol	Meaning
$\times$	: cross product of two vectors, e.g. $\mathbf{x} \times \mathbf{y}$
$\cdot$	: dot/inner/scalar product, e.g. $\mathbf{x} \cdot \mathbf{y} = \mathbf{x}^T\mathbf{y}$
^{T}	: superscript, the transposition operator of a vector or a matrix
$^{-1}$	: superscript, the inverse operator of a matrix
$^{-T}$	: superscript, the transpose of the inverse of a matrix
$^+$	: superscript, pseudo inverse of a matrix
$\| \|$	: norm-2 of a vector, or the Frobenius norm of a matrix
Λ	: covariance matrix of a random vector
$\mathbf{0}$	: zero vector, i.e. all elements of the vector are equal to 0
$\mathbf{A}$	: 3×3 matrix, the camera intrinsic matrix
$\mathbf{D}$	: 4×4 matrix, the 3D Euclidean transformation
$\mathbf{e}$	: 2D vector, the epipole in an image
$\mathbf{E}$	: 3×3 matrix, the essential matrix
$E[\mathbf{x}]$	: mean/expectation of the random vector $\mathbf{x}$
f	: scalar, the focal length of a camera
$\mathbf{f}$	: vector, composed of the elements of fundamental matrix $\mathbf{F}$
$\mathbf{F}$	: 3×3 matrix, the fundamental matrix
$\mathbf{F}_A$	: 3×3 matrix, the affine fundamental matrix
$\mathbf{H}$	: a 2D or 3D homography, or an Hessian matrix, depending on the context
$\mathbf{I}$	: Identity matrix
$\mathbf{l}$	: 3D vector, homogeneous coordinates of a 2D line
$\mathbf{m}$	: 2D vector, an image point
$\mathtt{M}$	: 3D vector, a space point
$P(e)$	: probability of an event e
$\mathbf{P}$	: 3×4 matrix, the camera perspective projection matrix
$\mathbf{P}_A$	: 3×4 matrix, the camera affine projection matrix
$\mathbf{R}$	: 3×3 matrix, the rotation matrix
$\mathbf{S}$	: diagonal matrix of the singular values in a singular value decomposition
$\mathbf{t}$	: 3D vector, the translation vector
$\mathbf{U}$	: left orthonormal matrix of a singular value decomposition
$\mathbf{V}$	: right orthonormal matrix of a singular value decomposition

2

CAMERA MODELS AND EPIPOLAR GEOMETRY

In this chapter we present a detailed description of the epipolar geometry between two images taken by two different cameras or by a single camera at two different time instants. Since it is intimately related to camera models, we will start from modeling cameras, followed by several approximations to the full perspective projection.

2.1 MODELING CAMERAS

In this section, we will proceed from introducing the pinhole camera which can be ideally modeled as the *perspective projection.* Then we will explore how changing the coordinate system of the image plane, or retina, affects the projection. This is important if one really wants to compute the epipolar geometry and motion parameters accurately. The same terminology is used as in [41] if possible, while avoiding to assume any knowledge of projective geometry.

2.1.1 Pinhole Camera and Perspective Projection

Figure 2.1 shows a pinhole camera model. There is a plane $\mathcal{F}$ at a fixed distance f in front of an *image plane* $\mathcal{I}$. The image plane is also called the *retinal plane*. An ideal pinhole C is found in the plane $\mathcal{F}$. Assume that an enclosure is provided so that only light coming through the pinhole can reach the image plane. The rays of light emitted or reflected by an object pass through the pinhole and form an inverted image of that object on the image plane. Each

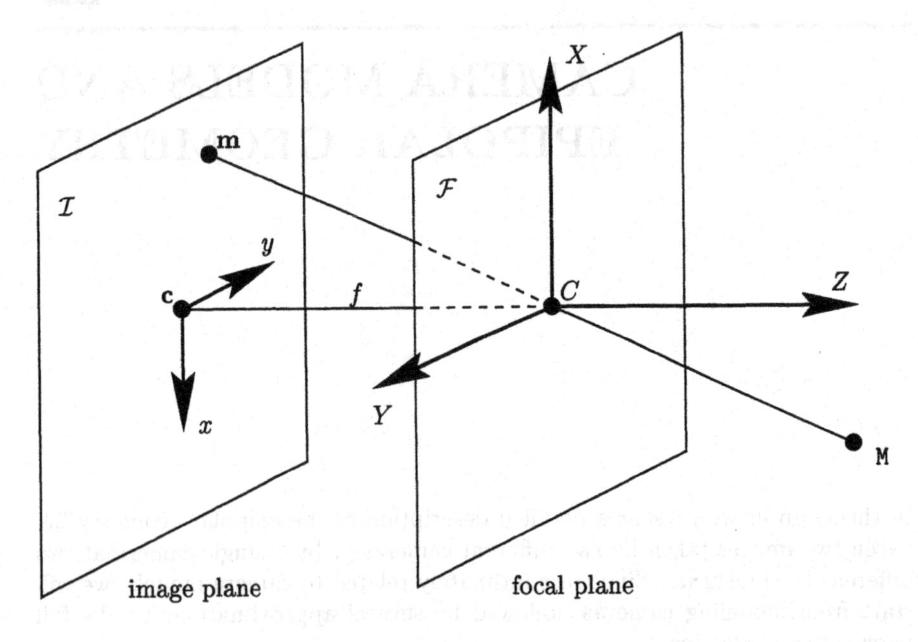

Figure 2.1 The pinhole camera model

point in the object, its corresponding image point and the pinhole constitute a straight line. This kind of projection from 3D space to a plane is called *perspective projection*.

The geometric model of a pinhole camera thus consists of an image plane $\mathcal{I}$ and a point C on the plane $\mathcal{F}$. The point C is called the *optical center*, or the *focus*. The plane $\mathcal{F}$ going through C and parallel to $\mathcal{I}$ is called the *focal plane*. The distance between the optical center and the image plane is the *focal length* of the optical system. The line going through the optical center C and perpendicular to the image plane $\mathcal{I}$ is called the *optical axis*, and it intersects $\mathcal{I}$ at a point c, called the *principal point*. It is clear that the focal plane is also perpendicular to the optical axis. Experiences have shown that such a simple system can accurately model the geometry and optics of most of the modern Vidicon and CCD cameras [41].

Now let us derive the equations for the perspective projection. The coordinate system (c, x, y) for the image plane is defined such that the origin is at the point c (intersection of the optical axis with the image plane) and that the axes are determined by the camera scanning and sampling system. We choose the coordinate system (C, X, Y, Z) for the three-dimensional space as indicated in

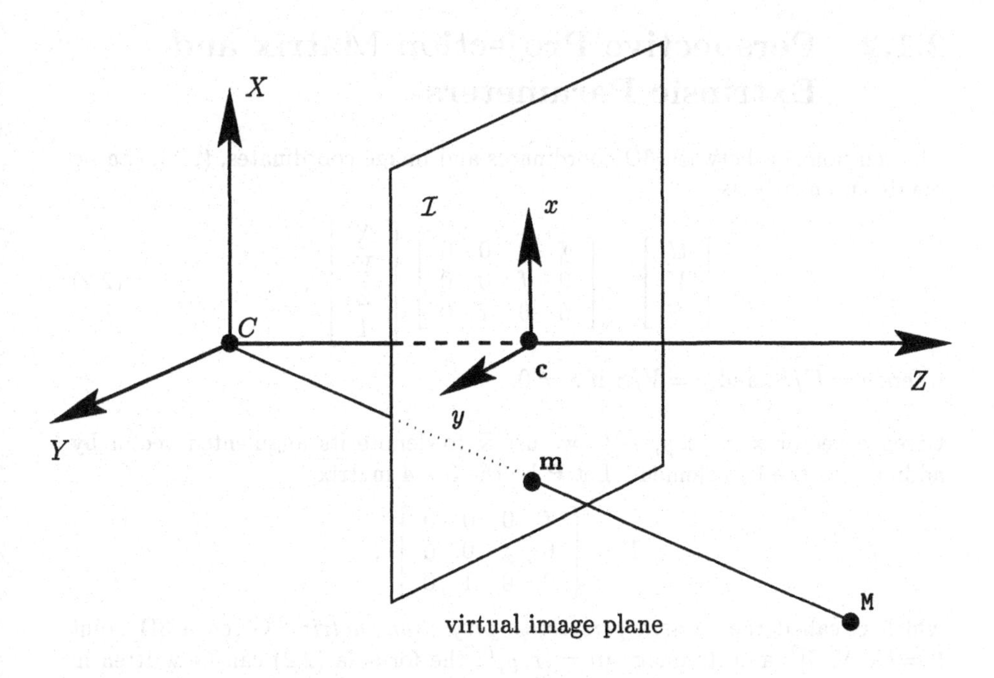

Figure 2.2 The pinhole camera model with a virtual image plane

Figure 2.1, where the origin is at the optical center and the Z-axis coincides the optical axis of the camera. The X- and Y-axes are parallel, but opposite in direction, to the image x- and y-axes. The coordinate system (C, X, Y, Z) is called the *standard coordinate system* of the camera, or simply *camera coordinate system*. From the above definition of the camera and image coordinate system, it is clear that the relationship between 2D image coordinates and 3D space coordinates can be written as

$$\frac{x}{X} = \frac{y}{Y} = \frac{f}{Z} . \tag{2.1}$$

It should be noted that, from the geometric viewpoint, there is no difference to replace the image plane by a virtual image plane located on the other side of the focal plane (Fig. 2.2). Actually this new system is what people usually use. In the new coordinate system, an image point (x, y) has 3D coordinates (x, y, f), if the scale of the image coordinate system is the same as that of the 3D coordinate system.

2.1.2 Perspective Projection Matrix and Extrinsic Parameters

The relationship between 3D coordinates and image coordinates, (2.1), can be rewritten linearly as

$$\begin{bmatrix} U \\ V \\ S \end{bmatrix} = \begin{bmatrix} f & 0 & 0 & 0 \\ 0 & f & 0 & 0 \\ 0 & 0 & 1 & 0 \end{bmatrix} \begin{bmatrix} X \\ Y \\ Z \\ 1 \end{bmatrix}, \tag{2.2}$$

where $x = U/S$, and $y = V/S$ if $S \neq 0$.

Given a vector $\mathbf{x} = [x, y, \cdots]^T$, we use $\widetilde{\mathbf{x}}$ to denote its augmented vector by adding 1 as the last element. Let $\mathbf{P}$ be the 3×4 matrix

$$\mathbf{P} = \begin{bmatrix} f & 0 & 0 & 0 \\ 0 & f & 0 & 0 \\ 0 & 0 & 1 & 0 \end{bmatrix},$$

which is called the camera *perspective projection matrix*. Given a 3D point $\mathtt{M} = [X, Y, Z]^T$ and its image $\mathbf{m} = [x, y]^T$, the formula (2.2) can be written in matrix form as

$$s\widetilde{\mathbf{m}} = \mathbf{P}\widetilde{\mathtt{M}}, \tag{2.3}$$

where $s = S$ is an arbitrary nonzero scalar.

For a real image point, S should not be 0. We now make an extension to include the case $S = 0$. If $S = 0$, then $Z = 0$, i.e. the 3D point is in the focal plane of the camera, and the image coordinates x and y are not defined. For all points in the focal plane but the optical center, their corresponding points in the image plane are at infinity. For the optical center C, we have $U = V = S = 0$ (i.e. $s = 0$) since $X = Y = Z = 0$.

So far, we assume that 3D points are expressed in the camera coordinate system. In practice, they can be expressed in any 3D coordinate system, which is sometimes referred as the *world coordinate system*. As shown in Fig. 2.3, we go from the old coordinate system centered at the optical center C (camera coordinate system) to the new coordinate system centered at point O (world coordinate system) by a rotation $\mathbf{R}$ followed by a translation $\mathbf{t} = CO$. Then, for a single point, its coordinates expressed in the camera coordinate system, M_c, and those expressed in the world coordinate system, M_w, are related by

$$\mathtt{M}_c = \mathbf{R}\mathtt{M}_w + \mathbf{t},$$

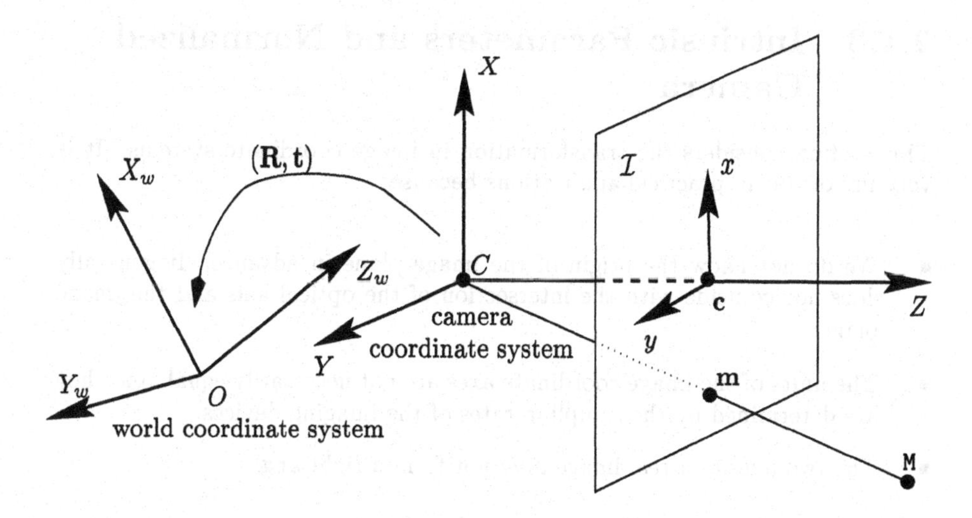

Figure 2.3 World coordinate system and camera extrinsic parameters

or more compactly

$$\widetilde{\mathtt{M}}_c = \mathbf{D}\widetilde{\mathtt{M}}_w , \tag{2.4}$$

where $\mathbf{D}$ is a Euclidean transformation of the three-dimensional space:

$$\mathbf{D} = \begin{bmatrix} \mathbf{R} & \mathbf{t} \\ \mathbf{0}_3^T & 1 \end{bmatrix} \quad \text{with } \mathbf{0}_3 = [0,0,0]^T . \tag{2.5}$$

The matrix $\mathbf{R}$ and the vector $\mathbf{t}$ describe the orientation and position of the camera with respect to the new world coordinate system. They are called the *extrinsic parameters* of the camera.

From (2.3) and (2.4), we have

$$\widetilde{\mathbf{m}} = \mathbf{P}\widetilde{\mathtt{M}}_c = \mathbf{P}\mathbf{D}\widetilde{\mathtt{M}}_w .$$

Therefore the new perspective projection matrix is given by

$$\mathbf{P}_{new} = \mathbf{P}\mathbf{D} . \tag{2.6}$$

This tells us how the perspective projection matrix $\mathbf{P}$ changes when we change coordinate systems in the three-dimensional space: We simply multiply it on the right by the corresponding Euclidean transformation.

2.1.3 Intrinsic Parameters and Normalized Camera

This section considers the transformation in image coordinate systems. It is very important in practical applications because:

- We do not know the origin of the image plane in advance. It generally does not coincide with the intersection of the optical axis and the image plane.
- The units of the image coordinate axes are not necessarily equal, and they are determined by the sampling rates of the imaging devices.
- The two axes of a real image may not form a right angle.

To handle these effects, we introduce an affine transformation.

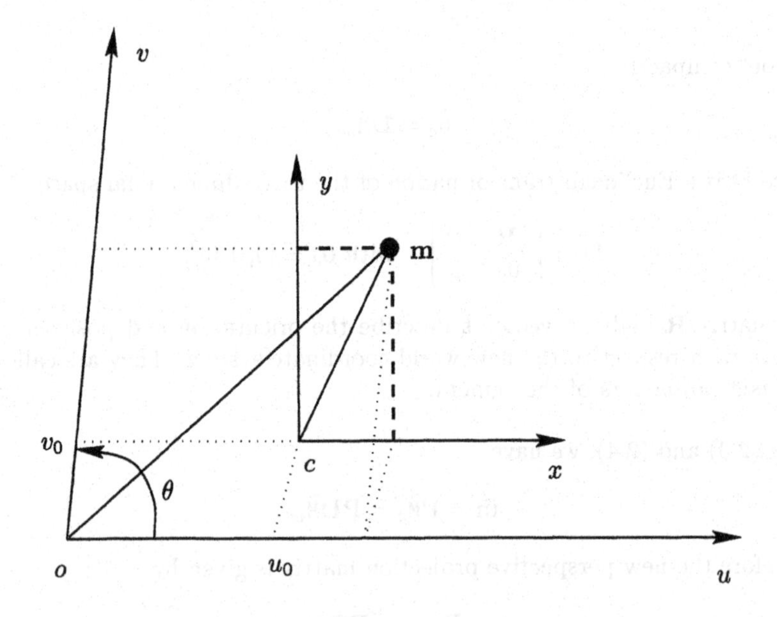

Figure 2.4 Camera intrinsic parameters

Consider Fig. 2.4. The original image coordinate system (c, x, y) is centered at the principal point c, and has the same units on both x- and y-axes. The coordinate system (o, u, v) is the coordinate system in which we address the pixels in an image. It is usually centered at the upper left corner of the image, which is usually not the principal point c. Due to the electronics of acquisition, the pixels are usually not square. Without loss of generality, the u-axis is assumed to be parallel to te x-axis. The units along the u- and v-axes are assumed to be k_u and k_v with respect to the unit used in (c, x, y). The u- and v-axes may not be exactly orthogonal, and we denote their angle by θ. Let the coordinates of the principal point c in (o, u, v) be $[u_0, v_0]^T$. These five parameters do not depend on the position and orientation of the cameras, and are thus called the camera *intrinsic parameters*.

For a given point, let $\mathbf{m}_{old} = [x, y]^T$ be the coordinates in the original coordinate system; let $\mathbf{m}_{new} = [u, v]^T$ be the pixel coordinates in the new coordinate system. It is easy to see that

$$\widetilde{\mathbf{m}}_{new} = \mathbf{H}\widetilde{\mathbf{m}}_{old} ,$$

where

$$\mathbf{H} = \begin{bmatrix} k_u & k_u \cot\theta & u_0 \\ 0 & k_v / \sin\theta & v_0 \\ 0 & 0 & 1 \end{bmatrix} .$$

Since, according to (2.3), we have

$$s\widetilde{\mathbf{m}}_{old} = \mathbf{P}_{old}\widetilde{\mathtt{M}} ,$$

we conclude that

$$s\widetilde{\mathbf{m}}_{new} = \mathbf{H}\mathbf{P}_{old}\widetilde{\mathtt{M}} ,$$

and thus

$$\mathbf{P}_{new} = \mathbf{H}\mathbf{P}_{old} = \begin{bmatrix} fk_u & fk_u \cot\theta & u_0 & 0 \\ 0 & fk_v / \sin\theta & v_0 & 0 \\ 0 & 0 & 1 & 0 \end{bmatrix} . \tag{2.7}$$

Note that it depends on the products fk_u and fk_v, which means that a change in the focal length and a change in the pixel units are indistinguishable. We thus introduce two parameters $\alpha_u = fk_u$ and $\alpha_v = fk_v$.

We will now define a special coordinate system that allows us to normalize the image coordinates [41]. This coordinate system is called the *normalized coordinate system* of the camera. In this "normalized" camera, the image plane is located at a unit distance from the optical center (i.e. $f = 1$). The perspective projection matrix of the normalized camera is given by

$$\mathbf{P}_N = \begin{bmatrix} 1 & 0 & 0 & 0 \\ 0 & 1 & 0 & 0 \\ 0 & 0 & 1 & 0 \end{bmatrix} . \tag{2.8}$$

For a world point $[X, Y, Z]^T$ expressed in the camera coordinate system, its normalized coordinates are

$$\begin{aligned} \hat{x} &= \frac{X}{Z} \\ \hat{y} &= \frac{Y}{Z} . \end{aligned} \tag{2.9}$$

A matrix $\mathbf{P}$ defined by (2.7) can be decomposed into the product of two matrices:

$$\mathbf{P}_{new} = \mathbf{A}\mathbf{P}_N , \tag{2.10}$$

where

$$\mathbf{A} = \begin{bmatrix} \alpha_u & \alpha_u \cot\theta & u_0 \\ 0 & \alpha_v / \sin\theta & v_0 \\ 0 & 0 & 1 \end{bmatrix} . \tag{2.11}$$

The matrix $\mathbf{A}$ contains only the intrinsic parameters, and is called *camera intrinsic matrix*. It is thus clear that the normalized image coordinates are given by

$$\begin{bmatrix} \hat{x} \\ \hat{y} \\ 1 \end{bmatrix} = \mathbf{A}^{-1} \begin{bmatrix} u \\ v \\ 1 \end{bmatrix} . \tag{2.12}$$

Through this transformation from the available pixel image coordinates, $[u, v]^T$, to the imaginary normalized image coordinates, $[\hat{x}, \hat{y}]^T$, the projection from the space onto the normalized image does not depend on the specific cameras. This frees us from thinking about characteristics of the specific cameras and allows us to think in terms of ideal systems in stereo, motion and object recognitions.

2.1.4 The General Form of Perspective Projection Matrix

The camera can be considered as a system that depends upon the intrinsic and the extrinsic parameters. There are five intrinsic parameters: the scale factors

α_u and α_v, the coordinates u_0 and v_0 of the principal point, and the angle θ between the two image axes. There are six extrinsic parameters, three for the rotation and three for the translation, which define the transformation from the world coordinate system, to the standard coordinate system of the camera.

Combining (2.6) and (2.10) yields the general form of the perspective projection matrix of the camera:

$$\mathbf{P} = \mathbf{A}\mathbf{P}_N\mathbf{D} = \mathbf{A}[\mathbf{R}\ \mathbf{t}] \,. \tag{2.13}$$

The projection of 3D world coordinates $\mathtt{M} = [X, Y, Z]^T$ to 2D pixel coordinates $\mathbf{m} = [u, v]^T$ is then described by

$$s\widetilde{\mathbf{m}} = \mathbf{P}\widetilde{\mathtt{M}} \,, \tag{2.14}$$

where s is an arbitrary scale factor. Matrix $\mathbf{P}$ has $3 \times 4 = 12$ elements, but has only 11 degrees of freedom because it is defined up to a scale factor.

Let p_{ij} be the (i, j) entry of matrix $\mathbf{P}$. Eliminating the scalar s in (2.14) yields two nonlinear equations:

$$u = \frac{p_{11}X + p_{12}Y + p_{13}Z + p_{14}}{p_{31}X + p_{32}Y + p_{33}Z + p_{34}} \tag{2.15}$$

$$v = \frac{p_{21}X + p_{22}Y + p_{23}Z + p_{24}}{p_{31}X + p_{32}Y + p_{33}Z + p_{34}} \,. \tag{2.16}$$

Camera *calibration* is the process of estimating the intrinsic and extrinsic parameters of a camera, or the process of first estimating the matrix $\mathbf{P}$ and then deducing the camera parameters from $\mathbf{P}$. A wealth of work has been carried out in this domain by researchers either in Photogrammetry [17, 33] or in Computer Vision and Robotics [159, 40, 87, 158, 169, 173] (see [160] for a review). The usual method of calibration is to compute camera parameters from one or more images of an object of *known size and shape*, for example, a flat plate with a regular pattern marked on it. From (2.14) or (2.15) and (2.16), we have two nonlinear equations relating 2D to 3D coordinates. This implies that each pair of an identified image point and its corresponding point on the calibration object provides two constraints on the intrinsic and extrinsic parameters of the camera. The number of unknowns is 11. It can be shown that, given N points ($N \geq 6$) in general position, the camera can be calibrated. The presentation of calibration techniques is beyond the scope of this book. The interested reader is referred to the above-mentioned references.

Once the perspective projection matrix $\mathbf{P}$ is given, we can compute the coordinates of the optical center C of the camera in the world coordinate system. We first decompose the 3×4 matrix $\mathbf{P}$ as the concatenation of a 3×3 submatrix $\mathbf{B}$ and a 3-vector $\mathbf{b}$, i.e. $\mathbf{P} = [\mathbf{B}\ \mathbf{b}]$. Assume that the rank of $\mathbf{B}$ is 3. In Sect. 2.1.2, we explained that, under the pinhole model, the optical center projects to $[0,0,0]^T$ (i.e. $s = 0$). Therefore, the optical center can be obtained by solving

$$\mathbf{P}\widetilde{C} = \mathbf{0} \quad \text{i.e.} \quad [\mathbf{B}\ \mathbf{b}]\begin{bmatrix} C \\ 1 \end{bmatrix} = \mathbf{0}\,.$$

The solution is

$$C = -\mathbf{B}^{-1}\mathbf{b}\,. \tag{2.17}$$

Given matrix $\mathbf{P}$ and an image point $\mathbf{m}$, we can obtain the equation of the 3-D semi-line defined by the optical center C and point $\mathbf{m}$. This line is called the *optical ray* defined by $\mathbf{m}$. Any point on it projects to the single point $\mathbf{m}$. We already know that C is on the optical ray. To define it, we need another point. Without loss of generality, we can choose the point D such that the scale factor $s = 1$, i.e.

$$\widetilde{\mathbf{m}} = [\mathbf{B}\ \mathbf{b}]\begin{bmatrix} D \\ 1 \end{bmatrix}\,.$$

This gives $D = \mathbf{B}^{-1}(-\mathbf{b} + \widetilde{\mathbf{m}})$. A point on the optical ray is thus given by

$$\mathtt{M} = C + \lambda(D - C) = \mathbf{B}^{-1}(-\mathbf{b} + \lambda\widetilde{\mathbf{m}})\,,$$

where λ varies from 0 to ∞.

2.2 PERSPECTIVE APPROXIMATIONS

As can be seen in the previous section, the perspective projection is a nonlinear mapping. This makes many vision problems difficult to solve, and more importantly, they can become ill-conditioned when the perspective effects are small. Sometimes, if certain conditions are satisfied, for example, when the camera field of view is small and the object size is small enough with respect to the distance from the viewer to the object, the projection can be approximated by a linear mapping [2]. As will be seen later, this simplification leads to great reduction of complexity in computation of the epipolar geometry from images.

2.2.1 Orthographic and Weak Perspective Projections

The simplest approximation is the so-called *orthographic projection* (Fig. 2.5a). It ignores completely the depth dimension. There is some evidence that rays near the fovea are projected orthographically (see for example [70]). Because of this, orthographic projection has been used in human vision research [164]. But the orthographic projection can lead to two identical objects having the same image, even if one is much further away from the camera than the other (*distance effect*), or if one is much more distant from the optical axis than the other (*position effect*) (see Fig. 2.5b). Therefore, methods that use orthographic projection are only valid in a limited domain where the distance and position effects can be ignored.

A much more reasonable approximation is the so-called *weak perspective projection*. When the object size is small enough with respect to the distance from the camera to the object, the depth, or Z, can be replaced by a common depth, Z_c. Then the equations of (2.1) become linear:

$$\begin{aligned} x &= \frac{X}{Z_c} \\ y &= \frac{Y}{Z_c} \end{aligned} \tag{2.18}$$

Here, we assume that the focal length f is normalized to 1.

Suppose the common depth is the depth of the centroid of the object. Then this approximation can be understood as a two-step projection (Figure 2.6). The first one is a parallel projection of all the object surface points onto a plane which goes through Z_c, the depth of the object centroid, and is parallel to the image plane or focal plane. The second step is a perspective projection of that plane onto the image plane, which is actually a uniform scaling of that plane. Sometimes, this projection is also called the *scaled orthographic projection*. When Z_c is unity, the projection becomes the *orthographic* one.

Let

$$\mathbf{P}_{wp} = \begin{bmatrix} 1 & 0 & 0 & 0 \\ 0 & 1 & 0 & 0 \\ 0 & 0 & 0 & Z_c \end{bmatrix} . \tag{2.19}$$

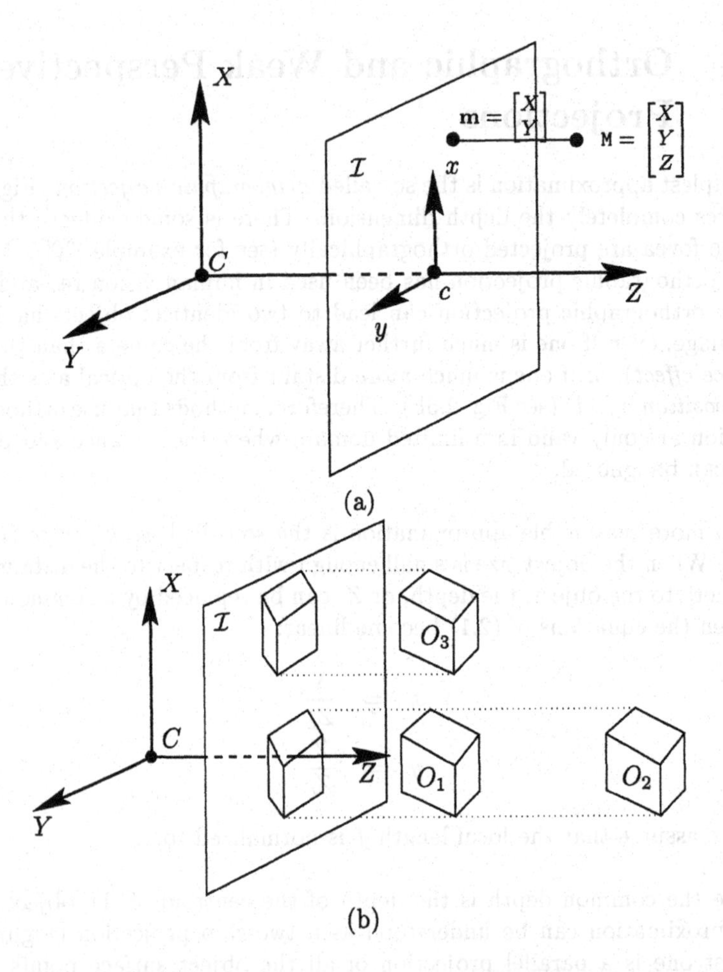

Figure 2.5 (a) Orthographic projection. (b) Three identical objects in different positions give the same image under orthographic projection: Object O_2 is much further away from the camera, and object O_3 is much distant from the optical axis, than object O_1

Equation (2.18) can then be written in matrix form, similar to (2.3), as

$$s\begin{bmatrix} x \\ y \\ 1 \end{bmatrix} = \mathbf{P}_{wp}\begin{bmatrix} X \\ Y \\ Z \\ 1 \end{bmatrix}$$

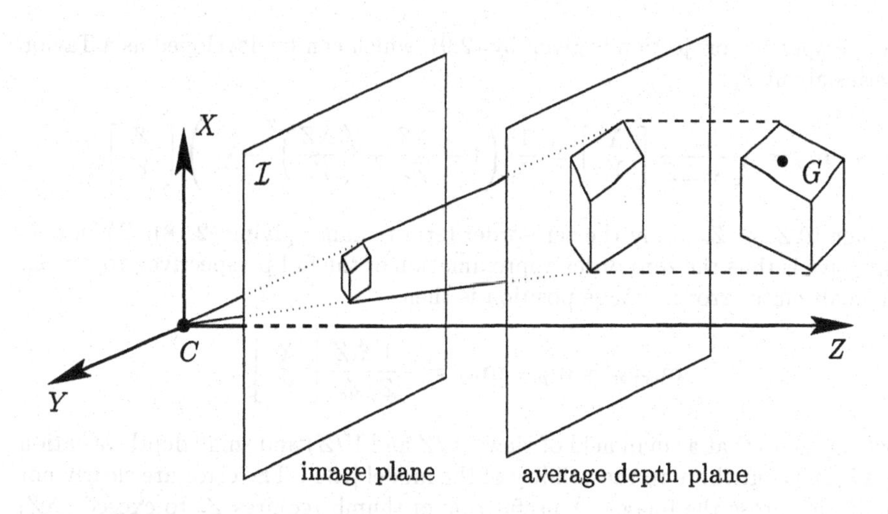

Figure 2.6 The weak perspective model. The average depth plane is drawn in front for clarity

Taking into account the intrinsic and extrinsic parameters of the camera yields:

$$s\widetilde{\mathbf{m}} = \mathbf{A}\mathbf{P}_{wp}\mathbf{D}\widetilde{\mathtt{M}} , \tag{2.20}$$

where $\mathbf{A}$ is the intrinsic matrix (2.11), and $\mathbf{D}$ is the rigid transformation (2.5). Expand it and eliminate the scale factor s, and we have

$$\mathbf{m} = \mathbf{T}_{wp}\mathtt{M} + \mathbf{t}_{wp} , \tag{2.21}$$

where

$$\begin{aligned} \mathbf{T}_{wp} &= \frac{1}{Z_c}\begin{bmatrix} \alpha_u(\mathbf{r}_1^T + \cot\theta\mathbf{r}_2^T) \\ \alpha_v\mathbf{r}_2^T / \sin\theta \end{bmatrix} \\ \mathbf{t}_{wp} &= \frac{1}{Z_c}\begin{bmatrix} \alpha_u(t_1 + t_2\cot\theta) \\ \alpha_v t_2 / \sin\theta \end{bmatrix} + \begin{bmatrix} u_0 \\ v_0 \end{bmatrix} . \end{aligned}$$

Here $\mathbf{r}_i$ denotes the ith row vector of the rotation matrix $\mathbf{R}$, and t_i is the ith element of the translation vector $\mathbf{t}$. It is clear that the relation between 3D coordinates and image coordinates is linear.

Let us examine the approximation error introduced by the weak perspective projection. For simplicity, we consider a normalized camera with the camera and world coordinate systems aligned. For a point $\mathtt{M}$ with depth $Z = Z_c + \Delta Z$, its perspective projection is given by (2.9), which can be developed as a Taylor series about Z_c:

$$\mathbf{m}_p = \frac{1}{Z_c + \Delta Z}\begin{bmatrix} X \\ Y \end{bmatrix} = \frac{1}{Z_c}\left(1 - \frac{\Delta Z}{Z_c} + \left(\frac{\Delta Z}{Z_c}\right)^2 - \cdots\right)\begin{bmatrix} X \\ Y \end{bmatrix} .$$

When $|\Delta Z| \ll Z_c$, only the zero-order term remains, giving (2.18). Weak perspective is thus the zero-order approximation of the full perspective projection. The absolute error in image position is then

$$\mathbf{m}_{error} = \mathbf{m}_p - \mathbf{m}_{wp} = -\frac{1}{Z}\frac{\Delta Z}{Z_c}\begin{bmatrix} X \\ Y \end{bmatrix} ,$$

which shows that a small field of view (X/Z and Y/Z) and small depth variation ($\Delta Z/Z_c$) contribute to the validity of the model [134]. The error are clearly not uniform across the image. A useful rule of thumb requires Z_c to exceed $|\Delta Z|$ by an order of magnitude [152], i.e. $Z_c \geq 10|\Delta Z|$.

2.2.2 Paraperspective Projection

Under weak perspective projection, a world point is first projected onto the average depth plane using the rays parallel to the optical axis. This causes a significant approximation error when an object is distant to the optical axis (large X/Z and Y/Z). The *paraperspective projection* generalizes this by projecting points using the rays parallel to the central projecting ray CG, where G is the centroid of the object [2]. As with weak perspective, the perspective projection is also approximated with a two-step process, but the projection in the first step is no more realized along rays parallel to the optical axis (see Fig. 2.7).

Let the coordinates of the centroid of the object G be $[X_c, Y_c, Z_c]^T$. A world point $\mathtt{M} = [X, Y, Z]^T$ is first projected, parallel to CG, to the average depth plane at $[X - \frac{X_c}{Z_c}Z + X_c, Y - \frac{Y_c}{Z_c}Z + Y_c, Z_c]^T$. Finally, this point is projected perspectively onto the image plane as

$$\begin{aligned} x &= \frac{1}{Z_c}\left(X - \frac{X_c}{Z_c}Z + X_c\right) \\ y &= \frac{1}{Z_c}\left(Y - \frac{Y_c}{Z_c}Z + Y_c\right) . \end{aligned} \tag{2.22}$$

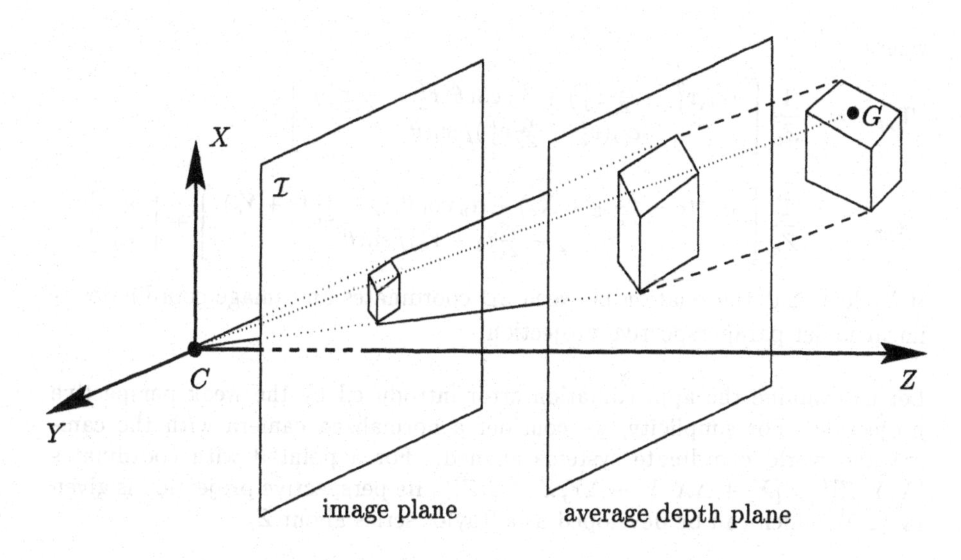

Figure 2.7 Paraperspective projection. The average depth plane is drawn in front for clarity

They can be written in matrix form as

$$s \begin{bmatrix} x \\ y \\ 1 \end{bmatrix} = \mathbf{P}_{pp} \begin{bmatrix} X \\ Y \\ Z \\ 1 \end{bmatrix},$$

with

$$\mathbf{P}_{pp} = \begin{bmatrix} 1 & 0 & -X_c/Z_c & X_c \\ 0 & 1 & -Y_c/Z_c & Y_c \\ 0 & 0 & 0 & Z_c \end{bmatrix}. \tag{2.23}$$

Taking into account the intrinsic and extrinsic parameters of the camera yields:

$$s\widetilde{\mathbf{m}} = \mathbf{A}\mathbf{P}_{pp}\mathbf{D}\widetilde{\mathtt{M}}. \tag{2.24}$$

Here, of course, X_c, Y_c and Z_c in $\mathbf{P}_{pp}$ must be still measured in the camera coordinate system. In terms of image and world coordinates, we have

$$\mathbf{m} = \mathbf{T}_{pp}\mathtt{M} + \mathbf{t}_{pp}, \tag{2.25}$$

where

$$\mathbf{T}_{pp} = \frac{1}{Z_c}\begin{bmatrix} \alpha_u(\mathbf{r}_1^T - \frac{X_c}{Z_c}\mathbf{r}_3^T) + \alpha_u \cot\theta(\mathbf{r}_2^T - \frac{Y_c}{Z_c}\mathbf{r}_3^T) \\ \alpha_v(\mathbf{r}_2^T - \frac{Y_c}{Z_c}\mathbf{r}_3^T)/\sin\theta \end{bmatrix}$$

$$\mathbf{t}_{pp} = \frac{1}{Z_c}\begin{bmatrix} \alpha_u(t_1 - \frac{X_c}{Z_c}t_3 + X_c) + \alpha_u \cot\theta(t_2 - \frac{Y_c}{Z_c}t_3 + Y_c) \\ \alpha_v(t_2 - \frac{Y_c}{Z_c}t_3 + Y_c)/\sin\theta \end{bmatrix} + \begin{bmatrix} u_0 \\ v_0 \end{bmatrix}.$$

It is clear that the relation between 3D coordinates and image coordinates is linear under paraperspective projection.

Let us examine the approximation error introduced by the weak perspective projection. For simplicity, we consider a normalized camera with the camera and world coordinate systems aligned. For a point M with coordinates $[X, Y, Z]^T = [X_c + \Delta X, Y_c + \Delta Y, Z_c + \Delta Z]^T$, its perspective projection is given by (2.9), which can be developed as a Taylor series about Z_c:

$$\mathbf{m}_p = \frac{1}{Z_c + \Delta Z}\begin{bmatrix} X_c + \Delta X \\ Y_c + \Delta Y \end{bmatrix} = \frac{1}{Z_c}\left(1 - \frac{\Delta Z}{Z_c} + \left(\frac{\Delta Z}{Z_c}\right)^2 - \mathcal{O}^3\right)\begin{bmatrix} X_c + \Delta X \\ Y_c + \Delta Y \end{bmatrix}$$

where $\mathcal{O}^3$ is used to denote terms of order higher than 2, i.e. $\mathcal{O}^3 = O(\frac{\Delta Z}{Z_c})^3$. The projection under paraperspective, $\mathbf{m}_{pp}$, is given by (2.22). The absolute error in image plane, after some algebra, is given by

$$\mathbf{m}_{error} = \mathbf{m}_p - \mathbf{m}_{pp} = \begin{bmatrix} -\frac{\Delta X \Delta Z}{Z_c^2} + \frac{X}{Z_c}\left(\frac{\Delta Z}{Z_c}\right)^2 - \mathcal{O}^3 \\ -\frac{\Delta Y \Delta Z}{Z_c^2} + \frac{Y}{Z_c}\left(\frac{\Delta Z}{Z_c}\right)^2 - \mathcal{O}^3 \end{bmatrix}.$$

As can be seen, the approximation error is of second order under paraperspective projection, and we say that it is the first-order approximation of the full perspective projection. The weak perspective projection, as was seen in the last subsection, is only the zero-order approximation. Figure 2.8 illustrates well the difference between all these projection models. Similar analysis is presented in [66] for pose determination.

An even better approximation is called *orthoperspective*. The difference from the paraperspective projection consists in replacing the average depth plane, which is parallel to the image plane, by another plane that is perpendicular to the central projecting ray (which connects the optical center C of the camera and the centroid G of the object (see Fig. 2.9 for a pictorial description). It can be understood as the result of the following sequence of operations [2]:

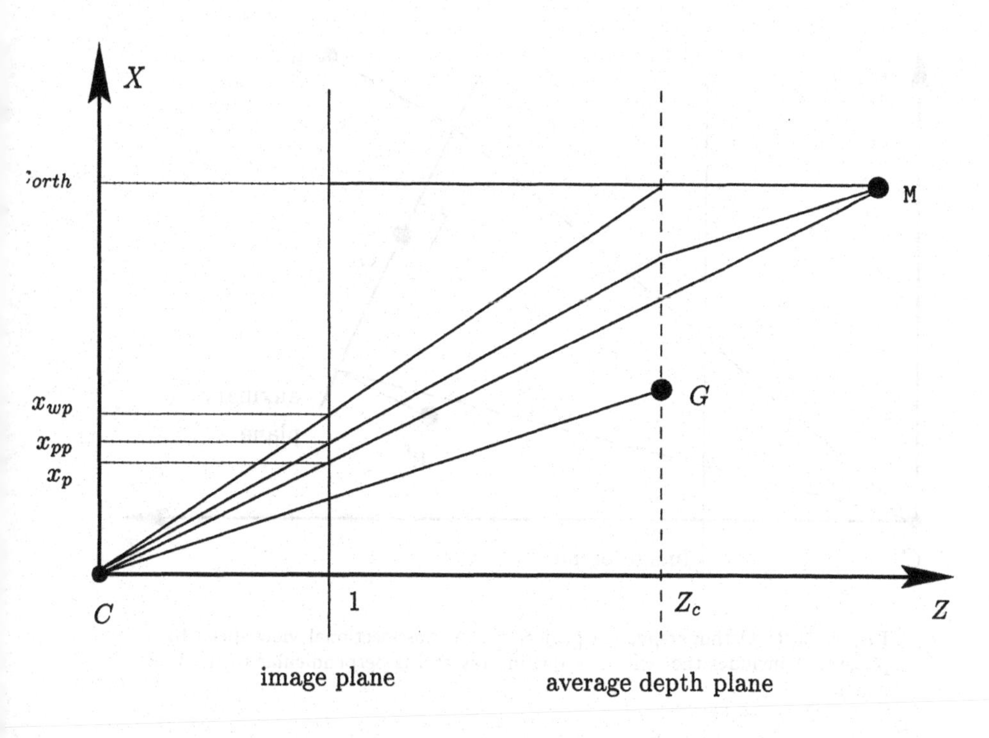

Figure 2.8 Comparison of different perspective approximations. Cross-sectional view sliced by a plane that includes the central projecting ray and is perpendicular to the Y-Z plane. The image plane is the line $Z = 1$. For x_p (full perspective), projection is along the ray connecting the world point M to the optical center C. For x_{orth} (orthographic), projection is perpendicular to the image. For x_{wp} (weak perspective), M is first orthographically projected onto the average depth plane, and then projected perspectively onto the image plane. For x_{pp} (paraperspective), M is first projected, parallel to the central projecting ray, onto the average depth plane, and then projected perspectively onto the image plane. (adapted from [134])

- a virtual rotation of the camera around C to make the central projecting ray CG coincide with the optical axis,
- a weak perspective projection of the object, and
- an inverse camera rotation to bring the camera back to its original position.

The orthoperspective projection involves more complicated formula, and will not be discussed anymore in this book.

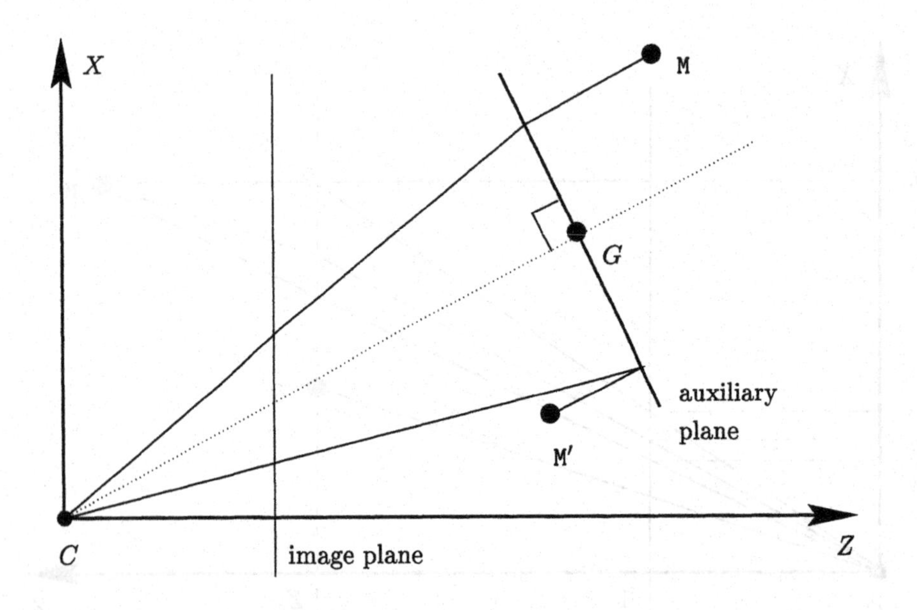

Figure 2.9 Orthoperspective projection. A cross-sectional view sliced by a plane that includes the central projecting ray and is perpendicular to the Y-Z plane

2.2.3 Affine Cameras

If we examine the camera projection matrices for orthographic, weak perspective, and paraperspective projections (see (2.19) and (2.23)), we find that they all have the same form:

$$\mathbf{P}_A = \begin{bmatrix} P_{11} & P_{12} & P_{13} & P_{14} \\ P_{21} & P_{22} & P_{23} & P_{24} \\ 0 & 0 & 0 & P_{34} \end{bmatrix} . \tag{2.26}$$

Depending on different projection models, some constraints exist on the elements of matrix $\mathbf{P}_A$ except P_{31}, P_{32}, and P_{33}, which are equal to 0. If we ignore the constraints on the matrix elements, $\mathbf{P}_A$ becomes the so-called *affine camera*, introduced by Mundy and Zisserman [105]. Unlike the general perspective projection matrix (2.13), an affine camera, which is also defined up to a scale factor, has only eight degrees of freedom. It can be determined from four points in general position.

In terms of image and world coordinates, the affine camera is written as

$$\mathbf{m} = \mathbf{T}_A \mathtt{M} + \mathbf{t}_A , \tag{2.27}$$

where $\mathbf{T}_A$ is a 2×3 matrix with elements $T_{ij} = P_{ij}/P_{34}$, and $\mathbf{t}_A$ is a 2D vector $[P_{14}/P_{34},\ P_{24}/P_{34}]^T$.

A key property of the affine camera is that it preserves parallelism [134]: lines that are parallel in the world remain parallel in the image, which is not the case with the perspective camera. The proof is simple. Consider two parametric parallel lines $\mathtt{M}_1(\lambda) = \mathtt{M}_a + \lambda\mathbf{u}$ and $\mathtt{M}_2(\mu) = \mathtt{M}_b + \mu\mathbf{u}$, where $\mathbf{u}$ is the 3D direction vector. They project onto the image as lines

$$\begin{aligned} \mathbf{m}_1(\lambda) &= (\mathbf{T}_A\mathtt{M}_a + \mathbf{t}_A) + \lambda\mathbf{T}_A\mathbf{u} \,, \\ \mathbf{m}_2(\mu) &= (\mathbf{T}_A\mathtt{M}_b + \mathbf{t}_A) + \mu\mathbf{T}_A\mathbf{u} \,. \end{aligned}$$

They are clearly parallel to the 2D direction vector $\mathbf{T}_A\mathbf{u}$.

Another attractive property of the affine camera is that the centroid of a set of 3D points and that of their image points correspond to each other. This is not the case for full perspective projections. Let $\mathtt{M}_i$ $(i = 1, \ldots, n)$ be the set of 3D points and $\mathbf{m}_i$, their corresponding image points. From (2.27), the centroid of the image points is given by

$$\bar{\mathbf{m}} = \frac{1}{n}\sum_{i=1}^{n}\mathbf{m}_i = \frac{1}{n}\sum_{i=1}^{n}(\mathbf{T}_A\mathtt{M}_i + \mathbf{t}_A) = \mathbf{T}_A\bar{\mathtt{M}} + \mathbf{t}_A \,,$$

where $\bar{\mathtt{M}}$ is the centroid of the 3D points. Thus, the 3D point centroid is projected by an affine camera to the image point centroid. It follows that, if 3D points are expressed with respect to their centroid, i.e. $\widehat{\mathtt{M}}_i = \mathtt{M}_i - \bar{\mathtt{M}}$, and if 2D points are also expressed with respect to their centroid, i.e. $\widehat{\mathbf{m}}_i = \mathbf{m}_i - \bar{\mathbf{m}}$, then we have

$$\widehat{\mathbf{m}}_i = \mathbf{T}_A\widehat{\mathtt{M}}_i \,.$$

Therefore, an affine camera has only 6 degrees of freedom (the 6 parameters of $\mathbf{T}_A$).

While the exact projection performed by an affine camera is not very clear, it is clearly the generalization of the orthographic, weak perspective, or paraperspective models, as can be seen from (2.26) or (2.27). The generalization can be understood in two ways (please refer to (2.13)):

- Some *non-rigid* deformation of the object is permitted. Actually, if we multiply $\mathbf{P}_A$ from the right by any 3D affine transformation of the object

$$\mathbf{D} = \begin{bmatrix} \mathbf{M} & \mathbf{t} \\ \mathbf{0}_3^T & 1 \end{bmatrix} ,$$

where $\mathbf{M}$ is a general 3×3 matrix and $\mathbf{t}$ is a 3-vector, the new projection matrix still has the form (2.26) of an affine camera.

- Camera calibration is unnecessary. Multiply $\mathbf{P}_A$ from the left by any 2D affine transformation of the image

$$\mathbf{A} = \begin{bmatrix} \mathbf{C} & \mathbf{c} \\ \mathbf{0}_2^T & 1 \end{bmatrix},$$

where $\mathbf{C}$ is a general 2×2 matrix and $\mathbf{c}$ is a 2-vector, the new projection matrix still has the form (2.26) of an affine camera.

Without camera calibration, various affine measurements, such as parallelism and ratios of lengths in parallel directions, can be extracted. For certain vision tasks, such properties may be sufficient. Affine cameras are extensively studies in particular by Oxford group [193, 24, 134, 11, 100].

A final note is that the affine camera is an approximation to the real camera, and is only valid when the depth variation of the object of interest is small compared to the distance from the object to the camera.

2.3 EPIPOLAR GEOMETRY UNDER FULL PERSPECTIVE PROJECTION

In this section, we introduce the epipolar geometry under full perspective projection, and describe its concepts and mathematical representations.

2.3.1 Concepts in Epipolar Geometry

The epipolar geometry exists between any two camera systems. Consider the case of two cameras as shown in Fig. 2.10. Let C and C' be the optical centers of the first and second cameras, respectively. Given a point $\mathbf{m}$ in the first image, its corresponding point in the second image is constrained to lie on a line called the *epipolar line* of $\mathbf{m}$, denoted by $\mathbf{l}'_\mathbf{m}$. The line $\mathbf{l}'_\mathbf{m}$ is the intersection of the plane Π, defined by $\mathbf{m}$, C and C' (known as the *epipolar plane*), with the second image plane $\mathcal{I}'$. This is because image point $\mathbf{m}$ may correspond to an arbitrary point on the semi-line $C\mathtt{M}$ ($\mathtt{M}$ may be at infinity) and that the projection of $C\mathtt{M}$ on $\mathcal{I}'$ is the line $\mathbf{l}'_\mathbf{m}$. Furthermore, one observes that all epipolar lines of the

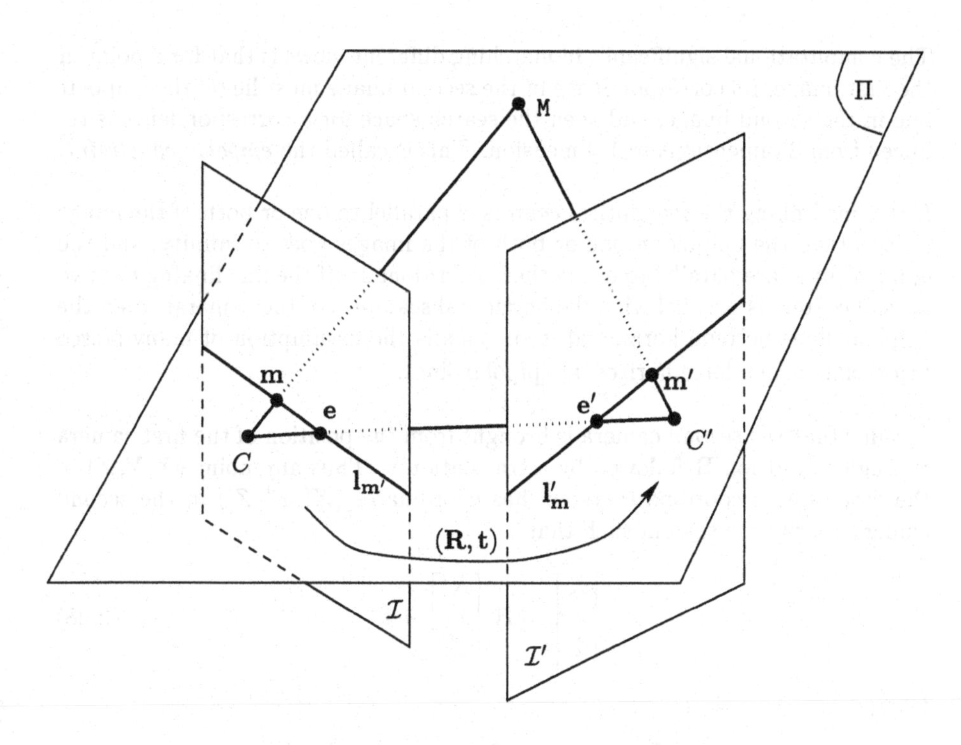

Figure 2.10 The epipolar geometry

points in the first image pass through a common point $\mathbf{e}'$, which is called the *epipole*. Epipole $\mathbf{e}'$ is the intersection of the line CC' with the image plane $\mathcal{I}'$. This can be easily understood as follows. For each point $\mathbf{m}_k$ in the first image $\mathcal{I}$, its epipolar line $\mathbf{l}'_{\mathbf{m}_k}$ in $\mathcal{I}'$ is the intersection of the plane Π^k, defined by $\mathbf{m}_k$, C and C', with image plane $\mathcal{I}'$. All epipolar planes Π^k thus form a pencil of planes containing the line CC'. They must intersect $\mathcal{I}'$ at a common point, which is $\mathbf{e}'$. Finally, one can easily see the symmetry of the epipolar geometry. The corresponding point in the first image of each point $\mathbf{m}'_k$ lying on $\mathbf{l}'_{\mathbf{m}_k}$ must lie on the epipolar line $\mathbf{l}_{\mathbf{m}'_k}$, which is the intersection of the same plane Π^k with the first image plane $\mathcal{I}$. All epipolar lines form a pencil containing the epipole $\mathbf{e}$, which is the intersection of the line CC' with the image plane $\mathcal{I}$. The symmetry leads to the following observation. If $\mathbf{m}$ (a point in $\mathcal{I}$) and $\mathbf{m}'$ (a point in $\mathcal{I}'$) correspond to a single physical point $\mathtt{M}$ in space, then $\mathbf{m}$, $\mathbf{m}'$, C and C' must lie in a single plane. This is the well-known *co-planarity constraint* in solving motion and structure from motion problems when the intrinsic parameters of the cameras are known [90].

The computational significance in matching different views is that for a point in the first image, its correspondence in the second image must lie on the epipolar line in the second image, and then the search space for a correspondence is reduced from 2 dimensions to 1 dimension. This is called the *epipolar constraint*.

If the line linking the two optical centers is parallel to one or both of the image planes, then the epipole in one or both of the images goes to infinity, and the epipolar lines are parallel to each other. Additionally, if the line linking the two optical centers is parallel with the horizontal scanlines of the cameras, then the epipolar lines become horizontal, too. This is the assumption of many stereo algorithms which have horizontal epipolar lines.

Assume that the second camera is brought from the position of the first camera through a rotation $\mathbf{R}$ followed by a translation $\mathbf{t}$. Thus any point (X, Y, Z) in the first camera coordinate system has coordinates (X', Y', Z') in the second camera coordinate system such that

$$\begin{bmatrix} X \\ Y \\ Z \end{bmatrix} = \mathbf{R} \begin{bmatrix} X' \\ Y' \\ Z' \end{bmatrix} + \mathbf{t} \tag{2.28}$$

where

$$\mathbf{R} = \begin{bmatrix} r_{11} & r_{12} & r_{13} \\ r_{21} & r_{22} & r_{23} \\ r_{31} & r_{32} & r_{33} \end{bmatrix} \quad \text{and} \quad \mathbf{t} = \begin{bmatrix} t_X \\ t_Y \\ t_Z \end{bmatrix} .$$

$\mathbf{R}$ has 9 components but there are only 3 degrees of freedom. There are 6 constraints on $\mathbf{R}$. Indeed, a rotation matrix $\mathbf{R}$ must satisfy

$$\mathbf{R}\mathbf{R}^T = \mathbf{I} , \tag{2.29}$$

and

$$\det(\mathbf{R}) = 1 . \tag{2.30}$$

See e.g. [190] for more details on the different representations of the rotation and its properties.

In the following, we first derive the epipolar equation with the normalized image coordinates, then extend it to include the pixel image coordinates, and finally formulate in terms of camera perspective projection matrices.

2.3.2 Working with Normalized Image Coordinates

The two images of a space point $\mathtt{X} = [X, Y, Z]^T$ are $[x, y, 1]^T$ and $[x', y', 1]^T$ in the first and second normalized images, respectively. They are denoted by $\tilde{\mathbf{x}}$ and $\tilde{\mathbf{x}}'$ with the notation introduced in Sect. 2.1.2. Let $\mathtt{X}' = [X', Y', Z']^T$ be the coordinates of the same space point in the second camera coordinate system. From the pinhole model (2.9), we have

$$\begin{aligned} \tilde{\mathbf{x}} &= \mathtt{X}/Z \,, \\ \tilde{\mathbf{x}}' &= \mathtt{X}'/Z' \,. \end{aligned}$$

Eliminating the structure parameters $\mathtt{X}$ and $\mathtt{X}'$ using (2.28), we obtain

$$\tilde{\mathbf{x}} = \frac{1}{Z}(Z'\mathbf{R}\tilde{\mathbf{x}}' + \mathbf{t}) \,,$$

which contains still two unknown structure parameters Z and Z'. The cross product of the above equation with vector $\mathbf{t}$ yields

$$\mathbf{t} \times \tilde{\mathbf{x}} = \frac{Z'}{Z}\mathbf{t} \times \mathbf{R}\tilde{\mathbf{x}}' \,.$$

Its dot product (or inner product) with $\tilde{\mathbf{x}}$ gives

$$\tilde{\mathbf{x}}^T\mathbf{t} \times (\mathbf{R}\tilde{\mathbf{x}}') = 0 \,. \tag{2.31}$$

Here, the quantity Z'/Z has been removed.

Equation (2.31) is very important in solving motion and structure from motion. Geometrically, it is very clear. The three vectors CC', $C\tilde{\mathbf{x}}$, and $C'\tilde{\mathbf{x}}'$ are coplanar. When expressed in the first camera coordinate system, they are equal to $\mathbf{t}$, $\tilde{\mathbf{x}}$, and $\mathbf{R}\tilde{\mathbf{x}}'$, respectively. The co-planarity of the three vectors implies that their mixed product should be equal to 0, which gives (2.31).

Let us define a mapping $[\cdot]_\times$ from a 3D vector to a 3×3 *antisymmetric matrix* (also called *skew symmetric matrix*:

$$\begin{bmatrix} x_1 \\ x_2 \\ x_3 \end{bmatrix}_\times = \begin{bmatrix} 0 & -x_3 & x_2 \\ x_3 & 0 & -x_1 \\ -x_2 & x_1 & 0 \end{bmatrix} \,. \tag{2.32}$$

It is clear that

$$[\mathbf{t}]_\times = -[\mathbf{t}]_\times^T \,. \tag{2.33}$$

Using this mapping we can express the cross product of two vectors by the matrix multiplication of a 3×3 matrix and a 3-vector: $\mathbf{t} \times \widetilde{\mathbf{x}} = [\mathbf{t}]_\times \widetilde{\mathbf{x}}$, $\forall \widetilde{\mathbf{x}}$. Equation (2.31) can then be rewritten as

$$\widetilde{\mathbf{x}}^T \mathbf{E} \widetilde{\mathbf{x}}' = 0 , \tag{2.34}$$

where

$$\mathbf{E} = [\mathbf{t}]_\times \mathbf{R} . \tag{2.35}$$

We call this equation the *epipolar equation.*

Matrix $\mathbf{E}$ is known under tha name of the *Essential Matrix*. It was first proposed by Longuet-Higgins (1981) for structure-from-motion. The essential matrix is determined completely by the rotation and translation between the two cameras. Because $[\mathbf{t}]_\times$ is antisymmetric, we have $\det([\mathbf{t}]_\times) = 0$. Thus we have

$$\det(\mathbf{E}) = \det([\mathbf{t}]_\times) \det(\mathbf{R}) = 0 . \tag{2.36}$$

For more properties of the Essential Matrix, see [98, 41].

Before gaining an insight of (2.34), we recall how to represent a line in a plane. Any line can be described by an equation of the form

$$ax + by + c = 0 . \tag{2.37}$$

Thus the line can be represented by a 3-vector $\mathbf{l} = [a, b, c]^T$ such that a point $\widetilde{\mathbf{x}} = [x, y, 1]^T$ on it must satisfy

$$\widetilde{\mathbf{x}}^T \mathbf{l} = 0 . \tag{2.38}$$

Of course, the 3-vector $\mathbf{l}$ is only defined up to a scale factor. Multiplying $\mathbf{l}$ by any non-zero scalar λ gives $\lambda \mathbf{l}$, which describes exactly the same line. If a line goes through two given points $\widetilde{\mathbf{x}}_1$ and $\widetilde{\mathbf{x}}_2$, we have

$$\widetilde{\mathbf{x}}_1^T \mathbf{l} = 0 \quad \text{and} \quad \widetilde{\mathbf{x}}_2^T \mathbf{l} = 0 ,$$

and it is easy to see that the line is represented by

$$\mathbf{l} = \widetilde{\mathbf{x}}_1 \times \widetilde{\mathbf{x}}_2 , \tag{2.39}$$

i.e. the cross product of the two point vectors.

For point $\widetilde{\mathbf{x}}' = [x', y', 1]^T$ in the second image, its corresponding point $\mathtt{X}'$ in space must be on the semi-line $C'\mathtt{X}'_\infty$ passing through $\widetilde{\mathbf{x}}'$, where $\mathtt{X}'_\infty$ is a point at infinity. From the pinhole model, point $\mathtt{X}'$ can be represented as

$$\mathtt{X}' = \lambda\widetilde{\mathbf{x}}' = \lambda \begin{bmatrix} x' \\ y' \\ 1 \end{bmatrix}, \quad \lambda \in (0, \infty) .$$

This is in fact the parametric representation of the semi-line $C'\mathtt{X}'_\infty$. If we express this point in the coordinate system of the first camera, we have

$$\mathtt{X} = \mathbf{R}\mathtt{X}' + \mathbf{t} = \lambda\mathbf{R}\widetilde{\mathbf{x}}' + \mathbf{t}\,, \quad \lambda \in (0, \infty) .$$

The projection of the semi-line $C'\mathtt{X}'_\infty$ on the first camera is still a line, denoted by $\mathbf{l}_{\mathbf{x}'}$, on which the corresponding point in the first image of point $\mathbf{x}$ must lie. The line $\mathbf{l}_{\mathbf{x}'}$ is known as the *epipolar line* of $\mathbf{x}'$. The epipolar line can be defined by two points. The first point can be obtained by projecting $\mathtt{X}$ with $\lambda = 0$, which gives $\widetilde{\mathbf{e}} = \frac{1}{t_Z}\mathbf{t}$, where t_Z is the Z-component of the translation vector $\mathbf{t}$. This is in fact the projection of the optical center C of the second camera on the first camera, and is called the *epipole*epipole in the first image. The second point can be obtained by projecting $\mathtt{X}$ with $\lambda = \infty$, which gives $\widetilde{\mathbf{x}}_\infty = \frac{1}{\mathbf{r}_3^T\widetilde{\mathbf{x}}'}\mathbf{R}\widetilde{\mathbf{x}}'$, where $\mathbf{r}_3$ is the third row of the rotation matrix $\mathbf{R}$. As described in the last paragraph, the epipolar line $\mathbf{l}_{\mathbf{x}'}$ is represented by

$$\mathbf{l}_{\mathbf{x}'} = \widetilde{\mathbf{e}} \times \widetilde{\mathbf{x}}_\infty = \mathbf{t} \times \mathbf{R}\widetilde{\mathbf{x}}' = \mathbf{E}\widetilde{\mathbf{x}}' . \tag{2.40}$$

Here we have multiplied the original vector by t_Z and $\mathbf{r}_3^T\widetilde{\mathbf{x}}'$ because, as we said, a 3-vector for a line is only defined up to a scalar factor.

If now we reverse the role of the two camera, we find that the epipolar geometry is symmetric for the two cameras. For a given point $\mathbf{x}$ in the first image, its corresponding epipolar line in the second image is

$$\mathbf{l}'_{\mathbf{x}} = \mathbf{E}^T\widetilde{\mathbf{x}} .$$

It is seen that the transpose of matrix $\mathbf{E}$, $\mathbf{E}^T$, defines the epipolar lines in the second image.

From the above discussion, (2.34) says nothing more than that point $\mathbf{x}$ is on the epipolar line $\mathbf{l}_{\mathbf{x}'}$, i.e.

$$\widetilde{\mathbf{x}}^T\mathbf{l}_{\mathbf{x}'} = 0 \quad \text{with } \mathbf{l}_{\mathbf{x}'} = \mathbf{E}\widetilde{\mathbf{x}}' ,$$

or that point $\widetilde{\mathbf{x}}'$ is on the epipolar line $\mathbf{l}'_{\mathbf{x}}$, i.e.

$$\mathbf{l}'^T_{\mathbf{x}}\widetilde{\mathbf{x}}' = 0 \quad \text{with } \mathbf{l}'_{\mathbf{x}} = \mathbf{E}^T\widetilde{\mathbf{x}} .$$

The epipoles are intersections of all epipolar lines. That is, epipoles satisfy all the epipolar line equations. Let the normalized coordinates of the epipole in the first image be **e**. Then, from (2.34), **e** satisfies

$$\tilde{\mathbf{e}}^T \mathbf{E} \tilde{\mathbf{x}}' = 0 \tag{2.41}$$

regardless of $\mathbf{x}'$. This means,

$$\tilde{\mathbf{e}}^T \mathbf{E} = \mathbf{0}^T \tag{2.42}$$

at anytime. That is,

$$\tilde{\mathbf{e}}^T \mathbf{E} = \tilde{\mathbf{e}}^T [\mathbf{t}]_\times \mathbf{R} = \mathbf{0}^T . \tag{2.43}$$

Since **R** is an orthonomal matrix, we have

$$\tilde{\mathbf{e}}^T [\mathbf{t}]_\times = \mathbf{0} . \tag{2.44}$$

The solution is

$$\tilde{\mathbf{e}} = [\frac{t_X}{t_Z}, \frac{t_Y}{t_Z}, 1]^T . \tag{2.45}$$

This is exactly the projection of the optical center of the second camera onto the first image plane, as we have already explained geometrically. For the second image, we have

$$\mathbf{E}\tilde{\mathbf{e}}' = [\mathbf{t}]_\times \mathbf{R} \tilde{\mathbf{e}}' = \mathbf{0} . \tag{2.46}$$

Thus,

$$\mathbf{R}\tilde{\mathbf{e}}' = \mathbf{t} . \tag{2.47}$$

The position of the epipole can then be determined as

$$\tilde{\mathbf{e}}' = \frac{1}{\mathbf{r}_3' \cdot \mathbf{t}} \mathbf{R}^T \mathbf{t} = \left[\frac{\mathbf{r}_1' \cdot \mathbf{t}}{\mathbf{r}_3' \cdot \mathbf{t}}, \frac{\mathbf{r}_1' \cdot \mathbf{t}}{\mathbf{r}_3' \cdot \mathbf{t}}, 1 \right]^T , \tag{2.48}$$

where $\mathbf{r}_i' = [r_{1i}, r_{2i}, r_{3i}]^T, i = 1, 2, 3$ are the column vectors of **R**.

For the epipole in the first image to go to infinity, we must have

$$t_Z = 0 . \tag{2.49}$$

This means that the translation of the camera has to be within the focal plane of the first camera. For both epipoles in the two images to go to infinity, then

$$\mathbf{r}_3' \cdot \mathbf{t} = 0 . \tag{2.50}$$

This implies that the optical center of the first camera lies in the focal plane of the second camera. Furthermore, if we require the two focal planes to be a single one, then besides $t_Z = 0$, r_{13} and r_{23} have to be 0. Since $\|\mathbf{r}'_i\| = 1$, we have

$$r_{33} = 1 \,. \tag{2.51}$$

Thus $\mathbf{R}$ is in the form of

$$\mathbf{R} = \begin{bmatrix} \cos\theta & \sin\theta & 0 \\ -\sin\theta & \cos\theta & 0 \\ 0 & 0 & 1 \end{bmatrix} . \tag{2.52}$$

This means that the rotation can be only around the optical axis of the cameras.

Substituting (2.49) and (2.52) for (2.35), we have

$$\mathbf{E} = [\mathbf{t}]_\times \mathbf{R} = \begin{bmatrix} 0 & 0 & t_Y \\ 0 & 0 & -t_X \\ -t_Y\cos\theta - t_X\sin\theta & -t_Y\sin\theta + t_X\cos\theta & 0 \end{bmatrix} . \tag{2.53}$$

If we expand the above equation, it is clear that there is only linear terms of the image coordinates, rather than quadric terms in the original form. That means the epipolar lines are parallel in both images and the orientations are independent of the image points.

2.3.3 Working with Pixel Image Coordinates

If two points $\mathbf{m}$ and $\mathbf{m}'$, expressed in pixel image coordinates in the first and second camera, are in correspondence, they must satisfy the following equation

$$\widetilde{\mathbf{m}}^T \mathbf{F} \widetilde{\mathbf{m}}' = 0 \,, \tag{2.54}$$

where

$$\mathbf{F} = \mathbf{A}^{-T} \mathbf{E} \mathbf{A}'^{-1} \,, \tag{2.55}$$

and $\mathbf{A}$ and $\mathbf{A}'$ are respectively the intrinsic matrix of the first and second camera. Equation (2.54) is easily verified. From (2.12), the normalized coordinates $\mathbf{x}$ are related to the pixel coordinates $\mathbf{m}$ by $\widetilde{\mathbf{x}} = \mathbf{A}^{-1}\widetilde{\mathbf{m}}$. Plunging it into (2.34) yields (2.54). This is a fundamental constraint for two pixels to be in correspondence between two images.

As with the normalized image coordinates, the above equation (2.54) can also be derived from the pinhole model (2.14). Without loss of generality, we assume that the world coordinate system coincides with the second camera coordinate system. From (2.13), we have

$$\begin{aligned} s\tilde{\mathbf{m}} &= \mathbf{A}\,[\mathbf{R}\ \mathbf{t}] \begin{bmatrix} \mathtt{M}' \\ 1 \end{bmatrix} \\ s'\tilde{\mathbf{m}}' &= \mathbf{A}'\,[\mathbf{I}\ \mathbf{0}] \begin{bmatrix} \mathtt{M}' \\ 1 \end{bmatrix} . \end{aligned}$$

Eliminating $\mathtt{M}'$, s and s' in the above two equations, we obtain, not at all surprising, equation (2.54).

The 3×3 matrix $\mathbf{F}$ is called the *fundamental matrix*. With this fundamental matrix, we can express the epipolar equation for two unnormalized images in the same form as for the normalized images. Since $\det(\mathbf{E}) = 0$,

$$\det(\mathbf{F}) = 0 . \tag{2.56}$$

$\mathbf{F}$ is of rank 2. Besides, it is only defined up to a scalar factor. If $\mathbf{F}$ is multiplied by an arbitray scalar, equation (2.54) still holds. Therefore, a fundamental matrix has only seven degrees of freedom. There are only 7 independent parameters among the 9 elements of the fundamental matrix.

We now derive the expression of the epipoles. The epipole $\mathbf{e}$ in the first image is the projection of the optical center C' of the second camera. Since $C' = \mathbf{0}$, from the pinhole model, we have

$$s_e\tilde{\mathbf{e}} = \mathbf{A}\,[\mathbf{R}\ \mathbf{t}] \begin{bmatrix} 0 \\ 1 \end{bmatrix} = \mathbf{A}\mathbf{t} , \tag{2.57}$$

where s_e is a scale factor. Thus, the epipole $\tilde{\mathbf{e}}$ is equal to $\mathbf{A}\mathbf{t}$ divided by its third element. Similarly, the epipole $\mathbf{e}'$ in the second image is the projection of the optical center C of the first camera. The optical center is determined, based on the discussion in Sect. 2.1.4, by

$$\mathbf{A}\,[\mathbf{R}\ \mathbf{t}] \begin{bmatrix} C \\ 1 \end{bmatrix} = \mathbf{0} ,$$

which gives

$$C = -\mathbf{R}^{-1}\mathbf{t} .$$

Therefore, the epipole $\mathbf{e}'$ is given by

$$s'_e\tilde{\mathbf{e}}' = \mathbf{A}'\,[\mathbf{I}\ \mathbf{0}] \begin{bmatrix} C \\ 1 \end{bmatrix} = -\mathbf{A}'\mathbf{R}^{-1}\mathbf{t} , \tag{2.58}$$

i.e. it is equal to $-\mathbf{A}'\mathbf{R}^{-1}\mathbf{t}$ divided by the third element of the vector.

Now we show, for a given point $\mathbf{m}'$ in the second image, how to compute the corresponding epipolar line $\mathbf{l}_{\mathbf{m}'}$ in the first image. It is determined by two points. Besides the epipole $\mathbf{e}$, we need another point. This point can be the projection of any point $\widehat{\mathtt{M}}'$ on the optical ray $\langle C', \widetilde{\mathbf{m}}' \rangle$. In particular, we can choose $\widehat{\mathtt{M}}'$ such that

$$\widetilde{\mathbf{m}}' = \mathbf{A}' \; [\mathbf{I} \; \mathbf{0}] \begin{bmatrix} \widehat{\mathtt{M}}' \\ 1 \end{bmatrix} = \mathbf{A}'\widehat{\mathtt{M}}' ,$$

i.e. the scale factor is equal to 1. This gives $\widehat{\mathtt{M}}' = \mathbf{A}'^{-1}\widetilde{\mathbf{m}}'$. The projection of this point in the first camera, $\widehat{\mathbf{m}}$, is given by

$$s_m \widetilde{\widehat{\mathbf{m}}} = \mathbf{A} \; [\mathbf{R} \; \mathbf{t}] \begin{bmatrix} \widehat{\mathtt{M}}' \\ 1 \end{bmatrix} = \mathbf{A}(\mathbf{R}\mathbf{A}'^{-1}\widetilde{\mathbf{m}}' + \mathbf{t}) ,$$

where s_m is the scale factor. As already described in (2.37) on page 30, a line can be represented by a 3-vector defined *up to a scale factor*. According to (2.39), the epipolar line $\mathbf{l}_{\mathbf{m}'}$ is given by

$$\begin{aligned} \mathbf{l}_{\mathbf{m}'} &= s_e s_m \widetilde{\mathbf{e}} \times \widetilde{\widehat{\mathbf{m}}} \\ &= (\mathbf{A}\mathbf{t}) \times [\mathbf{A}(\mathbf{R}\mathbf{A}'^{-1}\widetilde{\mathbf{m}}' + \mathbf{t})] \\ &= (\mathbf{A}\mathbf{t}) \times (\mathbf{A}\mathbf{R}\mathbf{A}'^{-1}\widetilde{\mathbf{m}}') . \end{aligned}$$

It can be shown that $(\mathbf{A}\mathbf{x}) \times (\mathbf{A}\mathbf{y}) = \det(\mathbf{A})\mathbf{A}^{-T}(\mathbf{x} \times \mathbf{y})$ for all vectors $\mathbf{x}$ and $\mathbf{y}$ if matrix $\mathbf{A}$ is invertible. Therefore, we have

$$\mathbf{l}_{\mathbf{m}'} = \mathbf{A}^{-T}[\mathbf{t} \times (\mathbf{R}\mathbf{A}'^{-1}\mathbf{m}')] = \mathbf{F}\widetilde{\mathbf{m}}'$$

with $\mathbf{F}$ given by (2.55). Then any point $\mathbf{m}$ on the epipolar line of $\mathbf{m}'$ satisfies $\widetilde{\mathbf{m}}^T\mathbf{F}\widetilde{\mathbf{m}}' = 0$, and this is exactly (2.61). Therefore, we obtain geometrically the same equation.

Now we reverse the role of the two images, and consider the epipolar line $\mathbf{l}'_{\mathbf{m}}$ in the second image for a given point $\mathbf{m}$ in the first image. Line $\mathbf{l}'_{\mathbf{m}}$ goes through the epipole $\mathbf{e}'$. We choose the projection of a point $\widehat{\mathtt{M}}'$ on the optical ray $\langle C, \widetilde{\mathbf{m}} \rangle$ such that

$$\widetilde{\mathbf{m}} = \mathbf{A} \, [\mathbf{R} \; \mathbf{t}] \begin{bmatrix} \widehat{\mathtt{M}}' \\ 1 \end{bmatrix} ,$$

i.e. the scale factor is chosen to be 1. This gives

$$\widehat{\mathtt{M}}' = (\mathbf{A}\mathbf{R})^{-1}(\widetilde{\mathbf{m}} - \mathbf{A}\mathbf{t}) .$$

Its projection in the second camera gives

$$\begin{aligned} s'_m \widetilde{\widehat{\mathbf{m}}}' &= \mathbf{A}'[\mathbf{I}\ \mathbf{0}] \begin{bmatrix} \widehat{\mathtt{M}}' \\ 1 \end{bmatrix} \\ &= \mathbf{A}'(\mathbf{A}\mathbf{R})^{-1}(\widetilde{\mathbf{m}} - \mathbf{A}\mathbf{t}) \\ &= \mathbf{A}'\mathbf{R}^{-1}\mathbf{A}^{-1}\widetilde{\mathbf{m}} - \mathbf{A}'\mathbf{R}^{-1}\mathbf{t}\,. \end{aligned}$$

The epipolar line $\mathbf{l}'_\mathbf{m}$ is thus represented by

$$\begin{aligned} \mathbf{l}'_\mathbf{m} &= s'_e s'_m \widetilde{\mathbf{e}}' \times \widetilde{\widehat{\mathbf{m}}}' \\ &= -(\mathbf{A}'\mathbf{R}^{-1}\mathbf{t}) \times (\mathbf{A}'\mathbf{R}^{-1}\mathbf{A}^{-1}\widetilde{\mathbf{m}}) \\ &= -(\mathbf{A}'\mathbf{R}^{-1})^{-T}(\mathbf{t} \times \mathbf{A}^{-1}\widetilde{\mathbf{m}}) \\ &= -\mathbf{A}'^{-T}\mathbf{R}^T[\mathbf{t}]_\times \mathbf{A}^{-1}\widetilde{\mathbf{m}} \\ &= \mathbf{F}^T\widetilde{\mathbf{m}}\,. \end{aligned}$$

In the above, we have used the following properties:

- $(\mathbf{A}\mathbf{x}) \times (\mathbf{A}\mathbf{y}) = \det(\mathbf{A})\mathbf{A}^{-T}(\mathbf{x} \times \mathbf{y}),\ \forall \mathbf{x}, \mathbf{y}$ if matrix $\mathbf{A}$ is invertible.
- $(\mathbf{A}\mathbf{B})^{-1} = \mathbf{B}^{-1}\mathbf{A}^{-1}$ if matrices $\mathbf{A}$ and $\mathbf{B}$ are invertible.
- $\mathbf{R}^T = \mathbf{R}^{-1}$ if $\mathbf{R}$ is a rotation matrix.
- $[\mathbf{t}]_\times^T = -[\mathbf{t}]_\times$ if $[\mathbf{t}]_\times$ is an antisymmetric matrix.

It is thus clear that if $\mathbf{F}$ describes epipolar lines in the first image for points given in the second image, then $\mathbf{F}^T$ describes epipolar lines in the second image for points given in the first image. The two images play a symmetric role in the epipolar geometry.

We now compute the epipoles from a different point of view. By definition, all epipolar lines in the first image go through the epipole $\mathbf{e}$. This implies

$$\widetilde{\mathbf{e}}^T\mathbf{F}\widetilde{\mathbf{m}}' = 0, \quad \forall \mathbf{m}'\,,$$

or in vector equation form

$$\mathbf{F}^T\widetilde{\mathbf{e}} = \mathbf{0}\,. \tag{2.59}$$

Pludging (2.55) into the above equation gives

$$\mathbf{A}'^{-T}\mathbf{R}^T[\mathbf{t}]_\times \mathbf{A}^{-1}\widetilde{\mathbf{e}} = \mathbf{0}\,.$$

Because $[\mathbf{t}]_\times \mathbf{t} = \mathbf{0}$, up to a scale factor, we have $s_e \mathbf{A}^{-1}\tilde{\mathbf{e}} = \mathbf{t}$, and thus $s_e \tilde{\mathbf{e}} = \mathbf{A}\mathbf{t}$. This is exactly (2.57). Similarly, for the epipole $\mathbf{e}'$ in the second image, we have

$$\mathbf{F}\tilde{\mathbf{e}}' = 0 \tag{2.60}$$

or

$$\mathbf{A}^{-T}[\mathbf{t}]_\times \mathbf{R}\mathbf{A}'^{-1}\tilde{\mathbf{e}}' = \mathbf{0} \, .$$

This gives $s'_e \mathbf{R}\mathbf{A}'^{-1}\tilde{\mathbf{e}}' = -\mathbf{t}$, or $s'_e \tilde{\mathbf{e}}' = -\mathbf{A}'\mathbf{R}^{-1}\mathbf{t}$. This is exactly (2.58).

2.3.4 Working with Camera Perspective Projection Matrices

In several applications, for example in the case of calibrated stereo, the camera perspective projection matrices are given, and we want to compute the epipolar geometry. Let $\mathbf{P}$ and $\mathbf{P}'$ be the projection matrices of the first and second camera. Furthermore, the 3×4 matrix $\mathbf{P}$ is decomposed as the concatenation of a 3×3 submatrix $\mathbf{B}$ and a 3-vector $\mathbf{b}$, i.e. $\mathbf{P} = [\mathbf{B}\ \mathbf{b}]$. Similarly, $\mathbf{P}' = [\mathbf{B}'\ \mathbf{b}']$.

From the pinhole model (2.14), we have

$$\begin{aligned} s\tilde{\mathbf{m}} &= [\mathbf{B}\ \mathbf{b}] \begin{bmatrix} \mathtt{M}' \\ 1 \end{bmatrix} \\ s'\tilde{\mathbf{m}}' &= [\mathbf{B}'\ \mathbf{b}'] \begin{bmatrix} \mathtt{M}' \\ 1 \end{bmatrix} . \end{aligned}$$

Assume that $\mathbf{B}$ and $\mathbf{B}'$ are invertible, we can compute $\mathtt{M}'$ from each of the above equations:

$$\begin{aligned} \mathtt{M}' &= s\mathbf{B}^{-1}\tilde{\mathbf{m}} - \mathbf{B}^{-1}\mathbf{b} \\ \mathtt{M}' &= s'\mathbf{B}'^{-1}\tilde{\mathbf{m}}' - \mathbf{B}'^{-1}\mathbf{b}' . \end{aligned}$$

The right sides of the above equations must be equal, which gives

$$s\mathbf{B}^{-1}\tilde{\mathbf{m}} = s'\mathbf{B}'^{-1}\tilde{\mathbf{m}}' + \mathbf{B}^{-1}\mathbf{b} - \mathbf{B}'^{-1}\mathbf{b}' .$$

Multiplying both sides by $\mathbf{B}$ gives

$$s\tilde{\mathbf{m}} = s'\mathbf{B}\mathbf{B}'^{-1}\tilde{\mathbf{m}}' + \mathbf{b} - \mathbf{B}\mathbf{B}'^{-1}\mathbf{b}' .$$

Performing a cross product with $\mathbf{b} - \mathbf{B}\mathbf{B}'^{-1}\mathbf{b}'$ yields

$$s(\mathbf{b} - \mathbf{B}\mathbf{B}'^{-1}\mathbf{b}') \times \widetilde{\mathbf{m}} = s'(\mathbf{b} - \mathbf{B}\mathbf{B}'^{-1}\mathbf{b}') \times \mathbf{B}\mathbf{B}'^{-1}\widetilde{\mathbf{m}}' .$$

Eliminating the arbitrary scalars s and s' by multiplying $\widetilde{\mathbf{m}}^T$ from the left (i.e. dot product) gives

$$\widetilde{\mathbf{m}}^T \mathbf{F} \widetilde{\mathbf{m}}' = 0 , \tag{2.61}$$

where

$$\mathbf{F} = [\mathbf{b} - \mathbf{B}\mathbf{B}'^{-1}\mathbf{b}']_\times \mathbf{B}\mathbf{B}'^{-1} . \tag{2.62}$$

We thus obtain the epipolar equation in terms of the perspective projection matrices. Again, it is clear that the roles of $\mathbf{m}$ and $\mathbf{m}'$ are symmetric, and we have $\widetilde{\mathbf{m}}'^T \mathbf{F}^T \widetilde{\mathbf{m}} = 0$.

Now let us show how to compute the epipoles. The epipole $\mathbf{e}$ in the first image is the projection of the optical center C' of the second camera. From (2.17), we have

$$C' = -\mathbf{B}'^{-1}\mathbf{b}' .$$

We thus have

$$s_e \widetilde{\mathbf{e}} = \mathbf{P} \begin{bmatrix} C' \\ 1 \end{bmatrix} = \mathbf{b} - \mathbf{B}\mathbf{B}'^{-1}\mathbf{b}' , \tag{2.63}$$

where s_e is a scale factor. Thus, epipole $\mathbf{e}$ is equal to $(\mathbf{b} - \mathbf{B}\mathbf{B}'^{-1}\mathbf{b}')$ divided by its third element. Similarly, the epipole in the second image, $\mathbf{e}'$, is equal to $(\mathbf{b}' - \mathbf{B}'\mathbf{B}^{-1}\mathbf{b})$ divided by its third element.

Next, we show how, for a given point $\mathbf{m}'$ in the second image, to compute the corresponding epipolar line $\mathbf{l}_{\mathbf{m}'}$ in the first image. The epipolar line must go through the epipole $\mathbf{e}$. We thus need another point to determine it. This point can be the projection of any point $\widehat{\mathtt{M}}'$ on the optical ray $\langle C', \widetilde{\mathbf{m}}' \rangle$. In particular, we can choose $\widehat{\mathtt{M}}'$ such that

$$\widetilde{\mathbf{m}}' = \mathbf{P}' \begin{bmatrix} \widehat{\mathtt{M}}' \\ 1 \end{bmatrix} = \mathbf{B}'\widehat{\mathtt{M}}' + \mathbf{b}' ,$$

i.e. the scale factor is equal to 1. This gives $\widehat{\mathtt{M}}' = \mathbf{B}'^{-1}(\widetilde{\mathbf{m}}' - \mathbf{b}')$. According to the pinhole model, the image $\widehat{\mathbf{m}}$ of this point is given by

$$s_m \widetilde{\widehat{\mathbf{m}}} = \mathbf{P} \begin{bmatrix} \widehat{\mathtt{M}}' \\ 1 \end{bmatrix} = \mathbf{B}\mathbf{B}'^{-1}\widetilde{\mathbf{m}}' + (\mathbf{b} - \mathbf{B}\mathbf{B}'^{-1}\mathbf{b}') ,$$

where s_m is the scale factor. As already described in (2.37) on page 30, a line can be represented by a 3-vector defined *up to a scale factor.* According to (2.39), the epipolar line $\mathbf{l}_{\mathbf{m}'}$ is given by

$$\begin{aligned} \mathbf{l}_{\mathbf{m}'} &= s_e s_m \widetilde{\mathbf{e}} \times \widetilde{\widetilde{\mathbf{m}}} \\ &= (\mathbf{b} - \mathbf{B}\mathbf{B}'^{-1}\mathbf{b}') \times [\mathbf{B}\mathbf{B}'^{-1}\widetilde{\mathbf{m}}' + (\mathbf{b} - \mathbf{B}\mathbf{B}'^{-1}\mathbf{b}')] \\ &= (\mathbf{b} - \mathbf{B}\mathbf{B}'^{-1}\mathbf{b}') \times (\mathbf{B}\mathbf{B}'^{-1}\widetilde{\mathbf{m}}') \,, \end{aligned}$$

or

$$\mathbf{l}_{\mathbf{m}'} = \mathbf{F}\widetilde{\mathbf{m}}' \,, \tag{2.64}$$

where $\mathbf{F}$ is given by (2.62). Then any point $\mathbf{m}$ on the epipolar line of $\mathbf{m}'$ satisfies $\widetilde{\mathbf{m}}^T\mathbf{F}\widetilde{\mathbf{m}}' = 0$, and this is exactly (2.61). Therefore, we obtain geometrically the same equation. Because of symmetry, for a given point $\mathbf{m}$ in the first image, its corresponding epipolar line in the second image is represented by the vector $\mathbf{F}^T\widetilde{\mathbf{m}}$.

Now, we show that if the images are calibrated, then the fundamental matrix $\mathbf{F}$ is reduced to the essential matrix $\mathbf{E}$. Since the images are calibrated, the points $\mathbf{m}$ can be expressed in normalized coordinates, i.e. $\mathbf{m} = \mathbf{x}$. Without loss of generality, the world coordinate system is assumed to coincide with the second camera coordinate system. From (2.13), we have the following camera projection matrices:

$$\mathbf{P} = [\mathbf{R}\ \mathbf{t}] \quad \text{and} \quad \mathbf{P}' = [\mathbf{I}\ \mathbf{0}] \,.$$

This implies that $\mathbf{B} = \mathbf{R}$, $\mathbf{b} = \mathbf{t}$, $\mathbf{B}' = \mathbf{I}$, and $\mathbf{b}' = \mathbf{0}$. Pludging them into (2.62) gives $\mathbf{F} = [\mathbf{t}]_\times\mathbf{R}$, which is exactly the essential matrix (2.35).

In the above derivation of the fundamental matrix, a camera perspective projection matrix $\mathbf{P}$ is decomposed into a 3×3 matrix $\mathbf{B}$ and a 3-vector $\mathbf{b}$, and $\mathbf{B}$ must be invertible. In Sect. 2.4, we provide a more general derivation directly in terms of the camera projection matrices $\mathbf{P}$ and $\mathbf{P}'$.

2.3.5 Fundamental Matrix and Epipolar Transformation

We examine the relationship between the fundamental matrix and the epipolar transformation (i.e. the transformation of the epipoles and the epipolar lines between the two images).

For any point $\mathbf{m}'$ in the second image, its epipolar line $\mathbf{l}_{\mathbf{m}'}$ in the first image is given by $\mathbf{l}_{\mathbf{m}'} = \mathbf{F}\widetilde{\mathbf{m}}'$. It must go through the epipole $\widetilde{\mathbf{e}} = [e_1, e_2, e_3]^T$ and a point $\widetilde{\mathbf{m}} = [u, v, s]^T$, i.e. $\mathbf{l}_{\mathbf{m}'} = \widetilde{\mathbf{e}} \times \widetilde{\mathbf{m}} = \mathbf{F}\widetilde{\mathbf{m}}'$. Here, we use the homogeneous coordinates for the image points. Symmetrically, the epipolar line in the second image $\mathbf{l}'_{\mathbf{m}}$ of point $\mathbf{m}$ is given by $\mathbf{l}'_{\mathbf{m}} = \mathbf{F}^T\widetilde{\mathbf{m}}$ and must go through the epipole $\widetilde{\mathbf{e}}' = [e'_1, e'_2, e'_3]^T$ and a point $\widetilde{\mathbf{m}}' = [u', v', s']^T$, i.e. $\mathbf{l}'_{\mathbf{m}} = \widetilde{\mathbf{e}}' \times \widetilde{\mathbf{m}}' = \mathbf{F}^T\widetilde{\mathbf{m}}$. In other words, the epipole $\mathbf{e}'$ is *on* the epipolar line $\mathbf{l}'_{\mathbf{m}}$ for any point $\mathbf{m}$; that is

$$\widetilde{\mathbf{e}}'^T\mathbf{l}'_{\mathbf{m}} = \widetilde{\mathbf{e}}'^T\mathbf{F}^T\widetilde{\mathbf{m}} = 0 \quad \forall \mathbf{m} \,,$$

which yields

$$\mathbf{F}\widetilde{\mathbf{e}}' = \mathbf{0} \,. \tag{2.65}$$

Let $\mathbf{c}_1$, $\mathbf{c}_2$ and $\mathbf{c}_3$ be the column vectors of $\mathbf{F}$, we have $e'_1\mathbf{c}_1 + e'_2\mathbf{c}_2 + e'_3\mathbf{c}_3 = \mathbf{0}$, thus the rank of $\mathbf{F}$ is at most two. The solution to the epipole $\widetilde{\mathbf{e}}'$ is given by

$$\begin{aligned} e'_1 &= F_{23}F_{12} - F_{22}F_{13} \\ e'_2 &= F_{13}F_{21} - F_{11}F_{23} \\ e'_3 &= F_{22}F_{11} - F_{21}F_{12} \,, \end{aligned} \tag{2.66}$$

up to, of course, a scale factor. Similarly, for the epipole in the first image, we have

$$\mathbf{F}^T\widetilde{\mathbf{e}} = \mathbf{0} \,, \tag{2.67}$$

which gives

$$\begin{aligned} e_1 &= F_{32}F_{21} - F_{22}F_{31} \\ e_2 &= F_{31}F_{12} - F_{11}F_{32} \\ e_3 &= F_{22}F_{11} - F_{21}F_{12} \,, \end{aligned} \tag{2.68}$$

also up to a scale factor.

Now let us examine the relationship between the epipolar lines. Once the epipole is known, an epipolar line can be parameterized by its direction vector. Consider $\mathbf{l}'_{\mathbf{m}} = \widetilde{\mathbf{e}}' \times \widetilde{\mathbf{m}}'$, its direction vector $\mathbf{u}'$ can be parameterized by one parameter τ' such that $\mathbf{u}' = [1, \tau', 0]^T$. A particular point on $\mathbf{l}'_{\mathbf{m}}$ is then given by $\widetilde{\mathbf{m}}' = \widetilde{\mathbf{e}}' + \lambda'\mathbf{u}'$, where λ' is a scalar. Its epipolar line in the first image is given by

$$\mathbf{l}_{\mathbf{m}'} = \mathbf{F}\widetilde{\mathbf{m}}' = \mathbf{F}\widetilde{\mathbf{e}}' + \lambda'\mathbf{F}\mathbf{u}' = \lambda'\mathbf{F}\mathbf{u}' = \lambda' \begin{bmatrix} F_{11} + F_{12}\tau' \\ F_{21} + F_{22}\tau' \\ F_{31} + F_{32}\tau' \end{bmatrix} . \tag{2.69}$$

This line can also be parameterized by its direction vector $\mathbf{u} = [1, \tau, 0]^T$ in the first image, which implies that

$$\mathbf{l}_{\mathbf{m}'} \cong \widetilde{\mathbf{e}} \times (\widetilde{\mathbf{e}} + \lambda \mathbf{u}) = \lambda \widetilde{\mathbf{e}} \times \mathbf{u} = \lambda \begin{bmatrix} -(F_{11}F_{22} - F_{21}F_{12})\tau \\ F_{11}F_{22} - F_{21}F_{12} \\ (F_{32}F_{21} - F_{22}F_{31})\tau - F_{31}F_{12} + F_{11}F_{32} \end{bmatrix}, \tag{2.70}$$

where $\cong$ means "equal" up to a scale factor and λ is a scalar. By requiring that (2.69) and (2.70) represent the same line, we have

$$\tau = \frac{a\tau' + b}{c\tau' + d}, \tag{2.71}$$

where

$$\begin{aligned} a &= F_{12} \\ b &= F_{11} \\ c &= -F_{22} \\ d &= -F_{21}\,. \end{aligned} \tag{2.72}$$

Writing in matrix form gives

$$\rho \begin{bmatrix} \tau \\ 1 \end{bmatrix} = \begin{bmatrix} a & b \\ c & d \end{bmatrix} \begin{bmatrix} \tau' \\ 1 \end{bmatrix},$$

where ρ is a scale factor. This relation is known as the *homography* between τ and τ', and we say that *there is a homography between the epipolar lines in the first image and those in the second image.* The above parameterization is, of course, only valid for the epipolar lines having the nonzero first element in the direction vector. If the first element is equal to zero, we should parameterize the direction vector as $[\tau, 1, 0]^T$, and similar results can be obtained.

At this point, we can see that the epipolar transformation is defined by the coordinates $\widetilde{\mathbf{e}} = [e_1, e_2, e_3]^T$ and $\widetilde{\mathbf{e}}' = [e'_1, e'_2, e'_3]^T$ of the epipoles, and the four parameters a, b, c and d of the homography between the two pencils of the epipolar lines. The coordinates of each epipole are defined up to a scale factor, and the parameters of the homography, as can be seen in (2.71), are also defined up to a scale factor. Thus, we have in total 7 free parameters. This is exactly the number of parameters of the fundamental matrix.

If we have identified the parameters of the epipolar transformation, i.e. the coordinates of the two epipoles and the coefficients of the homography, then we can construct the fundamental matrix, from (2.66), (2.68) and (2.72), as

$$\begin{aligned}
F_{11} &= be_3e'_3 \\
F_{12} &= ae_3e'_3 \\
F_{13} &= -(ae'_2 + be'_1)e_3 \\
F_{21} &= -de_3e'_3 \\
F_{22} &= -ce_3e'_3 \\
F_{23} &= (ce'_2 + de'_1)e_3 \\
F_{31} &= (de_2 - be_1)e'_3 \\
F_{32} &= (ce_2 - ae_1)e'_3 \\
F_{33} &= -(ce'_2 + de'_1)e_2 + (ae'_2 + be'_1)e_1 \,.
\end{aligned} \tag{2.73}$$

The determinant $ad - bc$ of the homography is equal to the determinant of the first 2×2 submatrix of $\mathbf{F}$, $F_{11}F_{22} - F_{12}F_{21}$, which is zero when the epipoles are at infinity.

2.4 A GENERAL FORM OF EPIPOLAR EQUATION FOR ANY PROJECTION MODEL

In this section we will derive a general form of epipolar equation which does not assume any particular projection model.

2.4.1 Intersecting Two Optical Rays

The projections for the first and second cameras are represented respectively as

$$s\tilde{\mathbf{m}} = \mathbf{P}\tilde{\mathtt{M}} \,, \tag{2.74}$$

and

$$s'\tilde{\mathbf{m}}' = \mathbf{P}'\tilde{\mathtt{M}}' \,, \tag{2.75}$$

where $\tilde{\mathbf{m}}$ and $\tilde{\mathbf{m}}'$ are augmented image coordinates, and $\tilde{\mathtt{M}}$ and $\tilde{\mathtt{M}}'$ are augmented space coordinates of a single point in the two camera coordinate systems. Here both projection matrices *do not include the extrinsic parameters*.

The same point in the two camera coordinate systems can be related by

$$\tilde{\mathtt{M}} = \mathbf{D}\tilde{\mathtt{M}}' \,, \tag{2.76}$$

where

$$\mathbf{D} = \begin{bmatrix} \mathbf{R} & \mathbf{t} \\ \mathbf{0}_3^T & 1 \end{bmatrix}$$

is the Euclidean transformation matrix compactly representing both rotation and translation. Now substituting (2.76) for (2.74), we have

$$s\tilde{\mathbf{m}} = \mathbf{P}\mathbf{D}\tilde{\mathtt{M}}' . \tag{2.77}$$

For an image point $\tilde{\mathbf{m}}'$, (2.75) actually defines an optical ray, on which every space point $\tilde{\mathtt{M}}'$ projects onto the second image at $\tilde{\mathbf{m}}'$. This optical ray can be written in pamametric form as

$$\tilde{\mathtt{M}}' = s'\mathbf{P}'^{+}\tilde{\mathbf{m}}' + \mathbf{p}'^{\perp} , \tag{2.78}$$

where $\mathbf{P}'^{+}$ is the pseudoinverse matrix of $\mathbf{P}'$:

$$\mathbf{P}'^{+} = \mathbf{P}'^{T}(\mathbf{P}'\mathbf{P}'^{T})^{-1} , \tag{2.79}$$

and $\mathbf{p}'^{\perp}$ is a 4-vector that is perpendicular to all the row vectors of $\mathbf{P}'$, i.e.,

$$\mathbf{P}'\mathbf{p}'^{\perp} = \mathbf{0} . \tag{2.80}$$

There are an infinite number of matrices that satisfy $\mathbf{P}'\mathbf{P}'^{+} = \mathbf{I}$. Thus $\mathbf{P}'^{+}$ is not unique. See Appendix (Sect. 2.A.2) for how to derive this particular pseudoinverse matrix.

It remains to determine $\mathbf{p}'^{\perp}$. First note that such a vector does exist because the difference between the row dimension and column dimension is one, and the row vectors are generally independent of each other. Actually, one way to obtain $\mathbf{p}'^{\perp}$ is

$$\mathbf{p}'^{\perp} = (\mathbf{I} - \mathbf{P}'^{+}\mathbf{P}')\boldsymbol{\omega} , \tag{2.81}$$

where $\boldsymbol{\omega}$ is an arbitrary 4-vector. To show that it is perpendicular to every row vector of $\mathbf{P}'$, we multiply $\mathbf{P}'$ and $\mathbf{p}'^{\perp}$:

$$\mathbf{P}'(\mathbf{I} - \mathbf{P}'^{+}\mathbf{P}')\boldsymbol{\omega} = (\mathbf{P}' - \mathbf{P}'\mathbf{P}'^{T}(\mathbf{P}'\mathbf{P}'^{T})^{-1}\mathbf{P}')\boldsymbol{\omega} = \mathbf{0}$$

which is indeed a zero vector.

Actually, the following equation always stands, as long as the rank of the 3×4 matrix $\mathbf{P}'$ is 3 (see Sect. 2.A.2 for a proof),

$$\mathbf{I} - \mathbf{P}'^{+}\mathbf{P}' = \mathbf{I} - \mathbf{P}'^{T}(\mathbf{P}'\mathbf{P}'^{T})^{-1}\mathbf{P}' = \frac{\mathbf{p}'^{\perp}\mathbf{p}'^{\perp T}}{\|\mathbf{p}'^{\perp}\|^{2}} . \tag{2.82}$$

The effect of matrix $\mathbf{I} - \mathbf{P}'^{+}\mathbf{P}'$ is to transform an arbitrary vector to a vector that is perpendicular to every row vector of $\mathbf{P}'$. If $\mathbf{P}'$ is of rank 3 (which is usually the case), then $\mathbf{p}^{\perp}$ is unique up to a scale factor.

Equation (2.78) is easily justified by projecting $\mathtt{M}'$ onto the image using (2.75), which indeed gives $\tilde{\mathbf{m}}'$. If we look closely at the equation, we can find that $\mathbf{p}'^{\perp}$ actually defines the optical center, which always projects onto the origin, and $\mathbf{P}'^{+}\tilde{\mathbf{m}}'$ defines the direction of the optical ray corresponding to image point $\tilde{\mathbf{m}}'$. For a particular value s', (2.78) corresponds to a point on the optical ray defined by $\mathbf{m}'$.

Similarly, an image point $\tilde{\mathbf{m}}$ in the first image also defines an optical ray. Requiring the two rays to intersect in space means that a point $\tilde{\mathtt{M}}'$ corresponding to a particular s' in (2.78) must project onto the first image at $\tilde{\mathbf{m}}$. That is,

$$s\tilde{\mathbf{m}} = s'\mathbf{P}\mathbf{D}\mathbf{P}'^{+}\tilde{\mathbf{m}}' + \mathbf{P}\mathbf{D}\mathbf{p}'^{\perp} , \tag{2.83}$$

where $\mathbf{PDp}'^{\perp}$ is the epipole $\mathbf{e}$ in the first image.

Performing a cross product with $\mathbf{PDp}'^{\perp}$ yields

$$s(\mathbf{PDp}'^{\perp}) \times \tilde{\mathbf{m}} = (\mathbf{PDp}'^{\perp}) \times (s'\mathbf{PDP}'^{+}\tilde{\mathbf{m}}') .$$

Eliminating s and s' by multiplying $\tilde{\mathbf{m}}^{T}$ from the left (equivalent to an inner product), we have

$$\tilde{\mathbf{m}}^{T}\mathbf{F}\tilde{\mathbf{m}}' = 0 , \tag{2.84}$$

where

$$\mathbf{F} = [\mathbf{PDp}'^{\perp}]_{\times} \mathbf{PDP}'^{+} \tag{2.85}$$

is the general form of fundamental matrix. It is evident that the roles that the two images play are symmetrical.

Note that (2.85) will be the essential matrix $\mathbf{E}$ if $\mathbf{P}$ and $\mathbf{P}'$ do not include the intrinsic parameters, i.e., if we work with normalized cameras.

We can also include all the intrinsic and extrinsic parameters in the two projection matrices $\mathbf{P}$ and $\mathbf{P}'$, so that for a 3D point $\tilde{\mathtt{M}}'$ in a world coordinate system we have

$$s\tilde{\mathbf{m}} = \mathbf{P}\tilde{\mathtt{M}}' , \tag{2.86}$$

$$s'\tilde{\mathbf{m}}' = \mathbf{P}'\tilde{\mathtt{M}}' . \tag{2.87}$$

Similarly we get

$$s\tilde{\mathbf{m}} = s'\mathbf{P}\mathbf{P}'^{+}\tilde{\mathbf{m}}' + \mathbf{P}\mathbf{p}'^{\perp} , \tag{2.88}$$

The same line of reasoning will lead to the general form of epipolar equation

$$\tilde{\mathbf{m}}^T\mathbf{F}\tilde{\mathbf{m}}' = 0 ,$$

where

$$\mathbf{F} = \left[\mathbf{P}\mathbf{p}'^{\perp}\right]_\times \mathbf{P}\mathbf{P}'^{+} . \tag{2.89}$$

It can also be shown that this expression is equivalent to (2.62) for the full perspective projection (see next subsection), but it is more general. Indeed, (2.62) assumes that the 3×3 matrix $\mathbf{B}'$ is invertible, which is the case for full perspective projection but not for affine cameras, while and (2.89) makes use of the pseudoinverse of the projection matrix, which is valid for both full perspective projection as well as affine cameras. Therefore the equation does not depend on any specific knowledge of projection model. Replacing the projection matrix in the equation by specific projection matrix for each specific projection model produces the epipolar equation for that specific projection model.

2.4.2 The Full Perspective Projection Case

Here we work with normalized cameras. Under the full perspective projection, the projection matrices for the two cameras are the same,

$$\mathbf{P}_p = \mathbf{P}'_p = \begin{bmatrix} 1 & 0 & 0 & 0 \\ 0 & 1 & 0 & 0 \\ 0 & 0 & 1 & 0 \end{bmatrix} . \tag{2.90}$$

It is not difficult to obtain

$$\mathbf{P}'^{+}_p = \mathbf{P}^T_p ,$$

and

$$\mathbf{p}_p'^{\perp} = \begin{bmatrix} 0 \\ 0 \\ 0 \\ 1 \end{bmatrix} .$$

Now substituting the above equations for (2.85), we have obtain the essential matrix

$$\mathbf{E}_p = \left[\mathbf{P}_p \mathbf{D} \mathbf{p}_p'^{\perp}\right]_\times \mathbf{P}_p \mathbf{D} \mathbf{P}_p'^{+} = [\mathbf{t}]_\times \mathbf{R} , \tag{2.91}$$

which is exactly the same as what we derived in the last section.

For the full perspective projection, we can prove that (2.89) is equivalent to (2.62). From definitions, we have

$$\begin{aligned} \left[\mathbf{B}' \quad \mathbf{b}'\right] \mathbf{p}_p'^{\perp} &= \mathbf{0} , \\ \left[\mathbf{B}' \quad \mathbf{b}'\right] \mathbf{P}_p'^{+} &= \mathbf{I} . \end{aligned}$$

It is easy to show

$$\begin{aligned} \mathbf{P} \mathbf{p}_p'^{\perp} &= \lambda(\mathbf{b} - \mathbf{B}\mathbf{B}'^{-1}\mathbf{b}') , \\ \mathbf{P}_p'^{+} &= \begin{bmatrix} \mathbf{B}'^{-1} - \mathbf{B}'^{-1}\mathbf{b}'\mathbf{q}^T \\ \mathbf{q}^T \end{bmatrix} , \end{aligned}$$

where λ is a non-zero scalar, and $\mathbf{q}$ is a non-zero arbitrary 3 vector. Substituting them for (2.89) yields

$$\begin{aligned} \mathbf{F} &= \lambda[\mathbf{b} - \mathbf{B}\mathbf{B}'^{-1}\mathbf{b}']_\times \left(\mathbf{B}\mathbf{B}'^{-1} - (\mathbf{b} - \mathbf{B}\mathbf{B}'^{-1}\mathbf{b}')\mathbf{q}^T)\right) \\ &= \lambda[\mathbf{b} - \mathbf{B}\mathbf{B}'^{-1}\mathbf{b}']_\times \mathbf{B}\mathbf{B}'^{-1} , \end{aligned} \tag{2.92}$$

which completes the proof as $\mathbf{F}$ is defined up to a scale factor.

2.5 EPIPOLAR GEOMETRY UNDER ORTHOGRAPHIC, WEAK PERSPECTIVE, PARAPERSPECTIVE AND GENERAL AFFINE PROJECTIONS

In this section we describe epipolar geometry under orthographic, weak perspective, paraperspective and general affine projections. As described in Sect. 2.2.3, the projection matrices of all these projective approximations have the same form:

$$\mathbf{P}_A = \begin{bmatrix} P_{11} & P_{12} & P_{13} & P_{14} \\ P_{21} & P_{22} & P_{23} & P_{24} \\ 0 & 0 & 0 & P_{34} \end{bmatrix} . \tag{2.93}$$

2.5.1 Orthographic and Weak Perspective Projections

Deriving the Epipolar Equation from the General Form

In the case of weak perspective projection, the two projection matrices are different, each having a different average depth plane. Without considering the intrinsic parameters, they are

$$\mathbf{P}_{wp} = \begin{bmatrix} 1 & 0 & 0 & 0 \\ 0 & 1 & 0 & 0 \\ 0 & 0 & 0 & Z_c \end{bmatrix} , \tag{2.94}$$

and

$$\mathbf{P}'_{wp} = \begin{bmatrix} 1 & 0 & 0 & 0 \\ 0 & 1 & 0 & 0 \\ 0 & 0 & 0 & Z'_c \end{bmatrix} . \tag{2.95}$$

A degenerate case is that both Z_c and Z'_c are 1, which is the orthographic projection. The projective matrices are then identical. In the following we will mainly treat weak perspective projection and mention orthographic projection as its special case.

Some algebra leads to

$$\mathbf{p}'^{\perp}_{wp} = \begin{bmatrix} 0 \\ 0 \\ 1 \\ 0 \end{bmatrix} ,$$

and

$$\mathbf{P}'^{+}_{wp} = \begin{bmatrix} 1 & 0 & 0 \\ 0 & 1 & 0 \\ 0 & 0 & 0 \\ 0 & 0 & \frac{1}{Z'_c} \end{bmatrix} .$$

Now substituting them for (2.85) yields

$$\begin{aligned} \mathbf{E}_{wp} &= \left[\mathbf{P}_{wp}\mathbf{D}\mathbf{p}^{\perp}_{wp}\right]_{\times} \mathbf{P}_{wp}\mathbf{D}\mathbf{P}'^{+}_{wp} = \begin{bmatrix} r_{13} \\ r_{23} \\ 0 \end{bmatrix}_{\times} \begin{bmatrix} r_{11} & r_{12} & t_X & \frac{1}{Z'_c} \\ r_{21} & r_{22} & t_Y & \frac{1}{Z'_c} \\ 0 & 0 & 0 & \frac{Z_c}{Z'_c} \end{bmatrix} \\ &= \frac{1}{Z'_c} \begin{bmatrix} 0 & 0 & -Z_c r_{23} \\ 0 & 0 & Z_c r_{13} \\ -Z'_c r_{32} & Z'_c r_{31} & t_X r_{23} - t_Y r_{13} \end{bmatrix} , \end{aligned} \tag{2.96}$$

which is exactly the same as derived in [109, 137], and can also be thought of as a generalization of the epipolar equation for the orthographic projection derived in [71].

Setting both Z_c and Z'_c to be 1, then (2.96) becomes

$$\mathbf{E}_o = \begin{bmatrix} 0 & 0 & -r_{23} \\ 0 & 0 & r_{13} \\ -r_{32} & r_{31} & t_X r_{23} - t_Y r_{13} \end{bmatrix} . \tag{2.97}$$

Deriving the Epipolar Equation from Rigid Motion Equation

To eliminate Z and Z' in the rigid motion equation

$$\mathbf{X} = \mathbf{R}\mathbf{X}' + \mathbf{t} , \tag{2.98}$$

we multiply $\mathbf{v}^T = \begin{bmatrix} r_{23} & -r_{13} & 0 \end{bmatrix}$ to both sides of it, yielding

$$\mathbf{v}^T\mathbf{X} = \mathbf{v}^T\mathbf{R}\mathbf{X}' + \mathbf{v}^T\mathbf{t} . \tag{2.99}$$

From $\mathbf{R}^T = \mathbf{R}^{-1}$, it is easy to show

$$\mathbf{v}^T\mathbf{R} = \begin{bmatrix} -r_{32} & r_{31} & 0 \end{bmatrix} .$$

Thus (2.99) leads to

$$-r_{23}X + r_{13}Y - r_{32}X' + r_{31}Y' + r_{23}t_X - r_{13}t_Y = 0 . \tag{2.100}$$

This equation represents a plane in space which is perpendicular to both the first image plane and the second image plane, because it is independent from both Z and Z'.

Assuming the weak perspective projection, we can rewrite the above equation as

$$-(r_{23}Z_c)x + (r_{13}Z_c)y - (r_{32}Z_c')x' + (r_{31}Z_c')y' + r_{23}t_X - r_{13}t_Y = 0 \,. \quad (2.101)$$

Note that Huang and Lee (1989) derived this equation under orthographic projection. In that case, both Z_c and Z_c' are 1, and (2.101) becomes

$$-r_{23}x + r_{13}y - r_{32}x' + r_{31}y' + r_{23}t_X - r_{13}t_Y = 0 \,. \quad (2.102)$$

If we put (2.101) into matrix form, we have

$$\tilde{\mathbf{x}}^T \mathbf{E}_{wp} \tilde{\mathbf{x}}' = 0 \,, \quad (2.103)$$

where $\tilde{\mathbf{x}}$ and $\tilde{\mathbf{x}}'$ are augmented vectors of the corresponding points in the first and second images, respectively, and

$$\mathbf{E}_{wp} = \begin{bmatrix} 0 & 0 & -r_{23}Z_c \\ 0 & 0 & r_{13}Z_c \\ -r_{32}Z_c' & r_{31}Z_c' & r_{23}t_X - r_{13}t_Y \end{bmatrix} . \quad (2.104)$$

The essential matrix for orthographic projection can be obtained by setting Z_c and Z_c' to be 1 in (2.104), which leads to the same formula as (2.97).

If we know the intrinsic matrices $\mathbf{A}$ and $\mathbf{A}'$ for the two cameras, we can change the normalized coordinates $\mathbf{x}$ to pixel coordinates $\mathbf{m}$ in the epipolar equation as

$$\tilde{\mathbf{m}}^T \mathbf{F}_{wp} \tilde{\mathbf{m}}' = 0 \,, \quad (2.105)$$

where fundamental matrix $\mathbf{F}_{wp}$ is determined by

$$\mathbf{F}_{wp} = (\mathbf{A}^{-1})^T \mathbf{E}_{wp} \mathbf{A}'^{-1} \,. \quad (2.106)$$

Since $\mathbf{A}$ and $\mathbf{A}'$ are both upper triangular matrices, the essential matrix $\mathbf{E}_{wp}$ and the fundamental matrix $\mathbf{F}_{wp}$ have the same form in the sense that the upper left 4 components are all zero. Expanding (2.105) also yields a linear equation

$$f_{13}u + f_{23}v + f_{31}u' + f_{32}v' + f_{33} = 0 \,, \quad (2.107)$$

which is also linear in the image coordinates as (2.101). In particular, if the optical axis is perpendicular to the image plane, and the vertical and horizontal spacings are in same unit, then the intrinsic matrix is simplified to be

$$\mathbf{A} = \begin{bmatrix} s & 0 & u_0 \\ 0 & s & v_0 \\ 0 & 0 & 1 \end{bmatrix}, \tag{2.108}$$

where s is product of the focal length f and the pixel unit k in both horizontal and vertical directions. In the case of CCD cameras, this model is close to reality. We will frequently use this model in the following discussions if otherwise mentioned. The inverse of the intrinsic matrix in (2.108) is

$$\mathbf{A}^{-1} = \begin{bmatrix} \frac{1}{s} & 0 & -\frac{u_0}{s} \\ 0 & \frac{1}{s} & -\frac{v_0}{s} \\ 0 & 0 & 1 \end{bmatrix}. \tag{2.109}$$

Substituting this for (2.105) yields

$$\mathbf{F} = \begin{bmatrix} 0 & 0 & -\frac{r_{23}Z_c}{s} \\ 0 & 0 & \frac{r_{13}Z_c}{s} \\ -\frac{r_{23}Z'_c}{s'} & \frac{r_{31}Z'_c}{s'} & r_{23}t_X - r_{13}t_Y + \frac{(r_{23}u_0 - r_{13}v_0)Z_c}{s} + \frac{(r_{32}u'_0 - r_{31}v'_0)Z'_c}{s'} \end{bmatrix},$$

where $s = s', u_0 = u'_0, v_0 = v'_0$ if the same camera is used. s can be understood as a uniform scaling of the image coordinates. Note that u_0, v_0, u'_0, v'_0 only appear in the lower right elements of the matrix, and they function in a similar manner as the translation components t_X and t_Y. This is understandable as changing the origin in the image plane amounts to a pure translation parallel to the image plane.

Geometric Interpretation

There are many different ways to define a 3D rotation. One of them is to define an arbitrary 3D rotation by three consecutive rotations around the coordinate axes, that is, a rotation by α around the z-axis first, then a rotation by β around the new y-axis, and finally a rotation by $-\gamma$ around the new z-axis.

$$\mathbf{R} = \mathbf{R}_z(\alpha)\mathbf{R}_y(\beta)\mathbf{R}_z(-\gamma) \tag{2.110}$$

$(\alpha, \beta, -\gamma)$ is the same as *Euler angles*, which is widely used in kinematics of robot manipulators [114]. Note that $-\gamma$ defined with respect to the first image is equivalent to rotating the second image by γ.

The first and third rotations are actually two rotations within the image planes, while only the second rotation is related to depth. Representing $\mathbf{R}$ by the three angles α, β and $-\gamma$, we have

$$\mathbf{R} = \begin{bmatrix} \cos\alpha & -\sin\alpha & 0 \\ \sin\alpha & \cos\alpha & 0 \\ 0 & 0 & 1 \end{bmatrix} \begin{bmatrix} \cos\beta & 0 & \sin\beta \\ 0 & 1 & 0 \\ -\sin\beta & 0 & \cos\beta \end{bmatrix} \begin{bmatrix} \cos\gamma & \sin\gamma & 0 \\ -\sin\gamma & \cos\gamma & 0 \\ 0 & 0 & 1 \end{bmatrix}$$

$$= \begin{bmatrix} \cos\alpha\cos\beta\cos\gamma + \sin\alpha\sin\gamma & \cos\alpha\cos\beta\sin\gamma - \sin\alpha\cos\gamma & \cos\alpha\sin\beta \\ \sin\alpha\cos\beta\cos\gamma - \cos\alpha\sin\gamma & \sin\alpha\cos\beta\sin\gamma + \cos\alpha\cos\gamma & \sin\alpha\sin\beta \\ -\sin\beta\cos\gamma & -\sin\beta\sin\gamma & \cos\beta \end{bmatrix} .$$

Substituting the components of $\mathbf{R}$ for $\mathbf{E}_{wp}$, we have

$$\mathbf{E}_{wp} = \sin\beta \begin{bmatrix} 0 & 0 & -Z_c\sin\alpha \\ 0 & 0 & Z_c\cos\alpha \\ Z'_c\sin\gamma & -Z'_c\cos\gamma & t_X\sin\alpha - t_Y\cos\alpha \end{bmatrix} .$$

Now we can rewrite the epipolar equation as

$$\sin\beta(-Z_c x\sin\alpha + Z_c y\cos\alpha + Z'_c x'\sin\gamma - Z'_c y'\cos\gamma + t_X\sin\alpha - t_Y\cos\alpha) = 0 . \tag{2.111}$$

If we work with pixel image coordinates, then the epipolar equation becomes

$$\sin\beta(-\frac{Z_c\sin\alpha}{s}u + \frac{Z_c\cos\alpha}{s}v + \frac{Z'_c\sin\gamma}{s'}u' - \frac{Z'_c\cos\gamma}{s'}v'$$
$$+t_X\sin\alpha - t_Y\cos\alpha + \frac{r_{23}Z_c u_0}{s} - \frac{r_{13}Z_c v_0}{s} + \frac{r_{32}Z'_c u'_0}{s'} - \frac{r_{31}Z'_c v'_0}{s'}) = 0 \tag{2.112}$$

This can be rewritten as

$$-u\sin\alpha + v\cos\alpha - \rho(-u'\sin\gamma + v'\cos\gamma) + \lambda = 0 , \tag{2.113}$$

where $\alpha, \gamma, \rho, \lambda$ have two sets of values:

$$\alpha_1 = atan2(-f_{13}, f_{23}) , \tag{2.114}$$

$$\gamma_1 = atan2(f_{31}, -f_{32}) , \tag{2.115}$$

$$\rho_1 = \sqrt{\frac{f_{31}^2 + f_{32}^2}{f_{13}^2 + f_{23}^2}} , \tag{2.116}$$

$$\lambda_1 = \frac{f_{33}}{\sqrt{f_{13}^2 + f_{23}^2}} . \tag{2.117}$$

or

$$\alpha_2 = atan2(f_{13}, -f_{23}), \quad (2.118)$$

$$\gamma_2 = atan2(-f_{31}, f_{32}), \quad (2.119)$$

$$\rho_2 = \sqrt{\frac{f_{31}^2 + f_{32}^2}{f_{13}^2 + f_{23}^2}}, \quad (2.120)$$

$$\lambda_2 = \frac{-f_{33}}{\sqrt{f_{13}^2 + f_{23}^2}}. \quad (2.121)$$

where $atan2(x, y)$ is the function for *arctangent* in C. It is easy to get the second set of parameters by multiplying -1 to the two sides of (2.113). It is noted that

$$\alpha_1 - \alpha_2 = \pm\pi,$$
$$\gamma_1 - \gamma_2 = \pm\pi,$$
$$\rho_1 = \rho_2,$$
$$\lambda_1 = -\lambda_2.$$

We further define

$$\theta = \alpha - \gamma. \quad (2.122)$$

α, γ, θ, ρ and λ are called *motion parameters.* They are the only motion information that can be determined from the epipolar equation.

Here, ρ stands for the scale change between the two images caused by different depths Z_c and Z_c' and possible different pixel scales, and λ stands for the translation along the direction perpendicular to the epipolar lines.

The first two terms of (2.113) can be thought of as the new vertical coordinate after a rotation of the first image by α, and the next two terms as the new vertical coordinate after a rotation of the second image by γ. Then the equation can be understood as saying that the new vertical coordinates are the same after a vertical translation of the first image by λ (see Fig.2.11.) Similar interpretation is independently developed by Shapiro *et al.* in [137], but they did not consider λ and the ambiguity in α and γ.

Next we show that the above representation can also be understood as a variant of the Koendrink and van Doorn representation [79].

From (2.122), Eq. (2.110) can be rewritten as

$$\mathbf{R} = \mathbf{R}_z(\alpha)\mathbf{R}_y(\beta)\mathbf{R}_z(-\alpha)\mathbf{R}_z(\theta) = \mathbf{R}_z(\theta)\mathbf{R}_z(\gamma)\mathbf{R}_z(\beta)\mathbf{R}_y(-\gamma). \quad (2.123)$$

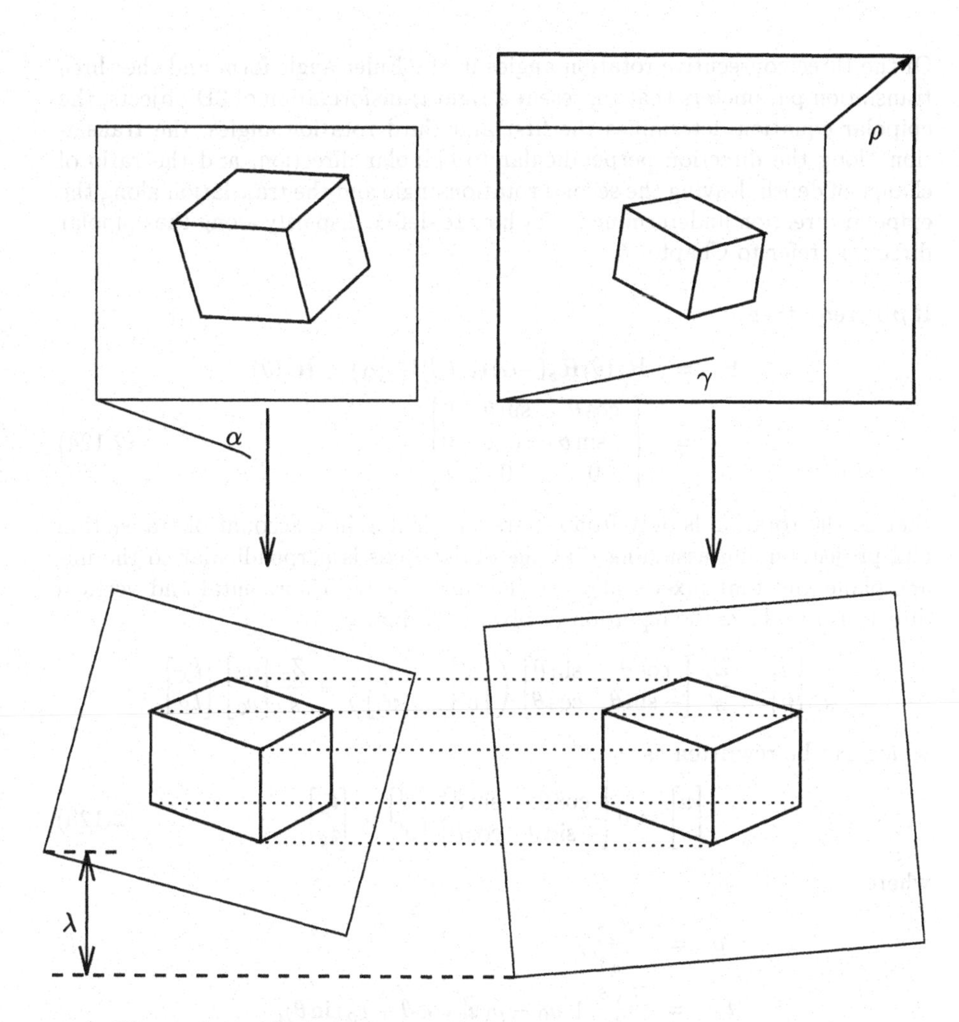

Figure 2.11 Corresponding points have same horizontal coordinates after the rotations, scaling and translation.

This means that a rotation can be represented as a rotation of β around an axis parallel to the image plane and angled at $\alpha + \frac{\pi}{2}$ to the positive horizontal axis, followed by another fronto-parallel rotation in the image plane of θ. Or it can be represented as a fronto-parallel rotation of θ in the image plane, followed by a rotation of β around an axis parallel to the image plane and angled at $\gamma + \frac{\pi}{2}$ to the new positive horizontal axis. This is first used in [79]. Similar interpretations can be found in [137].

Of the three consecutive rotation angles in the Euler angle form and the three translation parameters that represent a rigid transformation of 3D objects, the epipolar equation determines the first and third rotation angles, the translation along the direction perpendicular to epipolar direction, and the ratio of change of depth, leaving the second rotation angle and the translation along the epipolar direction undetermined. For how to define disparity along the epipolar direction, refer to Chapter 5.

If β is zero, then

$$\begin{aligned} \mathbf{R} &= \mathbf{R}_z(\theta)\mathbf{R}_z(-\alpha)\mathbf{R}_y(\beta)\mathbf{R}_z(\alpha) = \mathbf{R}_z(\theta) \\ &= \begin{bmatrix} \cos\theta & \sin\theta & 0 \\ -\sin\theta & \cos\theta & 0 \\ 0 & 0 & 1 \end{bmatrix}, \end{aligned} \tag{2.124}$$

that is, the rotation is only fronto-parallel. Taking into account of translation and projection and assuming that the optical axis is perpendicular to the image plane and that pixel scales are the same for both horizontal and vertical directions, we have the equation in pixel coordinates,

$$\frac{Z_c}{s}\begin{bmatrix} u \\ v \end{bmatrix} = \frac{Z_c'}{s'}\begin{bmatrix} \cos\theta & \sin\theta \\ -\sin\theta & \cos\theta \end{bmatrix}\left(\begin{bmatrix} u' \\ v' \end{bmatrix} - \begin{bmatrix} u_0' \\ v_0' \end{bmatrix}\right) + \frac{Z_c}{s}\begin{bmatrix} u_0 \\ v_0 \end{bmatrix}\begin{bmatrix} t_X \\ t_Y \end{bmatrix},$$

which can be rewritten as

$$\begin{bmatrix} u \\ v \end{bmatrix} = \rho\begin{bmatrix} \cos\theta & \sin\theta \\ -\sin\theta & \cos\theta \end{bmatrix}\begin{bmatrix} u' \\ v' \end{bmatrix} + \begin{bmatrix} t_x \\ t_y \end{bmatrix}, \tag{2.125}$$

where

$$\begin{aligned} \rho &= \frac{Z_c' s}{Z_c s'}, \\ t_x &= \frac{t_X s}{Z_c} + u_0 - \rho(u_0'\cos\theta + v_0'\sin\theta), \\ t_y &= \frac{t_Y s}{Z_c} + v_0 - \rho(-u_0'\sin\theta + v_0'\cos\theta). \end{aligned}$$

This means that the apparent motion is a combination of image rotation, scaling, and 2D translation. We call this *2D affine motion.* Note that Lee and Huang called this *degenerate motion* [86].

If we are given the equations in the form of

$$\begin{bmatrix} u \\ v \end{bmatrix} = \begin{bmatrix} a & b \\ -b & a \end{bmatrix}\begin{bmatrix} u' \\ v' \end{bmatrix} + \begin{bmatrix} t_u \\ t_v \end{bmatrix}, \tag{2.126}$$

then the parameters in (2.125) can be uniquely determined as

$$\rho = \sqrt{a^2 + b^2}\,, \tag{2.127}$$
$$\theta = atan2(b, a)\,, \tag{2.128}$$
$$x_0 = t_u\,, \tag{2.129}$$
$$y_0 = t_v\,. \tag{2.130}$$

In this case, given two pairs of matched points, all the motion information, a rotation angle θ, a scaling factor ρ and the translation vector $[x_0, y_0]^T$, can be uniquely determined. Motions with very slow rotation out of the image plane can often be modelled by 2D affine motion. Whether or not $\beta = 0$ can be detected by examining the rank of the matrix of coordinates of the points (see Sect. 3.2).

Epipolar Equation under Weak Perspective Projection as an Approximation of Epipolar Equation under Full Perspective Projection

In the following we show that the epipolar equation derived under the full perspective projection model can be approximated by the weak perspective epipolar equation given the condition that the projection can be approximated by the weak perspective projection.

The epipolar equation (2.34) can be rewritten as

$$\begin{bmatrix} X \\ Y \\ 1 \end{bmatrix}^T \begin{bmatrix} \frac{r_{31}t_X - r_{21}t_Z}{ZZ'} & \frac{r_{32}t_Y - r_{22}t_Z}{ZZ'} & \frac{r_{33}t_Y - r_{23}t_Z}{Z} \\ \frac{r_{11}t_Z - r_{31}t_X}{ZZ'} & \frac{r_{12}t_Z - r_{32}t_X}{ZZ'} & \frac{r_{13}t_Y - r_{33}t_Z}{Z} \\ \frac{r_{21}t_X - r_{11}t_Y}{Z'} & \frac{r_{22}t_X - r_{12}t_Y}{Z'} & r_{23}t_X - r_{13}t_Y \end{bmatrix} \begin{bmatrix} X' \\ Y' \\ 1 \end{bmatrix} = 0\,. \tag{2.131}$$

Using

$$\begin{bmatrix} X \\ Y \\ Z \end{bmatrix} = \mathbf{R} \begin{bmatrix} X' \\ Y' \\ Z' \end{bmatrix} + \mathbf{t}\,,$$

and

$$\mathbf{R}^T = \mathbf{R}^{-1} = \begin{bmatrix} r_{22}r_{33} - r_{23}r_{32} & r_{23}r_{31} - r_{21}r_{33} & r_{21}r_{32} - r_{22}r_{31} \\ r_{13}r_{32} - r_{12}r_{33} & r_{11}r_{33} - r_{13}r_{31} & r_{31}r_{12} - r_{11}r_{32} \\ r_{12}r_{23} - r_{13}r_{22} & r_{13}r_{21} - r_{11}r_{23} & r_{11}r_{22} - r_{12}r_{21} \end{bmatrix}\,, \tag{2.132}$$

we can represent $\mathbf{t}$ in terms of $\mathbf{R}$, $\mathtt{X}$ and $\mathtt{X}'$, and the central matrix in (2.131) can further be changed to

$$\begin{bmatrix} \frac{r_{31}Y-r_{21}Z+r_{13}Y'-r_{12}Z'}{ZZ'} & \frac{r_{32}Y-r_{22}Z-r_{13}X'+r_{11}Z'}{ZZ'} & \frac{r_{33}Y-r_{23}Z+r_{12}X'-r_{11}Y'}{Z} \\ \frac{-r_{31}X+r_{11}Z+r_{23}Y'-r_{22}Z'}{ZZ'} & \frac{-r_{32}X+r_{12}Z-r_{23}X'+r_{21}Z'}{ZZ'} & \frac{-r_{33}X+r_{13}Z+r_{22}X'-r_{21}Y'}{Z} \\ \frac{r_{21}X-r_{11}Y+r_{33}Y'-r_{32}Z'}{Z'} & \frac{r_{22}X-r_{12}Y-r_{33}X'+r_{31}Z'}{Z'} & r_{23}t_X - r_{13}t_Y \end{bmatrix}.$$

From the assumptions of the weak perspective projection, we know

$$Z, Z' \gg X, Y, X', Y' .$$

Provided that these conditions are satisfied, the above matrix can be approximated as

$$\begin{bmatrix} X \\ Y \\ 1 \end{bmatrix}^T \begin{bmatrix} 0 & 0 & -r_{23} \\ 0 & 0 & r_{13} \\ -r_{32} & r_{31} & r_{23}t_X - r_{13}t_Y \end{bmatrix} \begin{bmatrix} X' \\ Y' \\ 1 \end{bmatrix} = 0 ,$$

which is equivalent to

$$\begin{bmatrix} x \\ y \\ 1 \end{bmatrix}^T \begin{bmatrix} 0 & 0 & -r_{23}Z_c \\ 0 & 0 & r_{13}Z_c \\ -r_{32}Z'_c & r_{31}Z'_c & r_{23}t_X - r_{13}t_Y \end{bmatrix} \begin{bmatrix} x' \\ y' \\ 1 \end{bmatrix} = 0 .$$

This is exactly the weak perspective equation we derived in (2.104).

2.5.2 Paraperspective Projection

The Epipolar Equation

In the case of paraperspective projection, the two projection matrices are different, each having a different position of centroid. Without the intrinsic parameters, they are

$$\mathbf{P}_{pp} = \begin{bmatrix} 1 & 0 & -\frac{X_c}{Z_c} & X_c \\ 0 & 1 & -\frac{Y_c}{Z_c} & Y_c \\ 0 & 0 & 0 & Z_c \end{bmatrix} ,$$

and

$$\mathbf{P}'_{pp} = \begin{bmatrix} 1 & 0 & -\frac{X'_c}{Z'_c} & X'_c \\ 0 & 1 & -\frac{Y'_c}{Z'_c} & Y'_c \\ 0 & 0 & 0 & Z'_c \end{bmatrix} .$$

A sequence of algebraic operations lead to

$$\mathbf{p}'^{\perp}_{pp} = \frac{1}{\|\mathbf{X}'_c\|}\begin{bmatrix}\mathbf{X}'_c \\ 0\end{bmatrix}, \tag{2.133}$$

where $\mathbf{X}'_c = [X'_c, Y'_c, Z'_c]^T$, and

$$\mathbf{P}'^{+}_{pp} = \begin{bmatrix} & \mathbf{I}_3 & \\ 0 & 0 & \frac{1}{Z'_c}\end{bmatrix} - \frac{1}{\|\mathbf{X}'_c\|^2}\begin{bmatrix}\mathbf{X}'_c\mathbf{X}'^T_c \\ \mathbf{0}^T_3\end{bmatrix}, \tag{2.134}$$

where $\mathbf{I}_3$ is a 3×3 identity matrix, and $\mathbf{0}_3$ is a 3 zero vector.

Let us rewrite $\mathbf{P}_{pp}$ as $\begin{bmatrix}\mathbf{p}^T_{pp1} & X_c \\ \mathbf{p}^T_{pp2} & Y_c \\ \mathbf{0}^T_3 & Z_c\end{bmatrix}$. Simply, we have

$$\mathbf{P}_{pp}\mathbf{D} = \begin{bmatrix}\mathbf{p}^T_{pp1}\mathbf{R} & \mathbf{p}^T_{pp1}\mathbf{t} + X_c \\ \mathbf{p}^T_{pp2}\mathbf{R} & \mathbf{p}^T_{pp2}\mathbf{t} + Y_c \\ \mathbf{0}^T_3 & Z_c\end{bmatrix}.$$

Substituting these equations for (2.85) yields

$$\begin{aligned}
\mathbf{E}_{pp} &= [\mathbf{P}_{pp}\mathbf{D}\mathbf{p}'^{\perp}_{pp}]_\times \mathbf{P}_{pp}\mathbf{D}\mathbf{P}'^{+}_{pp} \\
&= \begin{bmatrix} 0 & 0 & \mathbf{p}^T_{pp2}\mathbf{R}\mathbf{X}'_c \\ 0 & 0 & -\mathbf{p}^T_{pp1}\mathbf{R}\mathbf{X}'_c \\ -\mathbf{p}^T_{pp2}\mathbf{R}\mathbf{X}'_c & \mathbf{p}^T_{pp1}\mathbf{R}\mathbf{X}'_c & 0\end{bmatrix}\mathbf{P}_{pp}\mathbf{D}\mathbf{P}'^{+}_{pp} \\
&= \left(\frac{1}{Z_c}\begin{bmatrix}\mathbf{0}^T_3 & 0 \\ \mathbf{0}^T_3 & 0 \\ (\mathbf{X}'_c \times (\mathbf{R}^T\mathbf{X}_c))^T & \mathbf{X}^T_c\mathbf{E}_p\mathbf{X}'_c\end{bmatrix} + [\mathbf{0}_3 \;\; \mathbf{0}_3 \;\; \mathbf{0}_3 \;\; \mathbf{X}_c \times (\mathbf{R}\mathbf{X}'_c)]\right)\mathbf{P}'^{+}_{pp} \\
&= \begin{bmatrix}\mathbf{0}^T_3 \\ \mathbf{0}^T_3 \\ \frac{(\mathbf{X}'_c \times (\mathbf{R}^T\mathbf{X}_c))^T}{Z_c}\end{bmatrix} + \begin{bmatrix}0 & 0 & 0 \\ 0 & 0 & 0 \\ 0 & 0 & \frac{\mathbf{X}^T_c\mathbf{E}_p\mathbf{X}'_c}{Z_c Z'_c}\end{bmatrix} + \begin{bmatrix}\mathbf{0}_3 & \mathbf{0}_3 & \frac{\mathbf{X}_c \times (\mathbf{R}\mathbf{X}'_c)}{Z'_c}\end{bmatrix}.
\end{aligned} \tag{2.135}$$

It is easy to verify that if $X_c = Y_c = X'_c = Y'_c = 0$, then the essential matrix for paraperspective projection becomes

$$\begin{bmatrix}0 & 0 & 0 \\ 0 & 0 & 0 \\ -Z'_c r_{32} & Z_c r_{31} & 0\end{bmatrix} + \begin{bmatrix}0 & 0 & 0 \\ 0 & 0 & 0 \\ 0 & 0 & r_{23}t_X - r_{13}t_Y\end{bmatrix} + \begin{bmatrix}0 & 0 & -r_{23}Z_c \\ 0 & 0 & r_{13}Z_c \\ 0 & 0 & 0\end{bmatrix},$$

where t_X and t_Y are the first and second components of $\mathbf{t}$, respectively. This is exactly identical to the essential matrix for weak perspective projection (2.96).

Now, as X_c and X'_c are the same point in space, we have

$$X_c^T \mathbf{E}_{pp} X'_c = 0 . \tag{2.136}$$

And though the values of Z_c and Z'_c are not available, the vectors to the centroids have the same orientations as the vectors to the centroid's images.

$$\begin{aligned} X_c &= Z_c \mathbf{x}_c , \\ X'_c &= Z'_c \mathbf{x}'_c , \end{aligned}$$

where $\mathbf{x}_c$ and $\mathbf{x}'_c$ are the 3D coordinates of the centroid's projections onto the two image planes, respectively. Taking into account of the above relations, we can write the essential matrix as

$$\mathbf{E}_{pp} = \begin{bmatrix} 0 & 0 & e_{13} \\ 0 & 0 & e_{23} \\ e_{31} & e_{32} & e_{33} \end{bmatrix} , \tag{2.137}$$

where

$$\begin{aligned} e_{13} &= Z_c \begin{bmatrix} 0 & -1 & y_c \end{bmatrix} \mathbf{R}\mathbf{x}'_c , \\ e_{23} &= Z_c \begin{bmatrix} 1 & 0 & -x_c \end{bmatrix} \mathbf{R}\mathbf{x}'_c , \\ e_{31} &= Z'_c \begin{bmatrix} 0 & -1 & y'_c \end{bmatrix} \mathbf{R}^T\mathbf{x}_c , \\ e_{32} &= Z'_c \begin{bmatrix} 1 & 0 & -x'_c \end{bmatrix} \mathbf{R}^T\mathbf{x}_c , \\ e_{33} &= Z_c \begin{bmatrix} -y_c & -x_c & 0 \end{bmatrix} \mathbf{R}\mathbf{x}'_c + Z'_c \begin{bmatrix} -y'_c & x'_c & 0 \end{bmatrix} \mathbf{R}^T\mathbf{x}_c . \end{aligned}$$

Similar to the essential matrix for the weak perspective projection, the upper left 4 components of the essential matrix for the paraperspective projection are also zero.

If we work with the pixel coordinates, as in the case of weak perspective projection, the upper left 4 components of the fundamental matrix for paraperspective projection are also zero:

$$\mathbf{F}_{pp} = \begin{bmatrix} 0 & 0 & \frac{e_{13}}{s} \\ 0 & 0 & \frac{e_{23}}{s} \\ \frac{e_{31}}{s'} & \frac{e_{32}}{s'} & e_{33} - \frac{e_{31}u_0}{s} - \frac{e_{32}v_0}{s} - \frac{e_{13}u'_0}{s'} - \frac{e_{23}v'_0}{s'} \end{bmatrix} . \tag{2.138}$$

Geometric Interpretation

In the following we assume that the camera's optical axis is perpendicular to the image plane, and the pixel scales are the same for both horizontal and

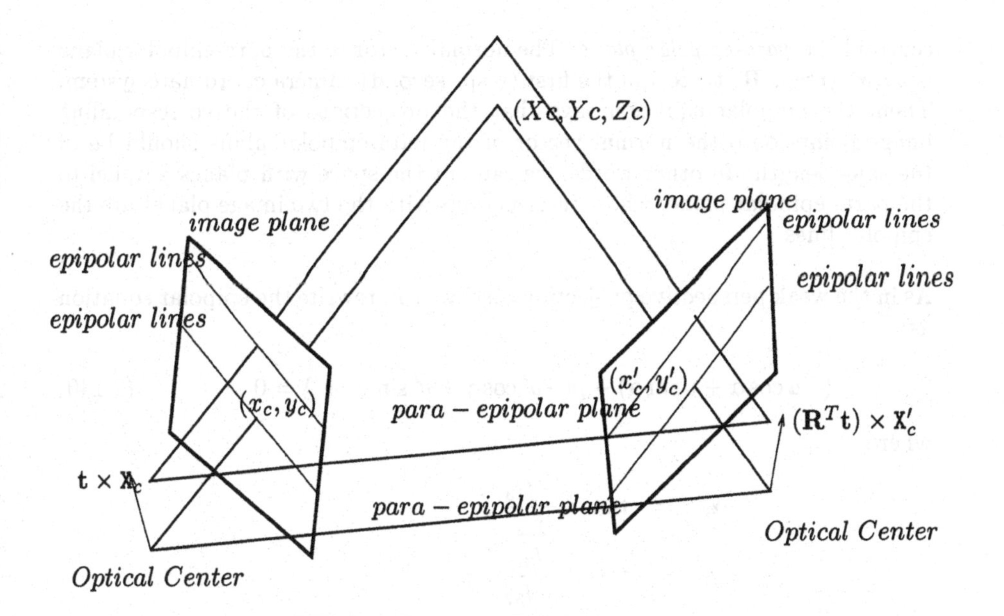

Figure 2.12 The epipolar geometry under paraperspective projection

vertical directions. Since

$$\mathbf{X}_c = \mathbf{R}\mathbf{X}'_c + \mathbf{t} \ ,$$

we can show

$$\begin{aligned} \mathbf{X}'_c \times (\mathbf{R}^T\mathbf{X}_c) &= -(\mathbf{R}^T\mathbf{t}) \times \mathbf{X}'_c \ , \\ \mathbf{X}_c \times (\mathbf{R}\mathbf{X}'_c) &= \mathbf{t} \times \mathbf{X}_c \ . \end{aligned}$$

Substituting these equations for (2.135), we have

$$\mathbf{E}_{pp} = \frac{Z'_c}{Z_c} \begin{bmatrix} \mathbf{0}_3^T \\ \mathbf{0}_3^T \\ -\left((\mathbf{R}^T\mathbf{t}) \times \mathbf{x}'_c\right)^T \end{bmatrix} + \frac{Z_c}{Z'_c} \left[\mathbf{0}_3 \quad \mathbf{0}_3 \quad (\mathbf{t} \times \mathbf{x}_c)\right] \ ,$$

and expanding the epipolar equation yields

$$Z_c^2 \tilde{\mathbf{m}}^T (\mathbf{t} \times \mathbf{x}_c) = Z_c'^2 \tilde{\mathbf{m}}'^T \left((\mathbf{R}^T\mathbf{t}) \times \mathbf{x}'_c\right) \ . \tag{2.139}$$

Clearly, the epipolar equation has an intuitive and nice interpretation. See Fig. 2.12. Let us call the plane that includes the two optical centers and object centroid the *para-epipolar plane.* The normal vector of the para-epipolar plane is $\mathbf{t} \times \mathbf{x}_c$ (resp. $\mathbf{R}^T\mathbf{t} \times \mathbf{x}'_c$) in the first (resp. second) camera coordinate system. Then, the epipolar equation says that the projections of the corresponding image points onto the normal vector of the para-epipolar plane should be of the same length. In other words, we can cut the space with planes parallel to the para- epipolar plane, whose intersections with the two image planes are the epipolar lines.

As in the weak perspective projection case, we can rewrite the epipolar equation as

$$(-u\cos\alpha + v\sin\alpha) - \rho(-u'\cos\gamma + v'\sin\gamma) + \lambda = 0\,, \tag{2.140}$$

where

$$\begin{aligned}
\alpha_1 &= \tan^{-1}\frac{-f_{13}}{f_{23}}\,,\\
\gamma_1 &= \tan^{-1}\frac{f_{31}}{-f_{32}}\,,\\
\rho_1 &= \sqrt{\frac{f_{31}^2 + f_{32}^2}{f_{13}^2 + f_{23}^2}} = \frac{s}{s'}\sqrt{\frac{pp_{31}^2 + pp_{32}^2}{pp_{13}^2 + pp_{23}^2}}\,,\\
\lambda_1 &= \frac{f_{33}}{\sqrt{f_{13}^2 + f_{23}^2}}\,,
\end{aligned}$$

or

$$\begin{aligned}
\alpha_2 &= atan2(-f_{13}, f_{23})\,,\\
\gamma_2 &= atan2(f_{31}, -f_{32})\,,\\
\rho_2 &= \sqrt{\frac{f_{31}^2 + f_{32}^2}{f_{13}^2 + f_{23}^2}} = \frac{s}{s'}\sqrt{\frac{pp_{31}^2 + pp_{32}^2}{pp_{13}^2 + pp_{23}^2}}\,,\\
\lambda_2 &= -\frac{f_{33}}{\sqrt{f_{13}^2 + f_{23}^2}}\,,
\end{aligned}$$

Once the epipolar equation is determined, these 4 parameters can be computed directly.

If we represent the arbitrary rotation between the two cameras by three consecutive rotations, the first one around the optical axis of the first camera, the second one around the axis perpendicular to the para-epipolar plane, and the third one around the optical axis of the second camera, then α and $-\gamma$ are exactly the first and third rotations. Their effects are two image torsions such that the parallel epipolar lines become horizontal in each image. The remaining second rotation cannot be computed directly from the epipolar equation, which apparently moves points only along the epipolar lines. This representation of arbitrary 3D rotation is slightly different from the Euler angle representation (the Koendrink and van Doorn representation can be thought of as its variant), in which the second angle is for rotation around the y-axis, while here the axis generally deviates from the y-axis, though within the $y - z$ plane.

ρ represents the scale change between the two images due to the change in distance from the camera to the object (centroid) and possible change of pixel scale. And λ accounts for the image translation perpendicular to the epipolar direction such that corresponding epipolar lines have the same vertical coordinates.

Comparing this with the interpretation for the weak perspective projection [109, 137], the difference is in epipolar plane. In the weak perspective case, the epipolar plane is always perpendicular to the image planes, while in the paraperspective projection case, the epipolar plane depends on the location of object centroid.

2.5.3 The General Affine Camera

The Epipolar Equation

In the case of general affine cameras [105, 137], the projection matrix, with or without the intrinsic parameters, can be rewritten as

$$\mathbf{P}_A = \begin{bmatrix} \mathbf{p}_1^T & \\ \mathbf{p}_2^T & \mathbf{p}_4 \\ \mathbf{0}_3^T & \end{bmatrix} , \tag{2.141}$$

where $\mathbf{p}_4 = [P_{14}, P_{24}, P_{34}]^T$.

For any affine camera, we can construct $\mathbf{p}_A'^{\perp}$ as

$$\mathbf{p}_A'^{\perp} = \frac{1}{\|\mathbf{p}_1' \times \mathbf{p}_2'\|} \begin{bmatrix} \mathbf{p}_1' \times \mathbf{p}_2' \\ 0 \end{bmatrix} . \tag{2.142}$$

Let us define $\mathbf{p}_3' = \mathbf{p}_1' \times \mathbf{p}_2'$. From

$$\begin{aligned} \mathbf{p}_1'^T \mathbf{p}_3' &= 0 , \\ \mathbf{p}_2'^T \mathbf{p}_3' &= 0 , \end{aligned}$$

we can verify that $\mathbf{p}_A'^{\perp}$ is indeed perpendicular to $\mathbf{P}_A'$:

$$\mathbf{P}_A'\mathbf{p}_A'^{\perp} = \frac{1}{\|\mathbf{p}_3'\|}\begin{bmatrix} \mathbf{p}_1'^T & \\ \mathbf{p}_2'^T & \mathbf{p}_4' \\ \mathbf{0}_3'^T & \end{bmatrix}\begin{bmatrix} \mathbf{p}_3' \\ 0 \end{bmatrix} = \mathbf{0}_3 .$$

Now, multiplying $\mathbf{P}_A\mathbf{D}$ with $\mathbf{p}_A'^{\perp}$ yields

$$\mathbf{P}_A\mathbf{D}\mathbf{p}_A'^{\perp} = \begin{bmatrix} \mathbf{p}_1^T\mathbf{R}\mathbf{p}_3' \\ \mathbf{p}_2^T\mathbf{R}\mathbf{p}_3' \\ 0 \end{bmatrix} . \tag{2.143}$$

Thus,

$$[\mathbf{P}_A\mathbf{D}\mathbf{p}_A'^{\perp}]_\times = \begin{bmatrix} 0 & 0 & \mathbf{p}_2^T\mathbf{R}\mathbf{p}_3' \\ 0 & 0 & -\mathbf{p}_1^T\mathbf{R}\mathbf{p}_3' \\ -\mathbf{p}_2^T\mathbf{R}\mathbf{p}_3' & \mathbf{p}_1^T\mathbf{R}\mathbf{p}_3' & 0 \end{bmatrix} .$$

Let us assume $\mathbf{P}_A'^{+} = \begin{bmatrix} \mathbf{Q} \\ \mathbf{q}_4^T \end{bmatrix}$, where $\mathbf{Q} = [\mathbf{q}_1 \quad \mathbf{q}_2 \quad \mathbf{q}_3]$ is a 3×3 matrix and $\mathbf{q}_4$ is a 3-vector. Since

$$\mathbf{P}_A'\mathbf{P}_A'^{+} = \begin{bmatrix} \mathbf{p}_1'^T\mathbf{Q} \\ \mathbf{p}_2'^T\mathbf{Q} \\ \mathbf{0}_3^T \end{bmatrix} + \mathbf{p}_4'\mathbf{q}_4^T = \mathbf{I}_3 ,$$

$\mathbf{q}_4$ can be uniquely determined as

$$\mathbf{q}_4 = \begin{bmatrix} 0 \\ 0 \\ \frac{1}{P_{34}'} \end{bmatrix} . \tag{2.144}$$

The constraint for matrix $\mathbf{Q}$ is then

$$\begin{bmatrix} \mathbf{p}_1'^T \\ \mathbf{p}_2'^T \end{bmatrix}\mathbf{Q} = \begin{bmatrix} 1 & 0 & -\frac{P_{14}'}{P_{34}'} \\ 0 & 1 & -\frac{P_{24}'}{P_{34}'} \end{bmatrix} . \tag{2.145}$$

It is evident that $\mathbf{Q}$ cannot be uniquely determined. In other words, any $\mathbf{Q}$ that satisfies the above equation suffices.

Now substituting these matrices for (2.85), we have the affine fundamental matrix

$$\mathbf{F}_A = \begin{bmatrix} 0 & 0 & a_{13} \\ 0 & 0 & a_{23} \\ a_{31} & a_{32} & a_{33} \end{bmatrix} , \tag{2.146}$$

where

$$\begin{aligned}
a_{13} &= \frac{P_{34}}{P'_{34}} \mathbf{p}_2^T \mathbf{R} \mathbf{p}'_3 , \\
a_{23} &= -\frac{P_{34}}{P'_{34}} \mathbf{p}_1^T \mathbf{R} \mathbf{p}'_3 , \\
a_{31} &= (-\mathbf{p}_2^T \mathbf{R} \mathbf{p}'_3 \mathbf{p}_1^T + \mathbf{p}_1^T \mathbf{R} \mathbf{p}'_3 \mathbf{p}_2^T) \mathbf{R} \mathbf{q}_1 , \\
a_{32} &= (-\mathbf{p}_2^T \mathbf{R} \mathbf{p}'_3 \mathbf{p}_1^T + \mathbf{p}_1^T \mathbf{R} \mathbf{p}'_3 \mathbf{p}_2^T) \mathbf{R} \mathbf{q}_2 , \\
a_{33} &= (-\mathbf{p}_2^T \mathbf{R} \mathbf{p}'_3 \mathbf{p}_1^T + \mathbf{p}_1^T \mathbf{R} \mathbf{p}'_3 \mathbf{p}_2^T)(\mathbf{R} \mathbf{q}_3 + \frac{1}{P_{34}} \mathbf{t}) .
\end{aligned}$$

Therefore, two matched points $\mathbf{m}$ and $\mathbf{m}'$ must satisfy the following equation:

$$\widetilde{\mathbf{m}}^T \mathbf{F}_A \widetilde{\mathbf{m}}' = 0 . \tag{2.147}$$

Expanding this epipolar equation, the lefthand side of the epipolar equation is a first-order polynomial of the image coordinates. It means that, as described in detail for the weak perspective projection and paraperspective projection, the epipolar lines are parallel everywhere in the image, and the orientations of the parallel epipolar lines are completely determined from the essential matrix.

Now, let us examine the constraints on the affine projection matrices if the affine fundamental matrix is given. Since $\mathbf{P}_A$ and $\mathbf{P}'_A$ are defined up to a scale factor, without loss of generality, we assume $P_{34} = P'_{34} = 1$. Then the relation between a 3D point and its 2D image is given by

$$\widetilde{\mathbf{m}} = \mathbf{P}_A \widetilde{\mathtt{M}} \quad \text{and} \quad \widetilde{\mathbf{m}}' = \mathbf{P}'_A \widetilde{\mathtt{M}} . \tag{2.148}$$

Note that there is no more scale factor in the above equations. From the affine epipolar equation (2.147), we have

$$\widetilde{\mathtt{M}}^T \underbrace{\mathbf{P}_A^T \mathbf{F}_A \mathbf{P}'_A}_{\mathbf{S}} \widetilde{\mathtt{M}} = 0 , \tag{2.149}$$

where

$$\begin{aligned}
\mathbf{S} &= \begin{bmatrix} P_{11} & P_{21} & 0 \\ P_{12} & P_{22} & 0 \\ P_{13} & P_{23} & 0 \\ P_{14} & P_{24} & 1 \end{bmatrix} \begin{bmatrix} 0 & 0 & a_{13} \\ 0 & 0 & a_{23} \\ a_{31} & a_{32} & a_{33} \end{bmatrix} \begin{bmatrix} P'_{11} & P'_{12} & P'_{13} & P'_{14} \\ P'_{21} & P'_{22} & P'_{23} & P'_{24} \\ 0 & 0 & 0 & 1 \end{bmatrix} \\
&= \begin{bmatrix} 0 & 0 & 0 & S_{14} \\ 0 & 0 & 0 & S_{24} \\ 0 & 0 & 0 & S_{34} \\ S_{41} & S_{42} & S_{43} & S_{44} \end{bmatrix} ,
\end{aligned}$$

with

$$
\begin{aligned}
S_{14} &= a_{13}P_{11} + a_{23}P_{21} \\
S_{24} &= a_{13}P_{12} + a_{23}P_{22} \\
S_{34} &= a_{13}P_{13} + a_{23}P_{23} \\
S_{41} &= a_{31}P'_{11} + a_{32}P'_{21} \\
S_{42} &= a_{31}P'_{12} + a_{32}P'_{22} \\
S_{43} &= a_{31}P'_{13} + a_{32}P'_{23} \\
S_{44} &= a_{13}P_{14} + a_{23}P_{24} + a_{31}P'_{14} + a_{32}P'_{24} + a_{33} .
\end{aligned}
$$

Equation (2.149) becomes

$$
(S_{14} + S_{41})X + (S_{24} + S_{42})Y + (S_{34} + S_{43})Z + S_{44} = 0 .
$$

Since this equation should be true for all points, the four coefficients must be all zero, which leads to

$$
\begin{aligned}
a_{13}P_{11} + a_{23}P_{21} + a_{31}P'_{11} + a_{32}P'_{21} &= 0 \\
a_{13}P_{12} + a_{23}P_{22} + a_{31}P'_{12} + a_{32}P'_{22} &= 0 \\
a_{13}P_{13} + a_{23}P_{23} + a_{31}P'_{13} + a_{32}P'_{23} &= 0 \\
a_{13}P_{14} + a_{23}P_{24} + a_{31}P'_{14} + a_{32}P'_{24} &= -a_{33} .
\end{aligned}
$$

We thus have 4 simple constraints on the coefficients of the projection matrices, which is consistent with the number of the degrees of freedom in an affine fundamental matrix. The above four equations can be written in the following vector form:

$$
a_{13}\mathbf{p}_1 + a_{23}\mathbf{p}_2 + a_{31}\mathbf{p}'_1 + a_{32}\mathbf{p}'_2 = \mathbf{0} , \tag{2.150}
$$

$$
a_{13}P_{14} + a_{23}P_{24} + a_{31}P'_{14} + a_{32}P'_{24} + a_{33} = 0 . \tag{2.151}
$$

Affine Camera as a Combination of Paraperspective Projection and Affine Transformation

In (2.145), the righthand side is the same as the left 3 vectors of the projection matrix for the paraperspective projection. If we construct another 4×4 matrix $\mathbf{Q}'$ as

$$
\mathbf{Q}' = \begin{bmatrix} \mathbf{Q} & \mathbf{0}_3 \\ \mathbf{0}_3^T & 1 \end{bmatrix} , \tag{2.152}
$$

then the projection matrix of an affine camera $\mathbf{P}_A$ can be transformed into that of the paraperspective projection by multiplying $\mathbf{Q}'$ from the right side,

$$\mathbf{P}_A\mathbf{Q}' = \mathbf{P}_{pp} . \tag{2.153}$$

Since there are only six constraints for the 9 unknowns in $\mathbf{Q}$, we can use only the six components in the upper triangle, and set the three other components to be zeros. The inverse matrix of $\mathbf{Q}$ is also an upper triangular matrix. This matrix resembles the camera intrinsic matrix. In the case of paraperspective projection, $\mathbf{Q}$ can be an identity matrix.

Substituting the above equation for (2.148), we have

$$s\tilde{\mathtt{m}} = \mathbf{P}_A\tilde{\mathtt{M}} = \mathbf{P}_{pp}\mathbf{Q}'^{-1}\tilde{\mathtt{M}} = \mathbf{P}_{pp}\tilde{\mathtt{M}}_{pp} , \tag{2.154}$$

where

$$\mathtt{M}_{pp} = \mathbf{Q}^{-1}\mathtt{M} . \tag{2.155}$$

The effect of $\mathbf{Q}$ can be understood as transforming a non-orthogonal coordinate system into an orthogonal system, with the constraints: (1) the two systems have the same origin; (2) the z-axes have the same direction; (3) the y-axis is rotated within the yz-plane. Given the projection matrix for a general affine camera, $\mathbf{Q}$ can be uniquely determined with the above 3 constraints.

In conclusion, a general affine camera can be thought of as a combination of an 3D affine transformation and a paraperspetive projection.

2.6 EPIPOLAR GEOMETRY BETWEEN TWO IMAGES WITH LENS DISTORTION

Up to now, and in almost all work on multiple-views problems in computer vision (with an exception of [170]), a camera is always modeled as being linearly projective, i.e. the homogeneous coordinates of a 3D point and those of an image point are related by a 3×4 matrix. This statement does not imply, though, that camera distortion has never been accounted for in such work. Indeed, distortion has usually been corrected off-line using classical methods by observing for example straight lines, if it is not weak enough to be neglected. Our treatment in this book intends to consider camera distorsion as an integral part of a camera.

2.6.1 Camera Distortion Modelling

Following [17, 33, 163], we can model the transformation from 3D world coordinates to camera pixel coordinates as a process of four steps:

Step 1: Rigid transformation from the object world coordinate system (X_w, Y_w, Z_w) to the camera 3D coordinate system (X, Y, Z):

$$\begin{bmatrix} X \\ Y \\ Z \end{bmatrix} = \mathbf{R} \begin{bmatrix} X_w \\ Y_w \\ Z_w \end{bmatrix} + \mathbf{t} .$$

Step 2: Perspective projection from 3D camera coordinates (X, Y, Z) to *ideal* image coordinates (x, y) under pinhole camera mdel:

$$x = f\frac{X}{Z} ,$$
$$y = f\frac{Y}{Z} ,$$

where f is the effective focal length.

Step 3: Lens distortion:

$$x = \hat{x} + \delta_x ,$$
$$y = \hat{y} + \delta_y ,$$

where $(\hat{x}, \hat{y})$ are the *distorted* or *true* image coordinates on the image plane, and (δ_x, δ_y) are the distortion corrections to (x, y). We will return back to this point later.

Step 4: Affine transformation from real image coordinates $(\hat{x}, \hat{y})$ to *frame buffer* (pixel) image coordinates (u, v):

$$u = d_x^{-1}\hat{x} + u_0 ,$$
$$v = d_y^{-1}\hat{y} + v_0 ,$$

where (u_0, v_0) are the coordinates of the image center (the principal point) in the frame buffer; d_x and d_y are the distances between adjacent pixels in the horizontal and vertical directions of the image plane, respectively.

Now let us examine how to model camera distortion. There are mainly two kinds of distortion: radial and decentering [143, 17]. The distortion corrections are expressed in the usual representation as power series in radial distance r:

$$\delta_x = \hat{x}(k_1 r^2 + k_2 r^4 + k_3 r^6 + \cdots) + [p_1(r^2 + 2\hat{x}^2) + 2p_2\hat{x}\hat{y}](1 + p_3 r^2 + \cdots) , \qquad (2.156)$$

$$\delta_y = \hat{y}(k_1 r^2 + k_2 r^4 + k_3 r^6 + \cdots) + [2p_1\hat{x}\hat{y} + p_2(r^2 + 2\hat{y}^2)](1 + p_3 r^2 + \cdots) , \qquad (2.157)$$

where $r = \sqrt{\hat{x}^2 + \hat{y}^2}$, k_1, k_2 and k_3 are coefficients of radial distortion, and p_1, p_2 and p_3 are coefficients of decentering distortion. Under radial (symmetric) distortion, which is caused by imperfect lens shape, the ideal image points are distorted along radial directions from the distortion center (here the same as the principal point). The radial distortion is symmetric (see Fig. 2.13). Under decentering distortion which is usually caused by improper lens assem-

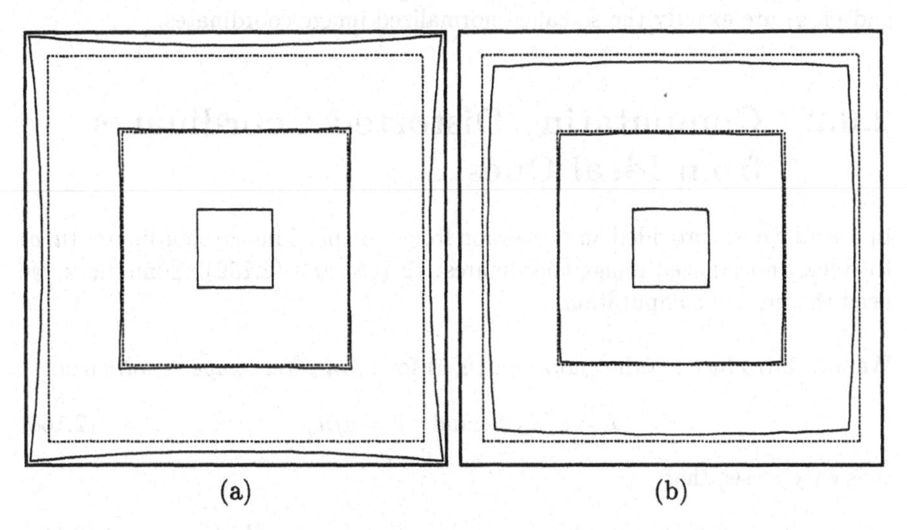

Figure 2.13 Image under radial distortion; image resolution 512 × 512; dashed curves are the ideal points. (a) $k_1 d_x^2 = -5e-07$ and $k_1 d_y^2 = -7e-07$; (b) $k_1 d_x^2 = 5e-07$ and $k_1 d_y^2 = 7e-07$

bly, the ideal image points are distorted in both radial and tangential directions. Depending on the lenses used, one needs to choose an appropriate distortion model. Based on the reports in the literature [17, 163], unless one is specifically concerned with the reduction of distortion to very low levels, it is likely that the distortion function is totally dominated by the radial components, and especially dominated by the k_1 term. Tsai [163], who used the first two radial terms, ever claimed that any more elaborate modelling not only would not help (negligible when compared with sensor quantization), but also would cause numerical instability. This has also been confirmed by Wei and Ma [170, section 3.4]. We thus consider only the first radial term in the sequel, although mathematically there is no reason to do this.

Combining Step 3 and Step 4 yields:

$$x = d_x(u - u_0)[1 + (u - u_0)^2 k_1 d_x^2 + (v - v_0)^2 k_1 d_y^2] , \tag{2.158}$$
$$y = d_y(v - v_0)[1 + (u - u_0)^2 k_1 d_x^2 + (v - v_0)^2 k_1 d_y^2] . \tag{2.159}$$

Examining the perspective model in Step 2, it is clear that we can not determine simultaneously f, d_x and d_y from visual information. We must specify either f, or one of d_x and d_y, or the ratio d_y/d_x using, for example, the information provided by camera manufacturers. For the problem at hand, we assume $f = 1$, and (x, y) are exactly the so-called normalized image coordinates.

2.6.2 Computating Distorted Coordinates from Ideal Ones

In the above, we provided an expression from true pixel image coordinates (u, v) to (ideal) normalized image coordinates: (2.158) and (2.159). Sometimes, we need the inverse computation.

We now introduce another notation $(\tilde{u}, \tilde{v})$ for *ideal pixel image coordinates*:

$$\tilde{u} = x/d_x , \quad \text{and} \quad \tilde{v} = y/d_y . \tag{2.160}$$

It is easy to see that

$$\tilde{u} = (u - u_0)[1 + (u - u_0)^2 k_1 d_x^2 + (v - v_0)^2 k_1 d_y^2] , \tag{2.161}$$
$$\tilde{v} = (v - v_0)[1 + (u - u_0)^2 k_1 d_x^2 + (v - v_0)^2 k_1 d_y^2] . \tag{2.162}$$

Divide the above two equations, and we have

$$\frac{\tilde{u}}{\tilde{v}} = \frac{u - u_0}{v - v_0} .$$

Substituting $(u - u_0)$ into (2.161) yields

$$\tilde{v} = (v - v_0)[1 + (v - v_0)^2 k_1 d_x^2 \frac{\tilde{u}^2}{\tilde{v}^2} + (v - v_0)^2 k_1 d_y^2] ,$$

i.e. :

$$(v - v_0)^3 + p(v - v_0) + q = 0\,, \tag{2.163}$$

where

$$p = \frac{\tilde{v}^2}{k_1(d_x^2\tilde{u}^2 + d_y^2\tilde{v}^2)}\,,$$

$$q = -\frac{\tilde{v}^3}{k_1(d_x^2\tilde{u}^2 + d_y^2\tilde{v}^2)} = -\tilde{v}p\,.$$

If $k_1 = 0$ (no distortion), there is only one solution: $v = \tilde{v} + v_0$. Let $\Delta = \left(\frac{q}{2}\right)^2 + \left(\frac{p}{3}\right)^2$. If $\Delta > 0$, then there is only one solution; if $\Delta = 0$, then $v = v_0$, which occurs when $\tilde{v} = 0$; if $\Delta < 0$, then there are three solutions, and in general the middle one is what we need.

Once v is solved, u coordinate is given by $u_0 + \tilde{u}(v - v_0)/\tilde{v}$.

2.6.3 Epipolar Constraint Between Two Images with Distortion

Given two points (u, v) and (u', v') in correspondence, there is a constraint on them even when the two images exhibit lens distortion. If we consider the ideal pixel coordinates $(\tilde{u}, \tilde{v})$ and $(\tilde{u}', \tilde{v}')$ based on (2.161) and (2.162), we see from the discussions in the previous sections that there exists a 3×3 so-called fundamental matrix $\mathbf{F}$ such that:

$$[\tilde{u}, \tilde{v}, 1]\mathbf{F}\begin{bmatrix}\tilde{u}'\\ \tilde{v}'\\ 1\end{bmatrix} = 0\,,$$

where

$$\begin{aligned}
\tilde{u} &= (u - u_0)[1 + (u - u_0)^2 k_1 d_x^2 + (v - v_0)^2 k_1 d_y^2]\,,\\
\tilde{v} &= (v - v_0)[1 + (u - u_0)^2 k_1 d_x^2 + (v - v_0)^2 k_1 d_y^2]\,,\\
\tilde{u}' &= (u' - u_0')[1 + (u' - u_0')^2 k_1' d_x'^2 + (v' - v_0')^2 k_1' d_y'^2]\,,\\
\tilde{v}' &= (v' - v_0')[1 + (u' - u_0')^2 k_1' d_x'^2 + (v' - v_0')^2 k_1' d_y'^2]\,.
\end{aligned}$$

If we go back to use the true pixel coordinates, we then have a constraint:

$$g(u, v; u', v') \equiv [\tilde{u}, \tilde{v}, 1]\mathbf{F}\begin{bmatrix}\tilde{u}'\\ \tilde{v}'\\ 1\end{bmatrix} = 0\,. \tag{2.164}$$

For a point (u', v') in the second image, unlike in the case of distortion-free images, its corresponding point in the first image does not lie on a line anymore. As a matter of fact, equation (2.164) describes a cubic curve in (u, v) on which the corresponding point must lie. We call the curve $g(u, v; u', v')$ the *epipolar curve* of point (u', v'). Symmetrically, for a point (u, v) in the first image, $g(u, v; u', v')$ describes the epipolar curve in (u', v') in the second image on which the corresponding point must lie.

2.7 SUMMARY

We began with cameras modeling. A camera is modeled as a pinhole, which performs a perspective projection. Intrinsic parameters and extrinsic parameters are introduced. The relationship between a space point and its image onto the image is nonlinear, and can be described by the projection matrix using homogeneous coordinates. We then presented several linear approximations to the nonlinear full perspective projection. Weak perspective and paraperspective projections are two well-known approximations, and all the linear approximations can be generalized as the affine camera.

We then provided a detailed description of epipolar geometry. We have described the geometric concepts of the epipolar geometry under full perspective projection. Then, it has been formulated as an essential matrix with normalized image coordinates (i.e. when the intrinsic parameters are known), and a fundamental matrix if pixel coordinates are used. To be complete, we have shown also how to compute the fundamental matrix if camera perspective projection matrices are known. The relationship between fundamental matrix and epipolar transformation was described.

A general form of epipolar equation has been derived using pseudoinverse of the camera projection matrices, which does not depend on any specific projection model. The epipolar equations for orthographic, weak perspective, paraperspective and general affine projections have then been specified and their geometric interpretations have been given.

Finally we have also included a discussion on the effects of lens distortion on the epipolar geometry.

2.A APPENDIX

2.A.1 Thin and Thick Lens Camera Models

Although pinhole cameras model quite well most of the cameras we use in computer vision community, they cannot be used physically in a real imaging system. This is for two reasons:

- An ideal pinhole, having an infinitesimal aperture, does not allow to gather enough amount of light to produce measurable image brightness (called *image irradiance*).
- Because of the wave nature of light, diffraction occurs at the edge of the pinhole and the light spread over the image [70]. As the pinhole is made smaller and smaller, a larger and larger fraction of the incoming light is deflected far from the direction of the incoming ray.

To avoid these problems, a real imaging system usually uses lenses with finite aperture. This appendix aims at having the reader know that there are other camera models available. One should choose an appropriate model for a particular imaging device.

For an ideal lens, which is known as the *thin lens*, all optical rays parallel to the optical axis converge to a point on the optical axis on the other side of the lens at a distance equal to the so-called *focal length* f (see Fig. 2.14). The light ray through the center of the lens is undeflected, thus a thin lens produces the same projection as the pinhole. However, it gathers also a finite amount of light reflected from (or emitted by) the object (see Fig. 2.15). By the familiar *thin lens law*, rays from points at a distance Z are focused by the lens at a distance $-F$, and Z and $-F$ satisfy

$$\frac{1}{Z} + \frac{1}{-F} = \frac{1}{-f}, \tag{2.165}$$

where f is the focal length.

If we put an image plane at the distance $-F$, then points at other distances than Z are imaged as small blur circles. This can be seen by considering the cone of light rays passing through the lens with apex at the point where they are correctly focused [70]. The size of the blur circle can be determined as follows. A point at distance $\hat{Z}$ is focused if it is imaged at a point $-\hat{F}$ from the

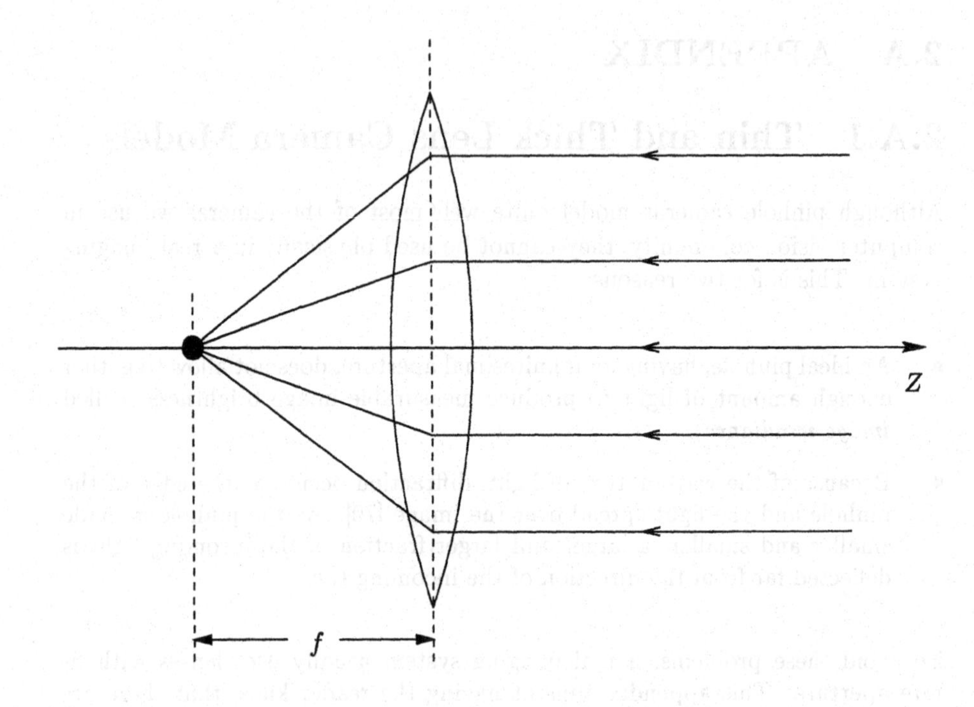

Figure 2.14 Cross-sectional view of a thin lens sliced by a plane containing the optical axis. All light rays parallel to the optical axis converge to a point at a distance equal to the focal length

lens (see Fig. 2.16), where

$$\frac{1}{\hat{Z}} + \frac{1}{-\hat{F}} = \frac{1}{-f}.$$

It gives rise to a blur circle on the image plane located at distance $-F$. The diameter of the blur circle, e, can be computed by triangle similarity

$$\frac{e}{d} = \frac{|F - \hat{F}|}{\hat{F}},$$

which gives

$$e = \frac{d}{\hat{F}}|F - \hat{F}| = \frac{fd}{\hat{Z}}\frac{|Z - \hat{Z}|}{Z - f},$$

where d is the diameter of the lens. If the diameter of blur circles, e, is less than the resolution of the image, then the object is well focused and its image is

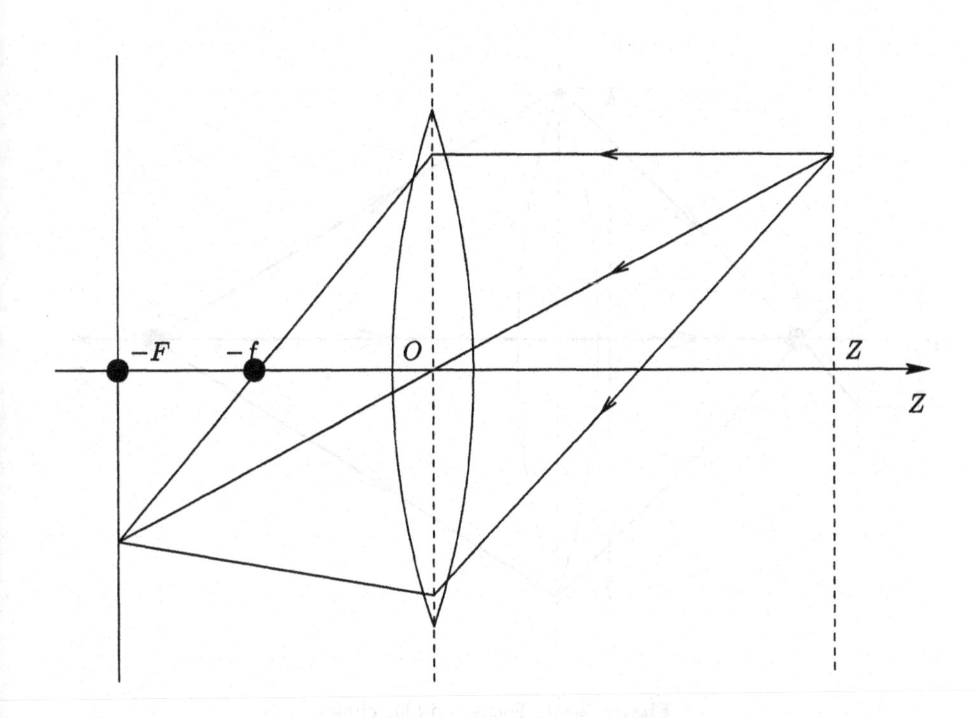

Figure 2.15 A thin lens gathers light from a finite area and produces a well-focused image at a particular distance

clean. The range of distances over which objects are focused "sufficiently well" is called the *depth of field*. It is clear that the larger the lens aperture d, the less the depth of field.

From (2.165), it is seen that for objects relatively distant from the lens (i.e. $Z \gg f$), we have $F = f$. If the image plane is located at distance f from the lens, then the camera can be modeled reasonably well by the pinhole.

It is difficult to manufacture a perfect lens. In practice, several simple lenses are carefully assembled to make a compound lens with better properties. In an imaging device with mechanism of focus and zoom, the lenses are allowed to move. It appears difficult to model such a device by a pinhole or thin lens. Another model, called the *thick lens*, is used by more and more researchers [113, 85].

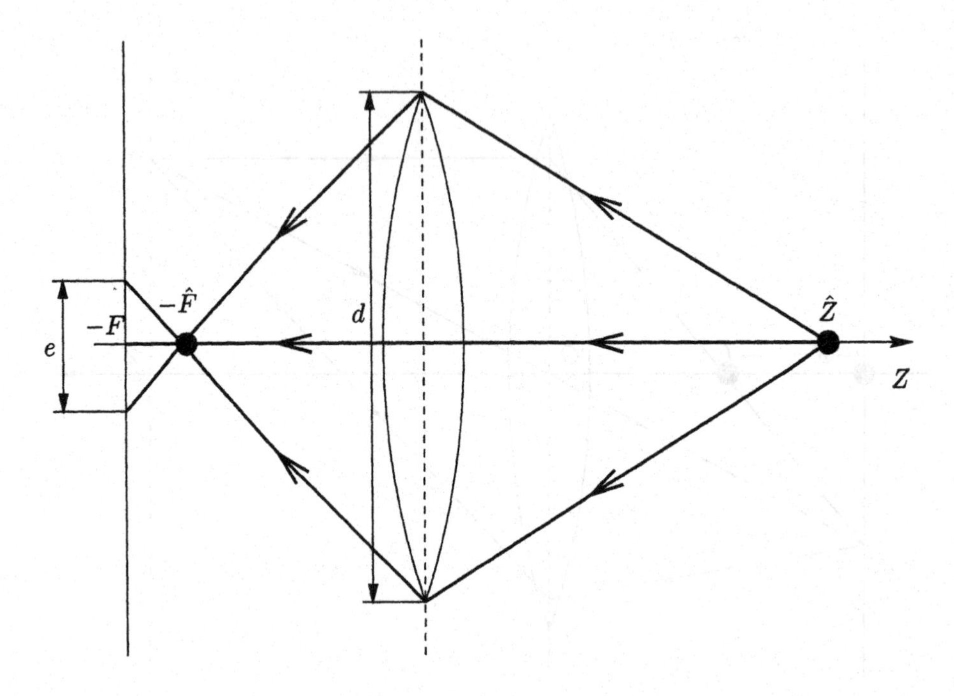

Figure 2.16 Focus and blur circles

An ideal thick lens is illustrated in Fig. 2.17. Is is composed of two lenses, each having two opposite surfaces one spherical and the other plane. These two planes p_1 and p_2, called the *principal planes*, are perpendicular to the optical axis, and are separated by a distance t, called the *thickness of the lens.* The principal planes intersect the optical axis at two points, called the *nodal points*. The thick lens produces the same perspective projection as the ideal thin lens, except for *an additional offset* equal to the lens thickness t along the optical axis. A light ray arriving at the first nodal point leaves the rear nodal point without changing direction. A thin lens can then be considered as a thick lens with $t = 0$.

It is thus clear that a thick lens can be considered as a thin lens if the object is relatively distant to the camera compared to the lens thickness (i.e. $\hat{Z} \gg t$). It can be further approximated by a pinhole only when the object is well focused (i.e. $F \approx \hat{F}$), and this is valid only locally.

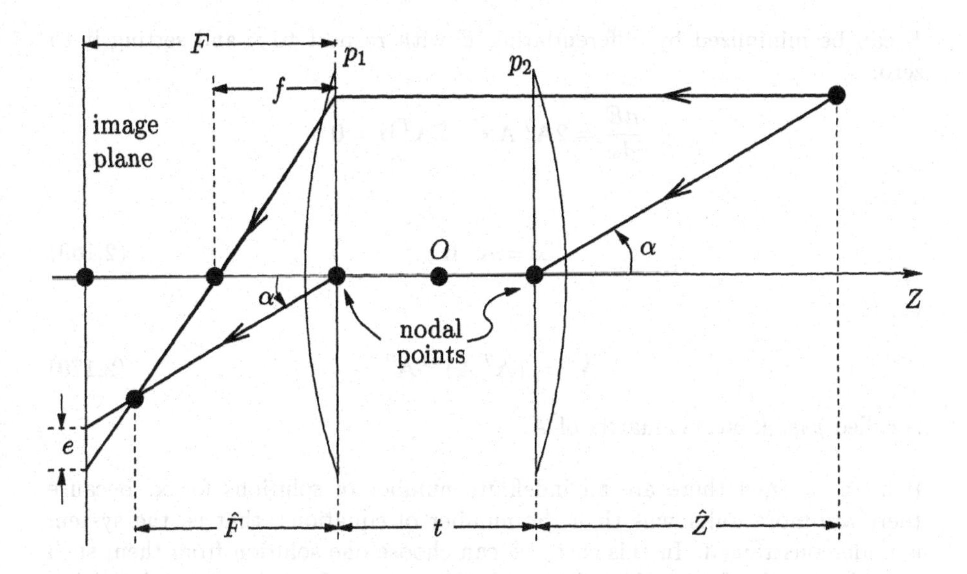

Figure 2.17 Cross-sectional view of a thick lens sliced by a plane containing the optical axis

2.A.2 Inverse and Pseudoinverse Matrices

For a group of linear equations, which we assume are independent of each other, using linear algebra, we can write them in the form of

$$\mathbf{A}\mathbf{x} = \mathbf{b}\,, \tag{2.166}$$

where $\mathbf{A}$ is a $m \times n$ matrix, $\mathbf{x}$ is a n vector of unknowns, and $\mathbf{b}$ is a m vector.

If $m = n$, then the inverse matrix of $\mathbf{A}$ exists (because we assume that the equations are linearly independent of each other.) $\mathbf{x}$ can be determined as

$$\mathbf{x} = \mathbf{A}^{-1}\mathbf{b}\,. \tag{2.167}$$

If $m > n$, then there does not exist a solution for $\mathbf{x}$ except for the noise-free case, as there are more equations than unknowns, that is, the system is overconstrained. In that case, one can make a compromise so as to find a solution of $\mathbf{x}$ such that the residual is minimized. Define the residual as

$$E = (\mathbf{A}\mathbf{x} - \mathbf{b})^T(\mathbf{A}\mathbf{x} - \mathbf{b})\,. \tag{2.168}$$

E can be minimized by differentiating E with respect to $\mathbf{x}$ and setting it to zero.

$$\frac{dE}{d\mathbf{x}} = 2\mathbf{A}^T\mathbf{A}\mathbf{x} - 2\mathbf{A}^T\mathbf{b} = \mathbf{0}$$

Thus,

$$\mathbf{x} = \mathbf{A}^+\mathbf{b} , \quad (2.169)$$

where

$$\mathbf{A}^+ = (\mathbf{A}^T\mathbf{A})^{-1}\mathbf{A}^T \quad (2.170)$$

is called *pseudoinverse* matrix of $\mathbf{A}$.

If $m < n$, then there are an indefinite number of solutions for $\mathbf{x}$, because there are more unknowns than the number of equations, that is, the system is underconstrained. In this case, we can choose one solution from them such that the length of $\mathbf{x}$ itself is minimized. This can be formulated as minimizing

$$E = \mathbf{x}^T\mathbf{x} + \boldsymbol{\lambda}^T(\mathbf{b} - \mathbf{A}\mathbf{x}) , \quad (2.171)$$

where $\boldsymbol{\lambda} = [\lambda_1, ..., \lambda_n]^T$ is the Lagrange multiplier. Differentiating E with respect to $\mathbf{x}$ and setting it to zero, we have

$$2\mathbf{x} - \mathbf{A}^T\boldsymbol{\lambda} = \mathbf{0} ,$$

or

$$\mathbf{x} = \frac{1}{2}\mathbf{A}^T\boldsymbol{\lambda} . \quad (2.172)$$

And, differentiating E with respect to $\boldsymbol{\lambda}$ and setting it to zero, we have

$$\mathbf{A}\mathbf{x} = \mathbf{b} ,$$

which is the original equation.

Substituting (2.172) for the original equation, we have

$$\boldsymbol{\lambda} = 2(\mathbf{A}\mathbf{A}^T)^{-1}\mathbf{b} . \quad (2.173)$$

Putting this back to (2.172) yields

$$\mathbf{x} = \mathbf{A}^+\mathbf{b} , \quad (2.174)$$

where

$$\mathbf{A}^{+} = \mathbf{A}^{T}(\mathbf{A}\mathbf{A}^{T})^{-1} \tag{2.175}$$

is another form of *pseudoinverse* matrix of $\mathbf{A}$.

Eq.(2.174) is only one solution out of an indefinite number of solutions. The class of solutions can be expressed as

$$\mathbf{x} = \mathbf{A}^{+}\mathbf{b} + \mathbf{y}\,, \tag{2.176}$$

where $\mathbf{y}$, which has the same dimension as $\mathbf{x}$, is an arbitrary n vector that satisfies

$$\mathbf{A}\mathbf{y} = \mathbf{0}\,.$$

It is easy to verify that $\mathbf{y}$ can be expressed as

$$\mathbf{y} = (\mathbf{I} - \mathbf{A}^{+}\mathbf{A})\boldsymbol{\omega} = (\mathbf{I} - \mathbf{A}(\mathbf{A}^{T}\mathbf{A})^{-1}\mathbf{A}^{T})\boldsymbol{\omega}\,, \tag{2.177}$$

where $\boldsymbol{\omega}$ is an arbitrary n vector.

The pseudoinverse matrices in (2.170) and (2.175) actually do the same operation. If we let $\mathbf{B} = \mathbf{A}^{T}$, and substitute this for, say, (2.175), we have

$$(\mathbf{A}^{+})^{T} = (\mathbf{B}^{T}\mathbf{B})^{-1}\mathbf{B}^{T}\,,$$

which has the same form as (2.170).

Using the pseudoinverse, the residual of (2.166) can be expressed as

$$(\mathbf{I} - \mathbf{A}^{T}(\mathbf{A}\mathbf{A}^{T})^{-1}\mathbf{A})\mathbf{b}\,, \quad \text{for } m < n\,,$$

or

$$(\mathbf{I} - \mathbf{A}(\mathbf{A}^{T}\mathbf{A})^{-1}\mathbf{A}^{T})\mathbf{b}\,, \quad \text{for } m > n\,.$$

Now let us discuss a special case of $\mathbf{A}$, that is, $n - m = 1$. The result for the case of $m - n = 1$ is the same.

Let us construct a $n \times n$ matrix $\mathbf{C}$,

$$\mathbf{C} = \begin{bmatrix} \mathbf{A} \\ \mathbf{y}^{T} \end{bmatrix}\,, \tag{2.178}$$

where $\mathbf{y}$ satisfies $\mathbf{Ay} = \mathbf{0}$.

As

$$\mathbf{I} = \mathbf{C}^T \left(\mathbf{C}\mathbf{C}^T\right)^{-1} \mathbf{C} = \mathbf{A}^T \left(\mathbf{A}\mathbf{A}^T\right)^{-1} \mathbf{A} + \frac{\mathbf{y}\mathbf{y}^T}{\|\mathbf{y}\|^2} ,$$

we have

$$\mathbf{I} - \mathbf{A}^T \left(\mathbf{A}\mathbf{A}^T\right)^{-1} \mathbf{A} = \frac{\mathbf{y}\mathbf{y}^T}{\|\mathbf{y}\|^2} . \qquad (2.179)$$

Substituting this for (2.177), it is easy to verify that $\frac{\mathbf{y}\mathbf{y}^T}{\|\mathbf{y}\|^2}\omega$ does produce $\mathbf{y}$ up to a scale factor.

3

RECOVERY OF EPIPOLAR GEOMETRY FROM POINTS

In this chapter, we present a number of methods to recover the epipolar geometry between two images from point matches. When the images are calibrated (i.e. their intrinsic parameters are known), the problem is equivalent to determining motion and structure. This has been widely studied in the last fifteen years. The minimum number of required point matches is five, which may give at most ten real solutions [39]. If more than 5 point matches are given, there usually exists a unique solution. When 8 or more point matches are available, a linear technique exists to solve the motion, which, however, usually yields an estimate sensitive to noise. Least-squares techniques are exploited to obtain a more robust estimate. Readers are referred to [41, Chap.7] for a complete exposition of the problem of motion and structure from motion.

3.1 DETERMINING FUNDAMENTAL MATRIX UNDER FULL PERSPECTIVE PROJECTION

We will describe different ways to estimate the fundamental matrix from point matches in pixel coordinates. The full perspective projection model is used. Neither the intrinsic parameters of the images nor the extrinsic parameters are assumed to be known. The only assumption is that the points undergo a rigid transformation between the two camera coordinate systems. This implies that all following problems are mathematically equivalent:

- The two images are taken by a moving camera at two different time instants in a static environment.
- The two images are taken by a fixed camera at two different time instants in a dynamic scene. We assume the two images are projections of a single moving rigid object, otherwise a pre-segmentation of images into different regions is necessary.
- The two images are taken by two cameras at the same time.
- The two images are taken by two cameras at two different instants. The scene is assumed to be static, or the images are assumed to be pre-segmented.

As one may remark, the segmentation of a scene into different moving objects is very important in practical applications, which will be discussed in Sect. 3.3.

Let a point $\mathbf{m}_i = [u_i, v_i]^T$ in the first image be matched to a point $\mathbf{m}'_i = [u'_i, v'_i]^T$ in the second image. They must satisfy the epipolar equation (2.54), i.e. $\widetilde{\mathbf{m}}_i^T \mathbf{F} \widetilde{\mathbf{m}}'_i = 0$. This equation can be written as a linear and homogeneous equation in the 9 unknown coefficients of matrix $\mathbf{F}$:

$$\mathbf{u}_i^T \mathbf{f} = 0 , \tag{3.1}$$

where

$$\begin{aligned}
\mathbf{u}_i &= [u_i u'_i, u_i v'_i, u_i, v_i u'_i, v_i v'_i, v_i, u'_i, v'_i, 1]^T \\
\mathbf{f} &= [F_{11}, F_{12}, F_{13}, F_{21}, F_{22}, F_{23}, F_{31}, F_{32}, F_{33}]^T ,
\end{aligned}$$

where F_{ij} is the element of $\mathbf{F}$ at row i and column j.

If we are given n point matches, by stacking (3.1), we have the following linear system to solve:

$$\mathbf{U}_n \mathbf{f} = \mathbf{0} ,$$

where

$$\mathbf{U}_n = \begin{bmatrix} \mathbf{u}_1^T \\ \vdots \\ \mathbf{u}_n^T \end{bmatrix} .$$

This set of linear homogeneous equations, together with the rank constraint of the matrix $\mathbf{F}$, allow to estimate the epipolar geometry.

3.1.1 Exact Solution with 7 Point Matches

As described in Sect. 2.3.3, a fundamental matrix $\mathbf{F}$ has only 7 degrees of freedom. Thus, 7 is the minimum number of point matches required for having a solution of the epipolar geometry.

In this case, $n = 7$ and $\mathrm{rank}(\mathbf{U}_7) = 7$. Through singular value decomposition, we obtain vectors $\mathbf{f}_1$ and $\mathbf{f}_2$ which span the null space of $\mathbf{U}_7$. The null space is a linear combination of $\mathbf{f}_1$ and $\mathbf{f}_2$, which correspond to matrices $\mathbf{F}_1$ and $\mathbf{F}_2$, respectively. Because of its homogeneity, the fundamental matrix is a one-parameter family of matrices $\alpha\mathbf{F}_1 + (1-\alpha)\mathbf{F}_2$. Since the determinant of $\mathbf{F}$ must be null, i.e.

$$\det[\alpha\mathbf{F}_1 + (1-\alpha)\mathbf{F}_2] = 0 ,$$

we obtain a cubic polynomial in α. The maximum number of real solutions is 3. For each solution α, the fundamental matrix is then given by

$$\mathbf{F} = \alpha\mathbf{F}_1 + (1-\alpha)\mathbf{F}_2 .$$

Actually, this technique has already been used in estimating the essential matrix when 7 point matches in normalized coordinates are available [72]. It is also mentioned in [156] for estimating the fundamental matrix.

As a matter of fact, this result, that there may have three solutions given 7 matches, has been known since 1800's [64, 148]. Sturm's algorithm [148] computes the epipoles and the epipolar transformation (see Sect. 2.3.5) from 7 point matches. It is based on the observation that the epipolar lines in the two images are related by a homography, and thus the cross-ratios of four epipolar lines is invariant. In each image, the 7 points define 7 lines going through the unknown epipole, thus providing 4 independent cross-ratios. Since these cross-ratios should remain the same in the two images, one obtains 4 cubic polynomial equations in the coordinates of the epipoles (4 independent parameters). It is shown that there may exist up to three solutions for the epipoles.

3.1.2 Analytic Method with 8 or More Point Matches

In practice, we are given more than 7 matches and we use a least-squares method to solve

$$\min_{\mathbf{F}} \sum_i (\tilde{\mathbf{m}}_i^T \mathbf{F} \tilde{\mathbf{m}}'_i)^2 , \tag{3.2}$$

which can be rewritten as:

$$\min_{\mathbf{f}} \|\mathbf{U}_n\mathbf{f}\|^2 \,. \tag{3.3}$$

The vector **f** is only defined up to an unknown scale factor. The trivial solution **f** to the above problem is $\mathbf{f} = \mathbf{0}$, which is not what we want. To avoid it, we need to impose some constraint on the coefficients of the fundamental matrix. Several methods are possible and are presented below. We will call them *the 8-point algorithm*, although more than 8 point matches can be used.

Linear Least-Squares Technique

The first method sets one of the coefficients of **F** to 1, and then solves the above problem using linear least-squares techniques. Without loss of generality, we assume that the last element of vector **f** (i.e. $f_9 = F_{33}$) is not equal to zero, and thus we can set $f_9 = -1$. This gives

$$\begin{aligned}\|\mathbf{U}_n\mathbf{f}\|^2 &= \|\mathbf{U}'_n\mathbf{f}' - \mathbf{c}_9\|^2 \\ &= \mathbf{f}'^T\mathbf{U}'^T_n\mathbf{U}'_n\mathbf{f}' - 2\mathbf{c}_9^T\mathbf{U}'_n\mathbf{f}' + \mathbf{c}_9^T\mathbf{c}_9 \,,\end{aligned}$$

where $\mathbf{U}'_n$ is the $n \times 8$ matrix composed of the first 8 columns of $\mathbf{U}_n$, and $\mathbf{c}_9$ is the ninth column of $\mathbf{U}_n$. The solution is obtained by requiring the first derivative to be zero, i.e.

$$\frac{\partial\|\mathbf{U}_n\mathbf{f}\|^2}{\partial\mathbf{f}'} = \mathbf{0} \,.$$

By definition of vector derivatives, $\partial(\mathbf{a}^T\mathbf{x})/\partial\mathbf{x} = \mathbf{a}$, for all vector **a**. We thus have

$$2\mathbf{U}'^T_n\mathbf{U}'_n\mathbf{f}' - 2\mathbf{U}'^T_n\mathbf{c}_9 = \mathbf{0} \,,$$

or

$$\mathbf{f}' = \left(\mathbf{U}'^T_n\mathbf{U}'_n\right)^{-1}\mathbf{U}'^T_n\mathbf{c}_9 \,.$$

The problem with this method is that we do not know a priori which coefficient is not zero. If we set an element to 1 which is actually zero or much smaller than the other elements, the result will be catastrophic. A remedy is to try all nine possibilities by setting one of the nine coefficients of **F** to 1 and retain the best estimation.

Eigen Analysis

The second method consists in imposing a constraint on the norm of **f**, and in particular we can set

$$\|\mathbf{f}\| = 1 .$$

Compared to the previous method, no coefficient of **F** prevails over the others. In this case, the problem (3.3) becomes a classical one:

$$\min_{\mathbf{f}} \|\mathbf{U}_n \mathbf{f}\|^2 \quad \text{subject to } \|\mathbf{f}\| = 1 . \tag{3.4}$$

It can be transformed into an unconstrained minimization problem through Lagrange multipliers:

$$\min_{\mathbf{f}} \mathcal{F}(\mathbf{f}, \lambda) , \tag{3.5}$$

where

$$\mathcal{F}(\mathbf{f}, \lambda) = \|\mathbf{U}_n \mathbf{f}\|^2 + \lambda(1 - \|\mathbf{f}\|^2) \tag{3.6}$$

and λ is the Lagrange multiplier. By requiring the first derivative of $\mathcal{F}(\mathbf{f}, \lambda)$ with respect to **f** to be zero, we have

$$\mathbf{U}_n^T \mathbf{U}_n \mathbf{f} = \lambda \mathbf{f} .$$

Thus, the solution **f** must be a unit eigenvector of the 9×9 matrix $\mathbf{U}_n^T \mathbf{U}_n$ and λ is the corresponding eigenvalue. Since matrix $\mathbf{U}_n^T \mathbf{U}_n$ is symmetric and positive semi-definite, all its eigenvalues are real and positive or zero. Without loss of generality, we assume the nine eigenvalues of $\mathbf{U}_n^T \mathbf{U}_n$ are in non-increasing order:

$$\lambda_1 \geq \cdots \geq \lambda_i \geq \cdots \geq \lambda_9 \geq 0 .$$

We therefore have 9 potential solutions: $\lambda = \lambda_i$ for $i = 1, \ldots, 9$. Back substituting the solution to (3.6) gives

$$\mathcal{F}(\mathbf{f}, \lambda_i) = \lambda_i .$$

Since we are seeking to minimize $\mathcal{F}(\mathbf{f}, \lambda)$, the solution to (3.4) is evidently the unit eigenvector of matrix $\mathbf{U}_n^T \mathbf{U}_n$ associated to the *smallest* eigenvalue, i.e. λ_9.

Imposing the Rank-2 Constraint

The advantage of the linear criterion is that it yields an analytic solution. However, we have found that it is quite sensitive to noise, even with a large set of data points. One reason is that the rank-2 constraint (i.e. $\det \mathbf{F} = 0$) is not satisfied. We can impose this constraint a posteriori. The most convenient way is to replace the matrix $\mathbf{F}$ estimated with any of the above methods by the matrix $\hat{\mathbf{F}}$ which minimizes the Frobenius norm (see Sect. 3.A.1) of $\mathbf{F} - \hat{\mathbf{F}}$ subject to the constraint $\det \hat{\mathbf{F}} = 0$. Let

$$\mathbf{F} = \mathbf{U}\mathbf{S}\mathbf{V}^T$$

be the singular value decomposition of matrix $\mathbf{F}$, where $\mathbf{S} = \mathrm{diag}(\sigma_1, \sigma_2, \sigma_3)$ is a diagonal matrix satisfying $\sigma_1 \geq \sigma_2 \geq \sigma_3$ (σ_i is the ith singular value), and $\mathbf{U}$ and $\mathbf{V}$ are orthogonal matrices. It can be shown that

$$\hat{\mathbf{F}} = \mathbf{U}\hat{\mathbf{S}}\mathbf{V}^T$$

with $\hat{\mathbf{S}} = \mathrm{diag}(\sigma_1, \sigma_2, 0)$ minimizes the Frobenius norm of $\mathbf{F} - \hat{\mathbf{F}}$ (see the appendix to this chapter, Sect. 3.A.1, for the proof). (This method was used by Tsai and Huang [162] in estimating the essential matrix.)

Geometric Interpretation of the Linear Criterion

Another problem with the linear criterion is that the quantity we are minimizing is not physically meaningful. A physically meaningful quantity should be something measured in the image plane, because the available information (2D points) are extracted from images. One such quantity is the distance from a point $\mathbf{m}_i$ to its corresponding epipolar line $\mathbf{l}_i = \mathbf{F}\widetilde{\mathbf{m}}'_i \equiv [l_1, l_2, l_3]^T$, which is given by[1]

$$d(\mathbf{m}_i, \mathbf{l}_i) = \frac{\widetilde{\mathbf{m}}_i^T \mathbf{l}_i}{\sqrt{l_1^2 + l_2^2}} = \frac{1}{c_i} \widetilde{\mathbf{m}}_i^T \mathbf{F} \widetilde{\mathbf{m}}'_i \,, \tag{3.7}$$

where $c_i = \sqrt{l_1^2 + l_2^2}$. Thus, the criterion (3.2) can be rewritten as

$$\min \sum_{i=1}^{n} c_i^2 d^2(\mathbf{m}_i, \mathbf{l}_i) \,.$$

[1] As described in (2.38), a 2D line in a plane oxy represented by a 3-vector $\mathbf{l} = [a, b, c]^T$ satisfies the equation $ax + by + c = 0$. Then, the distance of a point $\mathbf{m}_0 = [x_0, y_0]^T$ to this line is given by

$$d(\mathbf{m}_0, \mathbf{l}) = \frac{ax_0 + by_0 + c}{\sqrt{a^2 + b^2}} \,.$$

Note that this definition gives a *signed* distance.

This means that we are minimizing not only a physical quantity $d(\mathbf{m}_i, \mathbf{l}_i)$, but also c_i which is not physically meaningful. Luong [92] shows that the linear criterion introduces a bias and tends to bring the epipoles towards the image center.

Normalizing Input Data

Hartley [59] has analyzed, from numerical computation point of view, the high instability of this linear method if pixel coordinates are directly used, and proposed to perform a simple normalization of input data prior to running the 8-point algorithm. This technique indeed produces much better results, and is summarized below.

Suppose that coordinates $\mathbf{m}_i$ in one image are replaced by $\hat{\mathbf{m}}_i = \mathbf{T}\tilde{\mathbf{m}}_i$, and coordinates $\mathbf{m}'_i$ in the other image are replaced by $\hat{\mathbf{m}}'_i = \mathbf{T}'\tilde{\mathbf{m}}'_i$, where $\mathbf{T}$ and $\mathbf{T}'$ are any 3×3 matrices. Substituting in the equation $\tilde{\mathbf{m}}_i^T \mathbf{F} \tilde{\mathbf{m}}'_i = 0$, we derive the equation $\hat{\mathbf{m}}_i^T \mathbf{T}^{-T} \mathbf{F} \mathbf{T}'^{-1} \hat{\mathbf{m}}'_i = 0$. This relation implies that $\mathbf{T}^{-T} \mathbf{F} \mathbf{T}'^{-1}$ is the fundamental matrix corresponding to the point correspondences $\hat{\mathbf{m}}_i \leftrightarrow \hat{\mathbf{m}}'_i$. Thus, an alternative method of finding the fundamental matrix is as follows:

1. Transform the image coordinates according to transformations $\hat{\mathbf{m}}_i = \mathbf{T}\tilde{\mathbf{m}}_i$ and $\hat{\mathbf{m}}'_i = \mathbf{T}'\tilde{\mathbf{m}}'_i$.
2. Find the fundamental matrix $\hat{\mathbf{F}}$ corresponding to the matches $\hat{\mathbf{m}}_i \leftrightarrow \hat{\mathbf{m}}'_i$.
3. Retrieve the original fundamental matrix as $\mathbf{F} = \mathbf{T}^T \hat{\mathbf{F}} \mathbf{T}'$.

The question now is how to choose the transformations $\mathbf{T}$ and $\mathbf{T}'$.

Consider the second method described above, which consists in finding the eigenvector of the 9×9 matrix $\mathbf{U}_n^T \mathbf{U}_n$ associated with the least eigenvalue (for simplicity, this vector is called the *least eigenvector* in the sequel). This matrix can be expressed as $\mathbf{U}_n^T \mathbf{U}_n = \mathbf{U}\mathbf{D}\mathbf{U}^T$, where $\mathbf{U}$ is orthogonal and $\mathbf{D}$ is diagonal whose diagonal entries λ_i ($i = 1, \ldots, 9$) are assumed in non-increasing order. In this case, the least eigenvector of $\mathbf{U}_n^T \mathbf{U}_n$ is the last column of $\mathbf{U}$. Denote by κ the ratio λ_1/λ_8. The parameter κ is the *condition number*[2] of the matrix $\mathbf{U}_n^T \mathbf{U}_n$, well known to be an important factor in the analysis of stability of linear problems [46]. If κ is large, then very small changes to the data can

[2]Strictly speaking, λ_1/λ_9 is the condition number, but λ_1/λ_8 is the parameter of importance here.

cause large changes to the solution. The sensitivity of invariant subspaces is discussed in detail in [46, p.413].

The major reason for the poor condition of the matrix $\mathbf{U}_n^T\mathbf{U}_n \equiv \mathbf{X}$ is the lack of homogeneity in the image coordinates. In an image of dimension 200×200, a typical image point will be of the form $(100, 100, 1)$. If both $\widetilde{\mathbf{m}}_i$ and $\widetilde{\mathbf{m}}'_i$ are of this form, then $\mathbf{u}_i$ will be of the form $[10^4, 10^4, 10^2, 10^4, 10^4, 10^2, 10^2, 10^2, 1]^T$. The contribution to the matrix $\mathbf{X}$ is of the form $\mathbf{u}_i\mathbf{u}_i^T$, which will contain entries ranging between 10^8 and 1. The diagonal entries of $\mathbf{X}$ will be of the form $[10^8, 10^8, 10^4, 10^8, 10^8, 10^4, 10^4, 10^4, 1]^T$. Summing over all point matches will result in a matrix $\mathbf{X}$ whose diagonal entries are approximately in this proportion. We denote by $\mathbf{X}_r$ the trailing $r \times r$ principal submatrix (that is the last r columns and rows) of $\mathbf{X}$, and by $\lambda_i(\mathbf{X}_r)$ its ith largest eigenvalue. Thus $\mathbf{X}_9 = \mathbf{X} = \mathbf{U}_n^T\mathbf{U}_n$ and $\kappa = \lambda_1(\mathbf{X}_9)/\lambda_8(\mathbf{X}_9)$. First, we consider the eigenvalues of $\mathbf{X}_2$. Since the sum of the two eigenvalues is equal to the trace, we see that $\lambda_1(\mathbf{X}_2) + \lambda_2(\mathbf{X}_2) = \text{trace}(\mathbf{X}_2) = 10^4 + 1$. Since eigenvalues are non-negative, we know that $\lambda_1(\mathbf{X}_2) \leq 10^4 + 1$. From the *interlacing property* [46, p.411], we arrive that

$$\lambda_8(\mathbf{X}_9) \leq \lambda_7(\mathbf{X}_8) \leq \cdots \leq \lambda_1(\mathbf{X}_2) \leq 10^4 + 1 .$$

On the other hand, also from the interlacing property, we know that the largest eigenvalue of $\mathbf{X}$ is not less than the largest diagonal entry, i.e. $\lambda_1(\mathbf{X}_9) \geq 10^8$. Therefore, the ratio $\kappa = \lambda_1(\mathbf{X}_9)/\lambda_8(\mathbf{X}_9) \geq 10^8/(10^4 + 1)$. In fact, $\lambda_8(\mathbf{X}_9)$ will usually be much smaller than $10^4 + 1$ and the condition number will be far greater. This analysis shows that *scaling the coordinates so that they are on the average equal to* unity *will improve the condition of the matrix* $\mathbf{U}_n^T\mathbf{U}_n$.

Now consider the effect of translation. A usual practice is to fix the origin of the image coordinates at the top left hand corner of the image, so that all the image coordinates are positive. In this case, *an improvement in the condition of the matrix may be achieved by translating the points so that the centroid of the points is at the origin.* Informally, if the first image coordinates (the u-coordinates) of a set of points are $\{1001.5, 1002.3, 998.7, \ldots\}$, then the significant values of the coordinates are obscured by the coordinate offset of 1000. By translating by 1000, these numbers are changed to $\{1.5, 2.3, -1.3, \ldots\}$. The significant values become now prominent.

Based on the above analysis, Hartley [59] propose an isotropic scaling of the input data:

1. As a first step, the points are translated so that their centroid is at the origin.
2. Then, the coordinates are scaled, so that on the average a point $\widetilde{\mathbf{m}}_i$ is of the form $\widetilde{\mathbf{m}}_i = [1, 1, 1]^T$. Such a point will lie at a distance $\sqrt{2}$ from the origin. Rather than choosing different scale factors for u and v coordinates, we choose to scale the points isotropically so that the average distance from the origin to these points is equal to $\sqrt{2}$.

Such a transformation is applied to each of the two images independently.

An alternative to the isotropic scaling is an affine transformation so that the two principal moments of the set of points are both equal to unity. However, Hartley [59] found that the results obtained were little different from those obtained using the isotropic scaling method.

Beardley et al. [11] mention a normalization scheme which assumes some knowledge of camera parameters. Actually, if approximate intrinsic parameters (i.e. the intrinsic matrix $\mathbf{A}$) of a camera are available, we can apply the transformation $\mathbf{T} = \mathbf{A}^{-1}$ to obtain a "quasi-Euclidean" frame.

Boufama and Mohr [13] use implicitly data normalization by selecting 4 points, which are largely spread in the image (i.e. most distant from each other), to form a projective basis.

3.1.3 Analytic Method with Rank-2 Constraint

The method described in this section is due to [35] which imposes the rank-2 constraint during the minimization but still yields an analytic solution. Without loss of generality, let $\mathbf{f} = [\mathbf{g}^T, f_8, f_9]^T$, where $\mathbf{g}$ is a vector containing the first seven components of $\mathbf{f}$. Let $\mathbf{c}_8$ and $\mathbf{c}_9$ be the last two column vectors of $\mathbf{U}_n$, and $\mathbf{B}$ be the $n \times 7$ matrix composed of the first seven columns of $\mathbf{U}_n$. From $\mathbf{U}_n\mathbf{f} = \mathbf{0}$, we have

$$\mathbf{B}\mathbf{g} = -f_8\mathbf{c}_8 - \mathbf{f}_9\mathbf{c}_9 \ .$$

Assume that the rank of $\mathbf{B}$ is 7, we can solve for $\mathbf{g}$ by least-squares as

$$\mathbf{g} = -f_8(\mathbf{B}^T\mathbf{B})^{-1}\mathbf{B}^T\mathbf{c}_8 - f_9(\mathbf{B}^T\mathbf{B})^{-1}\mathbf{B}^T\mathbf{c}_9 \ .$$

The solution depends on two free parameters f_8 and f_9. As in Sect. 3.1.1, we can use the constraint $\det(\mathbf{F}) = 0$, which gives a third-degree homogeneous equation in f_8 and f_9, and we can solve for their ratio. Because a third-degree equation has at least one real root, we are guaranteed to obtain at least one solution for $\mathbf{F}$. This solution is defined up to a scale factor, and we can normalize $\mathbf{f}$ such that its vector norm is equal to 1. If there are three real roots, we choose the one that minimizes the vector norm of $\mathbf{U}_n\mathbf{f}$ subject to $\|\mathbf{f}\| = 1$. In fact, we can do the same computation for any of the 36 choices of pairs of coordinates of $\mathbf{f}$ and choose, among the possibly 108 solutions, the one that minimizes the previous vector norm.

The difference between this method and those described in Sect. 3.1.2 is that the latter imposes the rank-2 constraint after the linear least-squares. We have experimented this method with a limited number of examples, and found the results comparable with those obtained by the previous one.

3.1.4 Nonlinear Method Minimizing Distances of Points to Epipolar Lines

As discussed in page 84, the linear method (3.3) does not minimize a physically meaningful quantity. A natural idea is then to minimize the distances between points and their corresponding epipolar lines:

$$\min_{\mathbf{F}} \sum_i d^2(\widetilde{\mathbf{m}}_i, \mathbf{F}\widetilde{\mathbf{m}}'_i) ,$$

where $d(\cdot,\cdot)$ is given by (3.7). However, unlike the case of the linear criterion, the two images do not play a symmetric role. This is because the above criterion determines only the epipolar lines in the first image, and thus it should not be used to compute the epipolar lines in the second image. To avoid the inconsistency of the epipolar geometry between the two images, it is necessary and sufficient to exchange the role of the two images. As we have seen in the last chapter, by exchanging the two images, the fundamental matrix is changed to its transpose. This leads to the following criterion

$$\min_{\mathbf{F}} \sum_i \left(d^2(\widetilde{\mathbf{m}}_i, \mathbf{F}\widetilde{\mathbf{m}}'_i) + d^2(\widetilde{\mathbf{m}}'_i, \mathbf{F}^T\widetilde{\mathbf{m}}_i)\right) , \tag{3.8}$$

which operates simultaneously in the two images.

Let $\mathbf{l}_i = \mathbf{F}\widetilde{\mathbf{m}}'_i \equiv [l_1, l_2, l_3]^T$ and $\mathbf{l}'_i = \mathbf{F}^T\widetilde{\mathbf{m}}_i \equiv [l'_1, l'_2, l'_3]^T$. Using (3.7) and the fact that $\widetilde{\mathbf{m}}_i^T\mathbf{F}\widetilde{\mathbf{m}}'_i = \widetilde{\mathbf{m}}_i'^T\mathbf{F}^T\widetilde{\mathbf{m}}_i$, the criterion (3.8) can be rewritten as:

$$\min_{\mathbf{F}} \sum_i w_i^2 (\widetilde{\mathbf{m}}_i^T \mathbf{F} \widetilde{\mathbf{m}}'_i)^2 , \tag{3.9}$$

where

$$w_i = \left(\frac{1}{l_1^2 + l_2^2} + \frac{1}{l_1'^2 + l_2'^2} \right)^{1/2} = \left(\frac{l_1^2 + l_2^2 + l_1'^2 + l_2'^2}{(l_1^2 + l_2^2)(l_1'^2 + l_2'^2)} \right)^{1/2} .$$

We now present two methods for solving this problem.

Iterative Linear Method

The similarity between (3.9) and (3.2) conducts us to solve the above problem by a *weighted* linear least-squares technique. Indeed, if we can compute the weight w_i for each point match, the corresponding linear equation can be multiplied by w_i (which is equivalent to replacing $\mathbf{u}_i$ in (3.1) by $w_i\mathbf{u}_i$), and exactly the same 8-point algorithm can be run to estimate the fundamental matrix, which minimizes (3.9).

The problem is that the weights w_i depends themselves on the fundamental matrix. To overcome this difficulty, we apply an iterative linear method. We first assume that all $w_i = 1$ and run the 8-point algorithm to obtain an initial estimation of the fundamental matrix. The weights w_i are then computed from this initial solution. The weighted linear least-squares is then run for an improved solution. This procedure can be repeated several times.

Although this algorithm is simple to implement and minimizes a physical quantity, our experience shows that there is no significant improvement compared to the original linear method. The main reason is that the rank-2 constraint of the fundamental matrix is not taken into account.

Nonlinear Minimization in Parameter Space

From the above discussions, it is clear that the right thing to do is to search for a matrix among the 3×3 matrices of rank 2 which minimizes (3.9). There are several possible parameterizations for the fundamental matrix [92], e.g. we can express one row (or column) of the fundamental matrix as the linear combination of the other two rows (or columns). The parameterization described below is based directly on the parameters of the epipolar transformation (see Sect. 2.3.5).

Parameterization of fundamental matrix. Denoting the columns of $\mathbf{F}$ by the vectors $\mathbf{c}_1$, $\mathbf{c}_2$ and $\mathbf{c}_3$, we have :

$$\begin{aligned} & \text{rank}(\mathbf{F}) = 2 \\ & \Longleftrightarrow \\ (\exists j_0, j_1, j_2 \in [1,3])\ (\exists \lambda_1, \lambda_2 \in \mathcal{R}), \quad & \mathbf{c}_{j_0} + \lambda_1 \mathbf{c}_{j_1} + \lambda_2 \mathbf{c}_{j_2} = 0 \quad (3.10) \\ (\nexists \lambda \in \mathcal{R}), \quad & \mathbf{c}_{j_1} + \lambda \mathbf{c}_{j_2} = 0\,. \quad (3.11) \end{aligned}$$

Condition (3.11), as a non-existence condition, cannot be expressed by a parameterization: we shall only keep condition (3.10) and so extend the parameterized set to all the 3×3-matrices of rank strictly less than 3. Indeed, the rank-2 matrices of, for example, the following forms :

$$\begin{bmatrix} \mathbf{c}_1 & \mathbf{c}_2 & \lambda \mathbf{c}_2 \end{bmatrix} \text{ and } \begin{bmatrix} \mathbf{c}_1 & \mathbf{0}_3 & \mathbf{c}_3 \end{bmatrix} \text{ and } \begin{bmatrix} \mathbf{c}_1 & \mathbf{c}_2 & \mathbf{0}_3 \end{bmatrix}$$

do not have any parameterization if we take $j_0 = 1$. A parameterization of $\mathbf{F}$ is then given by $(\mathbf{c}_{j_1}, \mathbf{c}_{j_2}, \lambda_1, \lambda_2)$. This parameterization implies to divide the parameterized set among three maps, corresponding to $j_0 = 1$, $j_0 = 2$ and $j_0 = 3$.

If we construct a 3-vector such that λ_1 and λ_2 are the $j_1{}^{\text{th}}$ and $j_2{}^{\text{th}}$ coordinates and 1 is the $j_0{}^{\text{th}}$ coordinate, then it is obvious that this vector is the eigenvector of $\mathbf{F}$, and is thus the epipole in the case of the fundamental matrix. Using such a parameterization implies to compute directly the epipole which is often a useful quantity, instead of the matrix itself.

To make the problem symmetrical and since the epipole in the other image is also worth being computed, the same decomposition as for the columns is used for the rows, which now divides the parameterized set into 9 maps, corresponding to the choice of a column and a row as linear combinations of the two columns and two rows left. A parameterization of the matrix is then formed by the two coordinates x and y of the first epipole, the two coordinates x' and y' of the second epipole and the four elements a, b, c and d left by $\mathbf{c}_{i_1}$, $\mathbf{c}_{i_2}$, $\mathbf{l}_{j_1}$ and $\mathbf{l}_{j_2}$, which in turn parameterize the epipolar transformation mapping an epipolar line of the second image to its corresponding epipolar line in the first image. In that way, the matrix is written, for example, for $i_0 = 3$ and $j_0 = 3$:

$$\mathbf{F} = \begin{bmatrix} a & b & -ax' - by' \\ c & d & -cx' - dy' \\ -ax - cy & -bx - dy & (ax' + by')x + (cx' + dy')y \end{bmatrix}. \quad (3.12)$$

At last, to take into account the fact that the fundamental matrix is defined only up to a scale factor, the matrix is normalized by dividing the four elements

(a, b, c, d) by the largest in absolute value. We have thus in total 36 maps to parameterize the fundamental matrix.

Choosing the best map. Giving a matrix $\mathbf{F}$ and the epipoles, or an approximation to it, we must be able to choose, among the different maps of the parameterization, the most suitable for $\mathbf{F}$. Denoting by $\mathbf{f}_{i_0 j_0}$ the vector of the elements of $\mathbf{F}$ once decomposed as in equation (3.12), i_0 and j_0 are chosen in order to maximize the rank of the 9×8 Jacobian matrix:

$$\mathbf{J} = \frac{d\mathbf{f}_{i_0 j_0}}{d\mathbf{p}} \quad \text{where} \quad \mathbf{p} = [x, y, x', y', a, b, c, d]^T . \tag{3.13}$$

This is done by maximizing the norm of the vector whose coordinates are the determinants of the nine 8×8 submatrices of $\mathbf{J}$. An easy calculation shows that this norm is equal to

$$(ad - bc)^2 \sqrt{x^2 + y^2 + 1} \sqrt{x'^2 + y'^2 + 1} .$$

At the expense of dealing with different maps, the above parameterization works equally well whether the epipoles are at infinity or not. This is not the case with the original proposition in [92]. More details can be found in [23].

Minimization. The minimization of (3.9) can now be performed by any minimization procedure. The Levenberg-Marquardt method (as implemented in `MINPACK` from `NETLIB` and in the Numeric Recipes in C [119]) is used in our program. During the process of minimization, the parametrization of $\mathbf{F}$ can change: The parametrization chosen for the matrix at the begining of the process is not necessarily the most suitable for the final matrix. The nonlinear minimization method demands an initial estimate of the fundamental matrix, which is obtained by running the 8-point algorithm.

3.1.5 Nonlinear Method Minimizing Distances Between Observation and Reprojection

Last subsection presented a method based on minimizing the distances between points and their corresponding epipolar lines, but it considers only the distances along the direction perpendicular to the epipolar lines. If we can assume that the coordinates of the observed points are corrupted by additive noise and that the noises in different points are independant but with equal standard deviation, then the maximum likelihood estimation of the fundamental matrix is obtained by minimizing the following criterion:

$$\mathcal{F}(\mathbf{f}, \mathtt{M}) = \sum_i (\|\mathbf{m}_i - \mathbf{h}(\mathbf{f}, \mathtt{M}_i)\|^2 + \|\mathbf{m}'_i - \mathbf{h}'(\mathbf{f}, \mathtt{M}_i)\|^2) , \tag{3.14}$$

where $\mathbf{f}$ represents the parameter vector of the fundamental matrix such as the one described in the last section, $\mathtt{M} = [\mathtt{M}_1^T, \ldots, \mathtt{M}_n^T]^T$ are the structure parameters of the n points in space, while $\mathbf{h}(\mathbf{f}, \mathtt{M}_i)$ and $\mathbf{h}'(\mathbf{f}, \mathtt{M}_i)$ are the projection functions in the first and second image for a given space coordinates M_i and a given fundamental matrix between the two images represented by vector $\mathbf{f}$. Simply speaking, $\mathcal{F}(\mathbf{f}, \mathtt{M})$ is the sum of squared distances between observed points and the *reprojections* of the corresponding points in space. This implies that we estimate not only the fundamental matrix but also the structure parameters of the points in space. The estimation of the structure parameters, or *3D reconstruction*, in the uncalibrated case is an important subject and needs a separate section to describe it in sufficient details (see Sect. 3.4). In the remaining subsection, we assume that there is a procedure available for 3D reconstruction.

A generalization to (3.14) is to take into account different uncertainties, if available, in the image points. If a point $\mathbf{m}_i$ is assumed to be corrupted by a Gaussian noise with mean zero and covariance matrix $\Lambda_{\mathbf{m}_i}$ (a 2×2 symmetric positive-definite matrix), then the maximum likelihood estimation of the fundamental matrix is obtained by minimizing the following criterion:

$$\mathcal{F}(\mathbf{f}, \mathtt{M}) = \sum_i \left(\Delta\mathbf{m}_i^T \Lambda_{\mathbf{m}_i}^{-1} \Delta\mathbf{m}_i + \Delta\mathbf{m}_i'^T \Lambda_{\mathbf{m}'_i}^{-1} \Delta\mathbf{m}'_i \right)$$

with

$$\Delta\mathbf{m}_i = \mathbf{m}_i - \mathbf{h}(\mathbf{f}, \mathtt{M}_i) \quad \text{and} \quad \Delta\mathbf{m}'_i = \mathbf{m}'_i - \mathbf{h}'(\mathbf{f}, \mathtt{M}_i) .$$

Here we still assume that the noises in different points are independent, which is quite reasonable.

When the number of points n is large, the nonlinear minimization of $\mathcal{F}(\mathbf{f}, \mathtt{M})$ should be carried out in a huge parameter space ($3n + 7$ dimensions because each space point has 3 degrees of freedom), and the computation is very expensive. As a matter of fact, we can separate the structure parameters from the fundamental matrix such that the optimization of the structure parameters is conducted in each optimization iteration for the parameters of the fundamental matrix, that is:

$$\min_{\mathbf{f}} \left\{ \sum_i \min_{\mathtt{M}_i} \left(\|\mathbf{m}_i - \mathbf{h}(\mathbf{f}, \mathtt{M}_i)\|^2 + \|\mathbf{m}'_i - \mathbf{h}'(\mathbf{f}, \mathtt{M}_i)\|^2 \right) \right\} . \tag{3.15}$$

Therefore, a problem of minimization over $3n + 7$-D space (3.14) becomes a problem of minimization over 7-D space, in the later each iteration contains n independent optimizations of 3 structure parameters. The computation is thus considerably reduced. As will be seen in Sect. 3.4, the optimization of structure parameters is nonlinear. In order to speed up still more the computation, it can be approximated by an analytic method; when this optimization procedure converges, we then restart it with the nonlinear optimization method.

The idea underlying this method is already well known in motion and structure from motion [41, 185] and camera calibration [41]. Similar techniques have also been reported for uncalibrated images [104, 62].

3.1.6 Robust Methods

Up to now, we assume that point matches are given. They can be obtained by some techniques similar to that described in Chap. 6. They all exploit some *heuristics* in one form or another, for example, intensity similarity or rigid/affine transformation in image plane, which are not applicable to most cases. Among the matches established, we may find two types of *outliers* due to

bad locations. In the estimation of the fundamental matrix, the location error of a point of interest is assumed to exhibit Gaussian behavior. This assumption is reasonable since the error in localization for most points of interest is small (within one or two pixels), but a few points are possibly incorrectly localized (more than three pixels). The latter points will severely degrade the accuracy of the estimation.

false matches. In the establishment of correspondences, only heuristics have been used. Because the only geometric constraint, i.e., the epipolar constraint in terms of the *fundamental matrix*, is not yet available, many matches are possibly false. These will completely spoil the estimation process, and the final estimate of the fundamental matrix will be useless.

The outliers will severely affect the precision of the fundamental matrix if we directly apply the methods described above, which are all least-squares techniques.

Least-squares estimators assume that the noise corrupting the data is of zero mean, which yields an *unbiased* parameter estimate. If the noise variance is known, a *minimum-variance* parameter estimate can be obtained by choosing appropriate weights on the data. Furthermore, least-squares estimators implicitly assume that the entire set of data can be interpreted by *only one parameter vector* of a given model. Numerous studies have been conducted, which clearly show that least-squares estimators are vulnerable to the violation of these assumptions. Sometimes even when the data contains only one bad datum, least-squares estimates may be completely perturbed. During the last three decades, many robust techniques have been proposed, which are not very sensitive to departure from the assumptions on which they depend.

Hampel [51] gives some justifications to the use of robustness (quoted in [124]):

> What are the reasons for using robust procedures? There are mainly two observations which combined give an answer. Often in statistics one is using a parametric model implying a very limited set of probability distributions though possible, such as the common model of normally distributed errors, or that of exponentially distributed observations. Classical (parametric) statistics derives results under the assumption that these models were strictly true. However, apart from some simple discrete models perhaps, such models are never exactly true. We may try to distinguish three main reasons for the derivations: (i) rounding and grouping and other "local inaccuracies"; (ii) the occurrence of "gross errors" such as blunders in measuring, wrong decimal points, errors in copying, inadvertent measurement of a member of a different population, or just "something went wrong"; (iii) the model may have been conceived only as an approximation anyway, e.g. by virtue of the central limit theorem.

Recently, computer vision researchers have paid much attention to the robustness of vision algorithms because the data are unavoidably error prone [53, 192]. Many the so-called *robust regression* methods have been proposed that are not so easily affected by outliers [74, 131]. The reader is referred to [131, Chap. 1] for a review of different robust methods. The two most popular robust methods are the *M-estimators* and the *least-median-of-squares* (LMedS) method, which will be presented below. Kumar and Hanson [82] compared different robust methods for pose refinement from 3D-2D line correspondences, while Meer et al. [102], for image smoothing. Haralick et al. [54] applied M-estimators to solve the pose problem from point correspondences. Thompson et al. [153] applied the LMedS estimator to detect moving objects using point correspondences between orthographic views. Other recent works on the application of robust techniques to motion segmentation include [157, 111, 7].

Regarding the robust recovery of the epipolar geometry, our work described below is closely related to that of Olsen [112] and that of Shapiro and Brady [135]. Olsen uses a linear method to estimate the epipolar geometry, which has already been shown to be insufficiently accurate. He further assumes that knowledge of the epipolar geometry, as in many practical cases, is available. In particular, he assumes the epipolar lines are almost aligned horizontally. This knowledge is then used to find matches between the stereo image pair, and a robust method (an M-estimator) is used to detect false matches and to obtain a better estimate of the epipolar geometry. Shapiro and Brady also use a linear method. The camera model is however a simplified one, namely an affine camera. Correspondences are established by tracking corner features over time. False matches are rejected through a *regression diagnostic*, which computes an initial estimate of the epipolar geometry over all matches, and sees how the estimate changes if a match is deleted. The match whose removal maximally reduces the residual is identified to be an *outlier* and is rejected. The procedure is then repeated with the reduced set of matches until all outliers have been removed. These two approaches (M-estimators and Regression diagnostics) work well when the percentage of outliers is small and more importantly when their derivations from the valid matches are not too large, as in the above two works. In our work, two images can be quite different. There may be a large percentage of false matches (usually around 20%, sometimes 40%) using heuristic matching techniques such as correlation, and a false match may be completely different from the valid matches. This is why we use the least-median-of-squares technique to deal with these issues, which can theoretically detect outliers when they make up as much as 50% of whole data.

M-Estimators

Let r_i be the *residual* of the ith datum, i.e. the difference between the ith observation and its fitted value. The standard least-squares method tries to minimize $\sum_i r_i^2$, which is unstable if there are outliers present in the data. Outlying data give an effect so strong in the minimization that the parameters thus estimated are distorted. The M-estimators try to reduce the effect of outliers by replacing the squared residuals r_i^2 by another function of the residuals, yielding

$$\min \sum_i \rho(r_i) , \tag{3.16}$$

where ρ is a symmetric, positive-definite function with a unique minimum at zero, and is chosen to be less increasing than square. Instead of solving directly this problem, we can implement it as an iterated reweighted least-squares one. Now let us see how.

Let $\mathbf{p} = [p_1, \ldots, p_m]^T$ be the parameter vector to be estimated. The M-estimator of $\mathbf{p}$ based on the function $\rho(r_i)$ is the vector $\mathbf{p}$ which is the solution of the following m equations:

$$\sum_i \psi(r_i) \frac{\partial r_i}{\partial p_j} = 0 , \quad \text{for } j = 1, \ldots, m, \tag{3.17}$$

where the derivative $\psi(x) = \mathrm{d}\rho(x)/\mathrm{d}x$ is called the *influence function*. If now we define a *weight function*

$$w(x) = \frac{\psi(x)}{x} , \tag{3.18}$$

then Equation (3.17) becomes

$$\sum_i w(r_i) r_i \frac{\partial r_i}{\partial p_j} = 0 , \quad \text{for } j = 1, \ldots, m. \tag{3.19}$$

This is exactly the system of equations that we obtain if we solve the following iterated reweighted least-squares problem

$$\min \sum_i w(r_i^{(k-1)}) r_i^2 , \tag{3.20}$$

where the superscript $^{(k)}$ indicates the iteration number. The weight $w(r_i^{(k-1)})$ should be recomputed after each iteration in order to be used in the next iteration.

The influence function $\psi(x)$ measures the influence of a datum on the value of the parameter estimate. For example, for the least-squares with $\rho(x) = x^2/2$, the influence function is $\psi(x) = x$, that is, the influence of a datum on the estimate increases linearly with the size of its error, which confirms the non-robusteness of the least-squares estimate. When an estimator is robust, it may be inferred that the influence of any single observation (datum) is insufficient to yield any significant offset [124]. There are several constraints that a robust M-estimator should meet:

- The first is of course to have a bounded influence function.

- The second is naturally the requirement of the robust estimator to be unique. This implies that the objective function of parameter vector $\mathbf{p}$ to be minimized should have a unique minimum. This requires that *the individual ρ-function is* convex *in variable* $\mathbf{p}$. This is necessary because only requiring a ρ-function to have a unique minimum is not sufficient. This is the case with maxima when considering mixture distribution; the sum of unimodal probability distributions is very often multimodal. The convexity constraint is equivalent to imposing that $\frac{\partial^2 \rho(.)}{\partial \mathbf{p}^2}$ is non-negative definite.

- The third one is a practical requirement. Whenever $\frac{\partial^2 \rho(.)}{\partial \mathbf{p}^2}$ is singular, the objective should have a gradient, i.e. $\frac{\partial \rho(.)}{\partial \mathbf{p}} \neq \mathbf{0}$. This avoids having to search through the complete parameter space.

Table 3.1 lists a few commonly used influence functions. They are graphically dipicted in Fig. 3.1. Note that not all these functions satisfy the above requirements.

Briefly we give a few indications of these functions:

- L_2 (i.e. least-squres) estimators are not robust because their influence function is not bounded.

- L_1 (i.e. absolute value) estimators are not stable because the ρ-function $|x|$ is not trictly convex in x. Indeed, the second derivative at $x = 0$ is unbounded, and an indeterminant solution may result.

- L_1 estimators reduce the influence of large errors, but they still have an influence because the influence function has no cut off point.

- $L_1 - L_2$ estimators take both the advantage of the L_1 estimators to reduce the influence of large errors and that of L_2 estimators to be convex.

- The L_p (*least-powers*) function represents a family of functions. It is L_2 with $\nu = 2$ and L_1 with $\nu = 1$. The smaller the value of ν is, the smaller is the incidence of large errors in the estimate $\mathbf{p}$. It appears that ν must be fairly moderate to provide a relatively robust estimator or, in other words, to provide an estimator scarcely perturbed by outlying data. The selection of an optimal ν has been investigated, and for ν around 1.2, a good estimate may be expected [124]. However, many difficulties are encountered in the

Table 3.1 A few commonly used M-estimators

type	$\rho(x)$	$\psi(x)$	$w(x)$
L_2	$x^2/2$	x	1
L_1	$\|x\|$	$\text{sgn}(x)$	$\frac{1}{\|x\|}$
$L_1 - L_2$	$2(\sqrt{1+x^2/2}-1)$	$\frac{x}{\sqrt{1+x^2/2}}$	$\frac{1}{\sqrt{1+x^2/2}}$
L_p	$\frac{\|x\|^\nu}{\nu}$	$\text{sgn}(x)\|x\|^{\nu-1}$	$\|x\|^{\nu-2}$
"Fair"	$c^2[\frac{\|x\|}{c} - \log(1+\frac{\|x\|}{c})]$	$\frac{x}{1+\|x\|/c}$	$\frac{1}{1+\|x\|/c}$
Huber	$\begin{cases} x^2/2 & \text{if } \|x\| \leq k \\ k(\|x\|-k/2) & \text{if } \|x\| > k \end{cases}$	$\begin{cases} x \\ k\text{sgn}(x) \end{cases}$	$\begin{cases} 1 \\ k/\|x\| \end{cases}$
Cauchy	$\frac{c^2}{2}\log(1+(x/c)^2)$	$\frac{x}{1+(x/c)^2}$	$\frac{1}{1+(x/c)^2}$
Geman-McClure	$\frac{x^2/2}{1+x^2}$	$\frac{x}{(1+x^2)^2}$	$\frac{1}{(1+x^2)^2}$
Welsch	$\frac{c^2}{2}[1-\exp(-(x/c)^2)]$	$x\exp(-(x/c)^2)$	$\exp(-(x/c)^2)$
Tukey	$\begin{cases} \frac{c^2}{6}(1-[1-(x/c)^2]^3) \\ (c^2/6) \quad \text{if } \|x\| > c \end{cases}$	$\begin{cases} x[1-(x/c)^2]^2 \\ 0 \end{cases}$	$\begin{cases} [1-(x/c)^2]^2 \\ 0 \end{cases}$

computation when parameter ν is in the range of interest $1 < \nu < 2$, because zero residuals are troublesome.

- The function "Fair" is among the possibilities offered by the Roepack package (see [124]). It has everywhere defined continuous derivatives of first three orders, and yields a unique solution. The 95% asymptotic efficiency on the standard normal distribution is obtained with the tuning constan $c = 1.3998$.

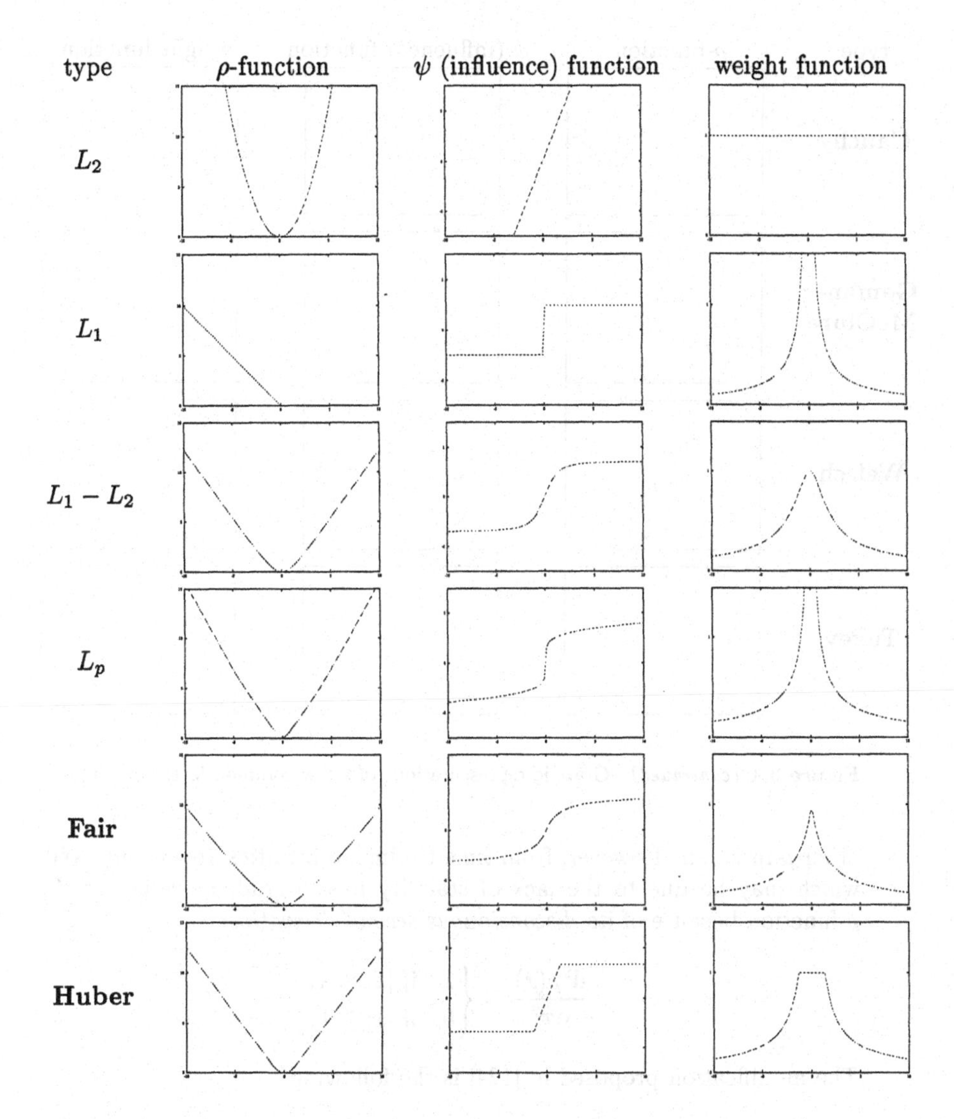

Figure 3.1 Graphic representations of a few common M-estimators (to be continued)

- Huber's function [74] is a parabola in the vicinity of zero, and increases linearly at a given level $|x| > k$. The 95% asymptotic efficiency on the standard normal distribution is obtained with the tuning constant $k = 1.345$. This estimator is so satisfactory that it has been recommended for almost all situations; very rarely it has been found to be inferior to some

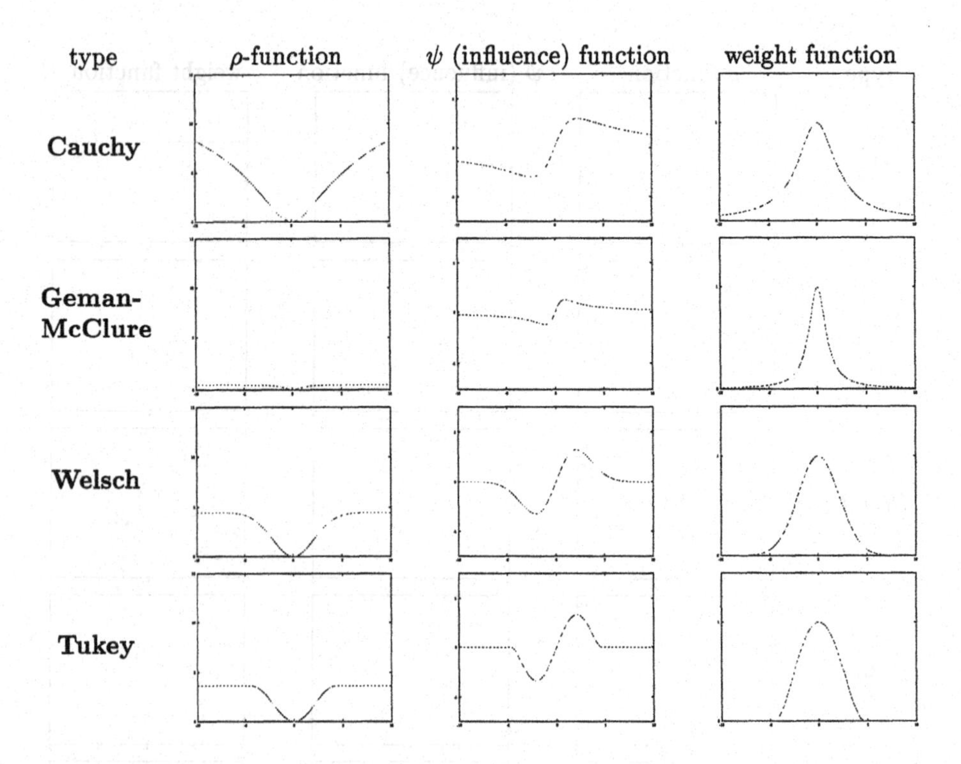

Figure 3.1 *(continued)* Graphic representations of a few common M-estimators

other ρ-function. However, from time to time, difficulties are encountered, which may be due to the lack of stability in the gradient values of the ρ-function because of its *discontinuous second derivative*:

$$\frac{d^2\rho(x)}{dx^2} = \begin{cases} 1 & \text{if } |x| \leq k, \\ 0 & \text{if } |x| \geq k. \end{cases}$$

The modification proposed in [124] is the following

$$\rho(x) = \begin{cases} c^2[1 - \cos(x/c)] & \text{if } |x|/c \leq \pi/2, \\ c|x| + c^2(1 - \pi/2) & \text{if } |x|/c \geq \pi/2. \end{cases}$$

The 95% asymptotic efficiency on the standard normal distribution is obtained with the tuning constan $c = 1.2107$.

- Cauchy's function, also known as the Lorentzian function, does not guarantee a unique solution. With a descending first derivative, such a function has a tendency to yield erroneous solutions in a way which cannot be observed. The 95% asymptotic efficiency on the standard normal distribution is obtained with the tuning constan $c = 2.3849$.

- The other remaining functions have the same problem as the Cauchy function. As can be seen from the influence function, the influence of large errors only decreases linearily with their size. The Geman-McClure function and Welsh function try to further reduce the effect of large errors, and the Tukey's biweight function even suppress the outliers. The 95% asymptotic efficiency on the standard normal distribution of the Tukey's biweight function is obtained with the tuning constant $c = 4.6851$; that of the Welsch function, with $c = 2.9846$.

There still exist many other ρ-functions, such as Andrew's cosine wave function. Another commonly used function is the following tri-weight one:

$$w_i = \begin{cases} 1 & |r_i| \leq \sigma \\ \sigma/|r_i| & \sigma < |r_i| \leq 3\sigma \\ 0 & 3\sigma < |r_i| \, , \end{cases}$$

where σ is some estimated standard deviation of errors. In [112, 92], this weight function was used for the estimation of the epipolar geometry.

It seems difficult to select a ρ-function for general use without being rather arbitrary. Following Rey [124], for the location (or regression) problems, the best choice is the L_p in spite of its theoretical non-robustness: they are quasi-robust. However, it suffers from its computational difficulties. The second best function is "Fair", which can yield nicely converging computational procedures. Eventually comes the Huber's function (either original or modified form). All these functions do not eliminate completely the influence of large gross errors.

The four last functions do not quarantee unicity, but reduce considerably, or even eliminate completely, the influence of large gross errors. As proposed by Huber [74], one can start the iteration process with a convex ρ-function, iterate until convergence, and then apply a few iterations with one of those non-convex functions to eliminate the effect of large errors.

Inherent in the different M-estimators is the simultaneous estimation of σ, the standard deviation of the residual errors. If we can make a good estimate of the standard deviation of the errors of good data (inliers), then datas whose error is larger than a certain number of standard deviations can be considered as outliers. Thus, the estimation of σ itself should be robust. The results of the M-estimators will depend on the method used to compute it. The *robust standard deviation* estimate is related to the median of the absolute values of the residuals, and is given by

$$\hat{\sigma} = 1.4826[1 + 5/(n-p)] \operatorname*{median}_i |r_i| \,. \tag{3.21}$$

The constant 1.4826 is a coefficient to achieve the same efficiency as a least-squares in the presence of only Gaussian noise (actually, the median of the absolute values of random numbers sampled from the Gaussian normal distribution $N(0,1)$ is equal to $\Phi^{-1}(\frac{3}{4}) \approx 1/1.4826$); $5/(n-p)$ (where n is the size of the data set and p is the dimension of the parameter vector) is to compensate the effect of a small set of data. The reader is referred to [131, page 202] for the details of these magic numbers.

Our experience shows that M-estimators are robust to outliers due to bad localization. They are, however, not robust to false matches. This leads us to use other more robust techniques.

Least Median of Squares (LMedS)

The LMedS method estimates the parameters by solving the nonlinear minimization problem:

$$\min \operatorname*{median}_i r_i^2 \,.$$

That is, the estimator must yield the smallest value for the median of squared residuals computed for the entire data set. It turns out that this method is very robust to false matches as well as outliers due to bad localization. Unlike the M-estimators, however, the LMedS problem cannot be reduced to a weighted least-squares problem. It is probably impossible to write down a straightforward formula for the LMedS estimator. It must be solved by a search in the space of possible estimates generated from the data. Since this space is too large, only a randomly chosen subset of data can be analyzed. The algorithm which we have implemented for robustly estimating the fundamental matrix follows the one structured in [131, Chap. 5], as outlined below.

Given n point correspondences: $\{(\mathbf{m}_i, \mathbf{m}'_i)\}$. A Monte Carlo type technique is used to draw m random subsamples of $p = 7$ different point correspondences (recall that 7 is the minimum number to determine the epipolar geometry). For each subsample, indexed by J, we determine the fundamental matrix $\mathbf{F}_J$. We may have at most 3 solutions. For each $\mathbf{F}_J$, we can determine the median of the squared residuals, denoted by M_J, with respect to the whole set of point correspondences, i.e.,

$$M_J = \operatorname*{median}_{i=1,\ldots,n}[d^2(\widetilde{\mathbf{m}}_i, \mathbf{F}_J\widetilde{\mathbf{m}}'_i) + d^2(\widetilde{\mathbf{m}}'_i, \mathbf{F}_J^T\widetilde{\mathbf{m}}_i)] .$$

We retain the estimate $\mathbf{F}_J$ for which M_J is minimal among all m M_J's. The question now is: *How do we determine m ?* A subsample is "good" if it consists of p good correspondences. Assuming that the whole set of correspondences may contain up to a fraction ε of outliers, the probability that at least one of the m subsamples is good is given by

$$P = 1 - [1 - (1 - \varepsilon)^p]^m . \tag{3.22}$$

By requiring that P must be near 1, one can determine m for given values of p and ε:

$$m = \frac{\log(1 - P)}{\log[1 - (1 - \varepsilon)^p]} .$$

In our implementation, we assume $\varepsilon = 40\%$ and require $P = 0.99$, thus $m = 163$. Note that the algorithm can be speeded up considerably by means of parallel computing, because the processing for each subsample can be done independently.

As noted in [131], the LMedS *efficiency* is poor in the presence of Gaussian noise. The efficiency of a method is defined as the ratio between the lowest achievable variance for the estimated parameters and the actual variance provided by the given method. To compensate for this deficiency, we further carry out a weighted least-squares procedure. The robust standard deviation estimate is given by (3.21), that is,

$$\hat{\sigma} = 1.4826[1 + 5/(n - p)]\sqrt{M_J} ,$$

where M_J is the minimal median estimated by the LMedS. Based on $\hat{\sigma}$, we can assign a weight for each correspondence:

$$w_i = \begin{cases} 1 & \text{if } r_i^2 \le (2.5\hat{\sigma})^2 \\ 0 & \text{otherwise ,} \end{cases}$$

where

$$r_i^2 = d^2(\widetilde{\mathbf{m}}_i, \mathbf{F}\widetilde{\mathbf{m}}'_i) + d^2(\widetilde{\mathbf{m}}_i, \mathbf{F}^T\widetilde{\mathbf{m}}'_i) .$$

The correspondences having $w_i = 0$ are outliers and should not be further taken into account. The fundamental matrix $\mathbf{F}$ is finally estimated by solving the weighted least-squares problem:

$$\min \sum_i w_i r_i^2 .$$

We have thus robustly estimated the fundamental matrix because outliers have been detected and discarded by the LMedS method.

As said previously, computational efficiency of the LMedS method can be achieved by applying a Monte-Carlo type technique. However, the seven points of a subsample thus generated may be very close to each other. Such a situation should be avoided because the estimation of the epipolar geometry from such points is highly instable and the result is useless. It is a waste of time to evaluate such a subsample. In order to achieve higher stability and efficiency, we develop a *regularly random selection method* based on bucketing techniques, which works as follows. We first calculate the min and max of the coordinates of the points in the first image. The region is then evenly divided into $b \times b$ buckets (see Fig. 3.2). In our implementation, $b = 8$. To each bucket is attached a set of points, and indirectly a set of matches, which fall in it. The buckets having no matches attached are excluded. To generate a subsample of 7 points, we first randomly select 7 mutually different buckets, and then randomly choose one match in each selected bucket.

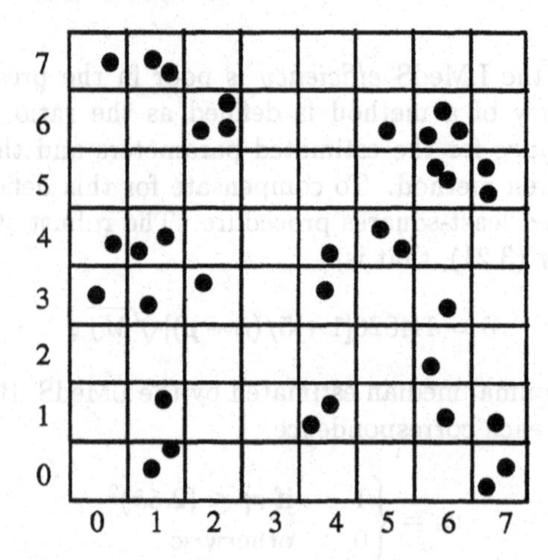

Figure 3.2 Illustration of a bucketing technique

One question remains: How many subsamples are required? If we assume that bad matches are uniformly distributed in space, and if each bucket has the same number of matches and the random selection is uniform, the formula (3.22) still holds. However, the number of matches in one bucket may be quite different from that in another. As a result, a match belonging to a bucket having fewer matches has a higher probability to be selected. It is thus preferred that a bucket having many matches has a higher probability to be selected than a bucket having few matches, in order for each match to have almost the same probability to be selected. This can be realized by the following procedure. If we have in total l buckets, we divide range [0 1] into l intervals such that the width of the ith interval is equal to $n_i / \sum_i n_i$, where n_i is the number of matches attached to the ith bucket (see Fig. 3.3). During the bucket selection procedure, a number, produced by a [0 1] uniform random generator, falling in the ith interval implies that the ith bucket is selected.

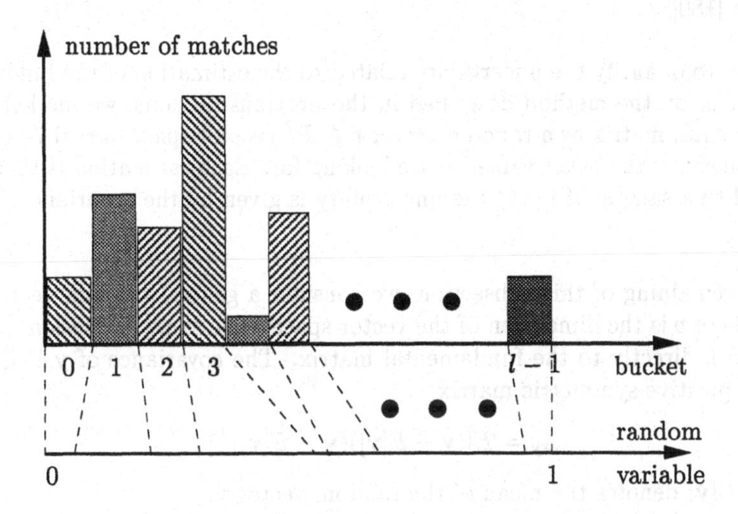

Figure 3.3 Interval and bucket mapping

Together with the matching technique described in Chap. 6, we have implemented this robust method and successfully solved, in an automatic way, the matching and epipolar geometry recovery problem for different types of scenes such as indoor, rocks, road, and textured dummy scenes [188].

3.1.7 Characterizing the Uncertainty of Fundamental Matrix

As described above, the fundamental matrix can be computed from a certain number of point correspondences obtained from the pair of images independently of any other knowledge about the world. Since the data points are always corrupted by noise, and sometimes the matches are even spurious or incorrect, one should model the uncertainty of the estimated fundamental matrix in order to exploit its underlying geometric information correctly and effectively. For example, one can use the covariance of the fundamental matrix to compute the uncertainty of the projective reconstruction or the projective invariants, or to improve the results of Kruppa's equation for a better self-calibration of a camera [180].

In order to quantify the uncertainty related to the estimation of the fundamental matrix by the method described in the previous sections, we modelize the fundamental matrix as a random vector $\mathbf{f} \in \mathbb{R}^7$ (vector space of real 7-vectors) whose mean is the exact value we are looking for. Each estimation is then considered to a sample of $\mathbf{f}$ and the uncertainty is given by the covariance matrix of $\mathbf{f}$.

In the remaining of this subsection, we consider a general random vector $\mathbf{y} \in \mathbb{R}^p$, where p is the dimension of the vector space. The same discussion applies, of course, directly to the fundamental matrix. The covariance of $\mathbf{y}$ is defined by the positive symmetric matrix

$$\Lambda_{\mathbf{y}} = E[(\mathbf{y} - E[\mathbf{y}])(\mathbf{y} - E[\mathbf{y}])^T] , \tag{3.23}$$

where $E[\mathbf{y}]$ denotes the mean of the random vector $\mathbf{y}$.

The Statistical Method

The statistical method consists in using the well-known large number law to approximate the mean: if we have a sufficiently large number N of samples $\mathbf{y}_i$ of a random vector $\mathbf{y}$, then $E[\mathbf{y}]$ can be approximated by the discrete mean

$$E_N[\mathbf{y}_i] = \frac{1}{N}\sum_{i=1}^{N} \mathbf{y}_i ,$$

and $\Lambda_{\mathbf{y}}$ is then approximated by

$$E_N[(\mathbf{y}_i - E_N[\mathbf{y}_i])(\mathbf{y}_i - E_N[\mathbf{y}_i])^T] . \tag{3.24}$$

A rule of thumb is that this method works reasonable well when $N > 30$. It is especially useful for simulation. For example, through simulation, we have found that the covariance of the fundamental matrix estimated by the analytical method through a first order approximation (see below) is quite good when the noise level in data points is moderate (the standard deviation is less than two pixels) [23].

The Analytical Method

The explicit case. We now consider the case that $\mathbf{y}$ is computed from another random vector $\mathbf{x}$ of $\mathbb{R}^m$ using a C^1 function φ:

$$\mathbf{y} = \varphi(\mathbf{x}) .$$

Writing the first order Taylor expansion of φ in the neighborhood of $E[\mathbf{x}]$ yields

$$\varphi(\mathbf{x}) = \varphi(E[\mathbf{x}]) + \mathbf{D}_\varphi(E[\mathbf{x}]) \cdot (\mathbf{x} - E[\mathbf{x}]) + \varepsilon(\|\mathbf{x} - E[\mathbf{x}]\|^2) , \tag{3.25}$$

where the function $t \rightarrow \varepsilon(t)$ from $\mathbb{R}$ into $\mathbb{R}^p$ is such that $\lim_{t\to 0} \varepsilon(t) = \mathbf{0}$, and $\mathbf{D}_\varphi(\mathbf{x}) = \partial\varphi(\mathbf{x})/\partial\mathbf{x}$ is the Jacobian matrix. By considering now that any sample of $\mathbf{x}$ is sufficiently close to $E[\mathbf{x}]$, we can approximate φ by the first order terms of equation (3.25) which yields:

$$\begin{aligned} E[\mathbf{y}] &\simeq \varphi(E[\mathbf{x}]) , \\ \varphi(\mathbf{x}) - \varphi(E[\mathbf{x}]) &\simeq \mathbf{D}_\varphi(E[\mathbf{x}]) \cdot (\mathbf{x} - E[\mathbf{x}]) . \end{aligned}$$

We have then

$$\begin{aligned} & E[(\varphi(\mathbf{x}) - \varphi(E[\mathbf{x}]))(\varphi(\mathbf{x}) - \varphi(E[\mathbf{x}]))^T] \\ &\simeq E[\mathbf{D}_\varphi(E[\mathbf{x}])(\mathbf{x} - E[\mathbf{x}])(\mathbf{x} - E[\mathbf{x}])^T \mathbf{D}_\varphi(E[\mathbf{x}])^T] \\ &= \mathbf{D}_\varphi(E[\mathbf{x}]) E[(\mathbf{x} - E[\mathbf{x}])(\mathbf{x} - E[\mathbf{x}])^T] \mathbf{D}_\varphi(E[\mathbf{x}])^T , \end{aligned}$$

which gives us a first order approximation of the covariance matrix of $\mathbf{y}$ in function of the covariance matrix of $\mathbf{x}$:

$$\Lambda_\mathbf{y} = \mathbf{D}_\varphi(E[\mathbf{x}]) \Lambda_\mathbf{x} \mathbf{D}_\varphi(E[\mathbf{x}])^T . \tag{3.26}$$

The case of an implicit function. In some cases, φ is implicit and we have to make use of the well-known implicit functions theorem to obtain the following result (see [41, chap.6]).

Proposition 3.1. *Let a criterion function $C : \mathbb{R}^m \times \mathbb{R}^p \rightarrow \mathbb{R}$ be a function of class C^∞, $\mathbf{x}_0 \in \mathbb{R}^m$ be the measurement vector and $\mathbf{y}_0 \in \mathbb{R}^p$ be a local minimum of $C(\mathbf{x}_0, \mathbf{z})$. If the Hessian $\mathbf{H}$ of C with respect to $\mathbf{z}$ is invertible at $(\mathbf{x}, \mathbf{z}) = (\mathbf{x}_0, \mathbf{y}_0)$ then there exists an open set U' of $\mathbb{R}^m$ containing $\mathbf{x}_0$ and an open set U'' of $\mathbb{R}^p$ containing $\mathbf{y}_0$ and a C^∞ mapping $\varphi : \mathbb{R}^m \rightarrow \mathbb{R}^p$ such that for $(\mathbf{x}, \mathbf{y})$ in $U' \times U''$ the two relations "$\mathbf{y}$ is a local minimum of $C(\mathbf{x}, \mathbf{z})$ with respect to $\mathbf{z}$" and $\mathbf{y} = \varphi(\mathbf{x})$ are equivalent. Furthermore, we have the following equation:*

$$\mathbf{D}_\varphi(\mathbf{x}) = -\mathbf{H}^{-1}\frac{\partial \Phi}{\partial \mathbf{x}}, \tag{3.27}$$

where

$$\Phi = \left(\frac{\partial C}{\partial \mathbf{z}}\right)^T \quad \textit{and} \quad \mathbf{H} = \frac{\partial \Phi}{\partial \mathbf{z}}.$$

Taking $\mathbf{x}_0 = E[\mathbf{x}]$ and $\mathbf{y}_0 = E[\mathbf{y}]$, equation (3.26) then becomes

$$\Lambda_\mathbf{y} = \mathbf{H}^{-1}\frac{\partial \Phi}{\partial \mathbf{x}}\Lambda_\mathbf{x}\frac{\partial \Phi}{\partial \mathbf{x}}^T \mathbf{H}^{-T}. \tag{3.28}$$

The case of a sum of squares of implicit functions. Here we study the case where C is of the form:

$$\sum_{i=1}^{n} C_i^2(\mathbf{x}_i, \mathbf{z})$$

with $\mathbf{x} = [\mathbf{x}_1^T, \ldots, \mathbf{x}_i^T, \ldots, \mathbf{x}_n^T]^T$. Then, we have

$$\begin{aligned} \Phi &= 2\sum_i C_i \frac{\partial C_i}{\partial \mathbf{z}}^T \\ \mathbf{H} = \frac{\partial \Phi}{\partial \mathbf{z}} &= 2\sum_i \frac{\partial C_i}{\partial \mathbf{z}}^T \frac{\partial C_i}{\partial \mathbf{z}} + 2\sum_i C_i \frac{\partial^2 C_i}{\partial \mathbf{z}^2}. \end{aligned}$$

Now, it is a usual practice to neglect the terms $C_i \frac{\partial^2 C_i}{\partial \mathbf{z}^2}$ with respect to the terms $\frac{\partial C_i}{\partial \mathbf{z}}^T \frac{\partial C_i}{\partial \mathbf{z}}$ (see classical books of numerical analysis [119]) and the numerical tests we did confirm that we can do this because the former is much smaller than the latter. We can then write:

$$\mathbf{H} = \frac{\partial \Phi}{\partial \mathbf{z}} \approx 2\sum_i \frac{\partial C_i}{\partial \mathbf{z}}^T \frac{\partial C_i}{\partial \mathbf{z}}.$$

In the same way we have:

$$\frac{\partial \Phi}{\partial \mathbf{x}} \approx 2 \sum_i \frac{\partial C_i}{\partial \mathbf{z}}^T \frac{\partial C_i}{\partial \mathbf{x}} .$$

Therefore, equation (3.28) becomes:

$$\Lambda_{\mathbf{y}} = 4\mathbf{H}^{-1} \sum_{i,j} \frac{\partial C_i}{\partial \mathbf{z}}^T \frac{\partial C_i}{\partial \mathbf{x}} \Lambda_{\mathbf{x}} \frac{\partial C_j}{\partial \mathbf{x}}^T \frac{\partial C_j}{\partial \mathbf{z}} \mathbf{H}^{-T} . \tag{3.29}$$

Assume that the noise in $\mathbf{x}_i$ and that in $\mathbf{x}_j$ ($j \neq i$) are independent (which is quite reasonable because the points are extracted independently), then $\Lambda_{\mathbf{x}_{i,j}} = E[(\mathbf{x}_i - \bar{\mathbf{x}}_i)(\mathbf{x}_i - \bar{\mathbf{x}}_i)^T] = \mathbf{0}$ and $\Lambda_{\mathbf{x}} = \mathrm{diag}(\Lambda_{\mathbf{x}_1}, \ldots, \Lambda_{\mathbf{x}_n})$. Equation (3.29) can then be written:

$$\Lambda_{\mathbf{y}} = 4\mathbf{H}^{-1} \sum_i \frac{\partial C_i}{\partial \mathbf{z}}^T \frac{\partial C_i}{\partial \mathbf{x}_i} \Lambda_{\mathbf{x}_i} \frac{\partial C_i}{\partial \mathbf{x}_i}^T \frac{\partial C_i}{\partial \mathbf{z}} \mathbf{H}^{-T} .$$

Since $\Lambda_{C_i} = \frac{\partial C_i}{\partial \mathbf{x}_i} \Lambda_{\mathbf{x}_i} \frac{\partial C_i}{\partial \mathbf{x}_i}^T$ by definition (up to the first order approximation), the above equation reduces to

$$\Lambda_{\mathbf{y}} = 4\mathbf{H}^{-1} \sum_i \frac{\partial C_i}{\partial \mathbf{z}}^T \Lambda_{C_i} \frac{\partial C_i}{\partial \mathbf{z}} \mathbf{H}^{-T} . \tag{3.30}$$

Considering that the mean of the value of C_i at the minimum is zero and under the somewhat strong assumption that the C_i's are independent and have identical distributed errors[3], we can then approximate Λ_{C_i} by its sample variance:

$$\Lambda_{C_i} = \frac{1}{n-p} \sum_i C_i^2 = \frac{S}{n-p} ,$$

where S is the value of the criterion C at the minimum, and p is the number of parameters, i.e. the dimension of $\mathbf{y}$. Although it has little influence when n is big, the inclusion of p in the formula above aims at correcting the effect of a small sample set. Indeed, for $n = p$, we can almost always find an estimate of $\mathbf{y}$ such that $C_i = 0$ for all i, which makes the estimation of the variance using this formula senseless. Equation (3.30) then finally becomes

$$\Lambda_{\mathbf{y}} = \frac{2S}{n-p} \mathbf{H}^{-1} \mathbf{H} \mathbf{H}^{-T} = \frac{2S}{n-p} \mathbf{H}^{-T} . \tag{3.31}$$

[3] It is under this assumption that the solution given by the least-squares technique is optimal.

The case of the fundamental matrix. As explained in Sect. 3.1.4, $\mathbf{F}$ is computed using a sum of squares of implicit functions of n point correspondences. Thus, referring to the previous paragraph, we have $p = 7$, and the criterion function $C(\hat{\mathbf{m}}, \mathbf{f}_7)$ (where $\hat{\mathbf{m}} = [\mathbf{m}_1, \mathbf{m}'_1, \cdots, \mathbf{m}_n, \mathbf{m}'_n]^T$ and $\mathbf{f}_7$ is the vector of the seven chosen parameters for $\mathbf{F}$) is given by (3.8). $\Lambda_{\mathbf{f}_7}$ is thus computed by (3.31) using the Hessian obtained as a by-product of the minimization of $C(\hat{\mathbf{m}}, \mathbf{f}_7)$.

According to (3.26), $\Lambda_{\mathbf{F}}$ is then computed from $\Lambda_{\mathbf{f}_7}$:

$$\Lambda_{\mathbf{F}} = \frac{\partial \mathbf{F}(\mathbf{f}_7)}{\partial \mathbf{f}_7} \Lambda_{\mathbf{f}_7} \frac{\partial \mathbf{F}(\mathbf{f}_7)}{\partial \mathbf{f}_7}^T . \tag{3.32}$$

Here, we actually consider the fundamental matrix $\mathbf{F}(\mathbf{f}_7)$ as a 9-vector composed of the 9 coefficients which are functions of the 7 parameters $\mathbf{f}_7$.

The Hyper-Ellipsoid of Uncertainty

If we define the random vector χ by

$$\chi = \Lambda_{\mathbf{y}}^{-\frac{1}{2}}(\mathbf{y} - E[\mathbf{y}])$$

and consider that $\mathbf{y}$ follows a Gaussian distribution, then χ follows a Gaussian distribution of mean zero and of covariance

$$E[\chi\chi^T] = E[\Lambda_{\mathbf{y}}^{-\frac{1}{2}}(\mathbf{y} - E[\mathbf{y}])(\mathbf{y} - E[\mathbf{y}])^T \Lambda_{\mathbf{y}}^{-\frac{1}{2}}] = \Lambda_{\mathbf{y}}^{-\frac{1}{2}} \Lambda_{\mathbf{y}} \Lambda_{\mathbf{y}}^{-\frac{1}{2}} = \mathbf{I} .$$

Consequently, the random variable $\delta_{\mathbf{y}}$ defined by

$$\delta_{\mathbf{y}} = \chi^T \chi = (\mathbf{y} - E[\mathbf{y}])^T \Lambda_{\mathbf{y}}^{-1}(\mathbf{y} - E[\mathbf{y}])$$

follows a χ^2 distribution of r degrees of freedom, where r is the rank of $\Lambda_{\mathbf{y}}$ (see [190]). Given a scalar k, we thus know the probability, equal to $P_{\chi^2}(k, r)$, that $\delta_{\mathbf{y}}$ appears between 0 and k^2. In other words, we have the following property:

Property 3.1. *If we consider that* $\mathbf{y}$ *follows a Gaussian distribution, the probability that* $\mathbf{y}$ *lies inside the k-hyper-ellipsoid defined by the equation*

$$(\mathbf{y} - E[\mathbf{y}])^T \Lambda_{\mathbf{y}}^{-1}(\mathbf{y} - E[\mathbf{y}]) = k^2$$

is equal to

$$P_{\chi^2}(k, r) ,$$

where k is any scalar, and r is the rank of $\Lambda_{\mathbf{y}}$.

The k-hyper-ellipsoid makes possible to graphically represent the uncertainty related to $\Lambda_{\mathbf{y}}$. For that, we usually come down to the two- or three-dimensional case, in order to draw an ellipse or an ellipsoid, by choosing two or three coordinates of $\mathbf{y}$ and extracting from $\Lambda_{\mathbf{y}}$ the corresponding submatrix.

The reader is referred to [190, chap.2] for a more detailed and comprehensive exposition on uncertainty manipulation.

3.1.8 An Example of Fundamental Matrix Estimation

The pair of images is a pair of calibrated stereo images (see Fig. 3.4). There are 241 point matches, which are established automatically by the technique described in Chap. 6. The calibrated parameters of the cameras are of course not used, but the fundamental matrix computed from these parameters serves as a ground truth. This is shown in Fig. 3.5, where the four epipolar lines are displayed, corresponding, from the left to the right, to the point matches 1, 220, 0 and 183, respectively. The intersection of these lines is the epipole, which is clearly very far from the image. This is because the two cameras are placed almost in the same plane.

Figure 3.4 Image pair used for comparing different estimation techniques of the fundamental matrix

Figure 3.5 Epipolar geometry estimated through classical stereo calibration, which serves as the ground truth

The epipolar geometry estimated with the linear method is shown in Fig. 3.6 for the same set of point matches. One can find that the epipolar is now in the image, which is completely different from what we have seen with the calibrated result. If we perform a data normalization before applying the linear method, the result is considerably improved, as shown in Fig. 3.7. This is very close to the calibrated one.

Figure 3.6 Epipolar geometry estimated with the linear method

Figure 3.7 Epipolar geometry estimated with the linear method with prior data normalization

The nonlinear method gives even better result, as shown in Fig. 3.8. A comparison with the "true" epipolar geometry is shown in Fig. 3.9. There is only a small difference in the orientation of the epipolar lines. We have also tried the normalization method followed by the nonlinear method, and the same result was obtained. Other methods have also been tested, and visually almost no difference is observed.

Figure 3.8 Epipolar geometry estimated with the nonlinear method

Figure 3.9 Comparison between the Epipolar geometry estimated through classical stereo calibration (shown in Red/Dark lines) and that estimated with the nonlinear method (shown in Green/Grey lines)

Quantitative results are provided in Table 3.2, where the elements in the first column indicates the methods used in estimating the fundamental matrix: they are respectively the classical stereo calibration (**Calib.**), the linear method with SVD (**linear**), the linear method with prior data normalization (**normal.**), the nonlinear method based on minimization of distances between points and epipolar lines (**nonlinear**), the M-estimator with Turkey function (**M-estim.**), the nonlinear method based on minimization of distances between observed points and reprojected ones (**reproj.**), and the LMedS technique (**LMedS**). The fundamental matrix of **Calib** is used as a reference. The second column shows the difference between the fundamental matrix estimated by each method with that of **Calib**. The difference is measured as the Frobenius norm: $\Delta \mathbf{F} = \|\mathbf{F} - \mathbf{F}_{\mathbf{Calib}}\| \times 100\%$. Since each $\mathbf{F}$ is normalized by its Frobenius norm, $\Delta \mathbf{F}$ is directly related to the angle between two unit vectors. It can be seen that although we have observed that Method **normal** has considerably improved the result of the linear method, its $\Delta \mathbf{F}$ is the largest. It seems that $\Delta \mathbf{F}$ is not appropriate to measure the difference between two fundamental matrix. We will describe another one in the next paragraph. The third and fourth columns show the positions of the two epipoles. The fifth column gives the root of the mean of squared distances between points and their epipolar lines. We can see that even with **Calib**, the RMS is as high as 1 pixel. There are two possibilities: either the stereo system is not very well calibrated, or the points are not well localized; and we think the latter is the major reason because the corner detector we

Table 3.2 Comparison of different methods for estimating the fundamental matrix

Method	ΔF	e		e′		RMS	CPU
Calib.		5138.18	−8875.85	1642.02	−2528.91	0.99	
linear	5.85%	304.018	124.039	256.219	230.306	3.40	0.13s
normal.	7.20%	−3920.6	7678.71	8489.07	−15393.5	0.89	0.15s
nonlinear	0.92%	8135.03	−14048.3	1896.19	−2917.11	0.87	0.38s
M-estim.	0.12%	4528.94	−7516.3	1581.19	−2313.72	0.87	1.05s
reproj.	0.92%	8165.05	−14102.3	1897.74	−2920.01	0.87	19.1s
LMedS	0.13%	3919.12	−6413.1	1500.21	−2159.65	0.75	2.40s

use only extracts points within pixel precision. The last column shows the approximate CPU time in seconds when the program is run on a Sparc 20 workstation. **Nonlinear** and **reproj** give essentially the same result (but the latter is much more time consuming). This is quite natural, because the closest point on the optimal epipolar line corresponds to the reprojection of the optimal reconstruction, "optimal" being always quantified in terms of image distances. The M-estimator and LMedS techniques give the best results. This is because the influence of poorly localized points has been reduced in M-estimator or they are simply discarded in LMedS. Actually, LMedS has detected five matches as outliers, which are 226, 94, 17, 78 and 100. Of course, these two methods are more time consuming than the nonlinear method.

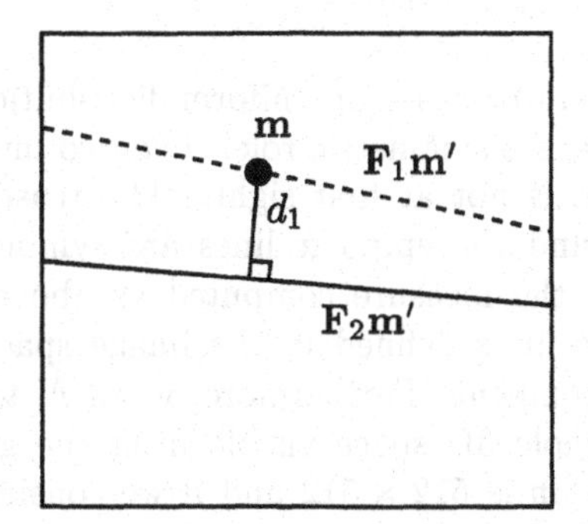

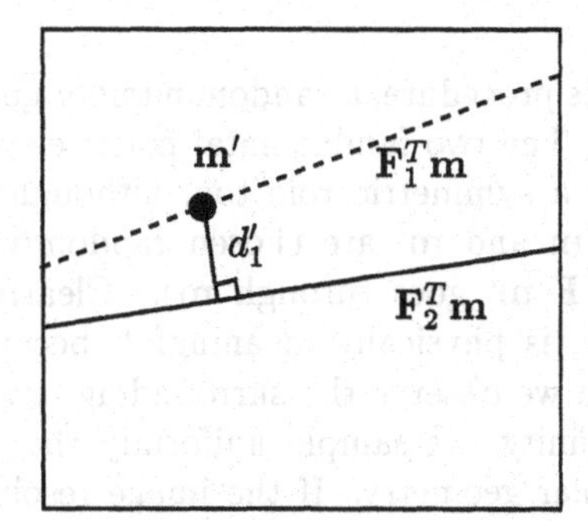

Figure 3.10 Definition of the difference between two fundamental matrices in terms of image distances

From the above discussion, the Frobenius norm of the difference between two normalized fundamental matrices is clearly not an appropriate measure. In the following, we describe a measure proposed by Stéphane Laveau from INRIA Sophia-Antipolis, which we think characterizes well the difference between two fundamental matrices. Let the two given fundamental matrices be $\mathbf{F}_1$ and $\mathbf{F}_2$. The measure is computed as follows (see Fig. 3.10):

Step 1: Choose *randomly* a point $\mathbf{m}$ in the first image.

Step 2: Draw the epipolar line of $\mathbf{m}$ in the second image using $\mathbf{F}_1$. The line is shown as a dashed line, and is defined by $\mathbf{F}_1^T\mathbf{m}$.

Step 3: If the epipolar line does not intersect the second image, go to **Step 1**.

Step 4: Choose *randomly* a point $\mathbf{m}'$ on the epipolar line. Note that $\mathbf{m}$ and $\mathbf{m}'$ correspond to each other exactly with respect to $\mathbf{F}_1$.

Step 5: Draw the epipolar line of $\mathbf{m}$ in the second image using $\mathbf{F}_2$, i.e. $\mathbf{F}_2^T\mathbf{m}$, and compute the distance, noted by d_1', between point $\mathbf{m}'$ and line $\mathbf{F}_2^T\mathbf{m}$.

Step 6: Draw the epipolar line of $\mathbf{m}'$ in the first image using $\mathbf{F}_2$, i.e. $\mathbf{F}_2\mathbf{m}'$, and compute the distance, noted by d_1, between point $\mathbf{m}$ and line $\mathbf{F}_2\mathbf{m}'$.

Step 7: Conduct the same procedure from **Step 2** through **Step 6**, but reversing the roles of $\mathbf{F}_1$ and $\mathbf{F}_2$, and compute d_2 and d_2'.

Step 8: Repeat N times **Step 1** through **Step 7**.

Step 9: Compute the average distance of d's, which is the measure of difference between the two fundamental matrices.

In this procedure, a random number generator based on uniform distribution is used. The two fundamental matrices plays a symmetric role. The two images plays a symmetric role too, although it is not at first sight. The reason is that $\mathbf{m}$ and $\mathbf{m}'$ are chosen randomly and the epipolar lines are symmetric (line $\mathbf{F}_1\mathbf{m}'$ goes through $\mathbf{m}$). Clearly, the measure computed as above, *in pixels*, is physically meaningful, because it is defined in the image space in which we observe the surrounding environment. Furthermore, when N tends to infinity, we sample uniformly the whole 3D space visible from the given epipolar geometry. If the image resolution is 512×512 and if we consider a pixel resolution, then the visible 3D space can be approximately sampled by 512^3 points. In our experiment, we set $N = 50000$. Using this method, we can compute the distance between each pair of fundamental matrices, and we obtain a symmetric matrix. The result is shown in Table 3.3, where only the upper triangle is displayed (because of symmetry). We arrive at the following conclusions:

- The linear method is very bad.
- The linear method with prior data normalization gives quite a reasonable result.
- The nonlinear method based on minimization of distances between points and epipolar lines gives the same result as that based on minimization of distances between observed points and reprojected ones. The latter should be avoided because it is too time consuming.
- M-estimators or the LMedS method give still better results because they try to limit or eliminate the effect of poorly localized points. The epipolar geometry estimated by LMedS is closer to the one computed through stereo calibration.

The LMedS method should be definitely used if the given set of matches contain false matches.

Table 3.3 Distances between the fundamental matrices estimated by different techniques

	linear	normal.	nonlinear	M-estim.	reproj.	LMedS
Calib.	116.4	5.97	2.66	2.27	2.66	1.33
linear		117.29	115.97	115.51	116.25	115.91
normal.			4.13	5.27	4.11	5.89
nonlinear				1.19	0.01	1.86
M-estim.					1.20	1.03
reproj.						1.88

3.1.9 Defining Epipolar Bands by Using the Estimated Uncertainty

Due to space limitation, the result on the uncertainty of the fundamental matrix is not shown here, and can be found in [23], together with its use in computing the uncertainty of the projective reconstruction and in improving the self-calibration based on Kruppa equations. We show in this section how to use the uncertainty to define the epipolar band for matching.

We only consider the epipolar lines in the second image (the same can be done for the first). For a given point $\mathbf{m}_0 = [u_0, v_0]^T$ in the first image together with its covariance matrix $\Lambda_{\mathbf{m}_0} = \left[\begin{smallmatrix}\sigma_{uu} & \sigma_{uv}\\ \sigma_{uv} & \sigma_{vv}\end{smallmatrix}\right]$, its epipolar line in the second image is given by $\mathbf{l}'_0 = \mathbf{F}^T\widetilde{\mathbf{m}}_0$. From (3.26), the covariance matrix of $\mathbf{l}'$ is computed by

$$\Lambda_{\mathbf{l}'_0} = \frac{\partial \mathbf{l}'_0}{\partial \mathbf{F}} \Lambda_{\mathbf{F}} \frac{\partial \mathbf{l}'_0}{\partial \mathbf{F}}^T + \mathbf{F}^T \left[\begin{smallmatrix}\Lambda_{\mathbf{m}_0} & \mathbf{0}_2\\ \mathbf{0}_2^T & 0\end{smallmatrix}\right] \mathbf{F}\,, \tag{3.33}$$

where $\mathbf{F}$ in the first term of the right hand is treated as a 9-vector, and $\mathbf{0}_2 = [0,0]^T$.

Any point $\mathbf{m}' = [u', v']^T$ on the epipolar line $\mathbf{l}'_0 \equiv [l'_1, l'_2, l'_3]^T$ must satisfy $\widetilde{\mathbf{m}}'^T\mathbf{l}'_0 = \mathbf{l}'^T_0\widetilde{\mathbf{m}}' = l'_1u' + l'_2v' + l'_3 = 0$ (we see the *duality* between points and lines). The vector $\mathbf{l}'_0$ is defined up to a scale factor. It is a projective point in the dual space of the image plane, and is the dual of the epipolar line. We consider the vector of parameters $\mathbf{x}_0 = (x_0, y_0) = (l'_1/l'_3, l'_2/l'_3)^T$ (if $l'_3 = 0$ we can choose $(l'_1/l'_2, l'_3/l'_2)$ or $(l'_2/l'_1, l'_3/l'_1)$). The covariance matrix of $\mathbf{x}_0$ is computed in the same way as (3.26): $\mathbf{C} = (\partial\mathbf{l}'_0/\partial\mathbf{x}_0)\Lambda_{\mathbf{l}'_0}(\partial\mathbf{l}'_0/\partial\mathbf{x}_0)^T$. The uncertainty of $\mathbf{x}_0$ can be represented in the usual way by an ellipse $\mathcal{C}$ in the dual space (denoted by $\mathbf{x}$) of the image plane:

$$(\mathbf{x} - \mathbf{x}_0)^T\mathbf{C}^{-1}(\mathbf{x} - \mathbf{x}_0) = k^2\,, \tag{3.34}$$

where k is a confidence factor determined by the χ^2 distribution of 2 degrees of freedom. The probability that $\mathbf{x}$ appears at the interior of the ellipse defined by (3.34) is equal to $P_{\chi^2}(k, 2)$. Equation (3.34) can be rewritten in projective form as

$$\widetilde{\mathbf{x}}^T\mathbf{A}\widetilde{\mathbf{x}} = 0 \quad \text{with} \quad \mathbf{A} = \begin{bmatrix} \mathbf{C}^{-1} & -\mathbf{C}^{-1}\mathbf{x}_0 \\ -\mathbf{x}_0^T\mathbf{C}^{-T} & \mathbf{x}_0^T\mathbf{C}^{-1}\mathbf{x}_0 - k^2 \end{bmatrix}.$$

The dual of this ellipse, denoted by $\mathcal{C}^*$, defines a conic in the image plane. It is given by

$$\widetilde{\mathbf{m}}^T\mathbf{A}^*\widetilde{\mathbf{m}} = 0 \tag{3.35}$$

where $\mathbf{A}^*$ is the adjoint of matrix $\mathbf{A}$ (i.e., $\mathbf{A}^*\mathbf{A} = \det(\mathbf{A})\mathbf{I}$). Because of the duality between the parameter space $\mathbf{x}$ and the image plane $\mathbf{m}$ (see Fig. 3.11), for a point $\mathbf{x}$ on $\mathcal{C}$, it defines an epipolar line in the image plane, line($\mathbf{x}$), which is tangent to conic $\mathcal{C}^*$ at a point $\mathbf{m}$, while the latter defines a line in the parameter space, line($\mathbf{m}$), which is tangent to $\mathcal{C}$ at $\mathbf{x}$. It can be shown [22] that, for a point in the interior of ellipse $\mathcal{C}$, the corresponding epipolar line lies outside of conic $\mathcal{C}^*$ (i.e., it does not cut the conic). Therefore, for a given k, the outside of this conic defines the region in which the epipolar line should lie with probability $P_{\chi^2}(k, 2)$. We call this region the *epipolar band*. For a given point in one

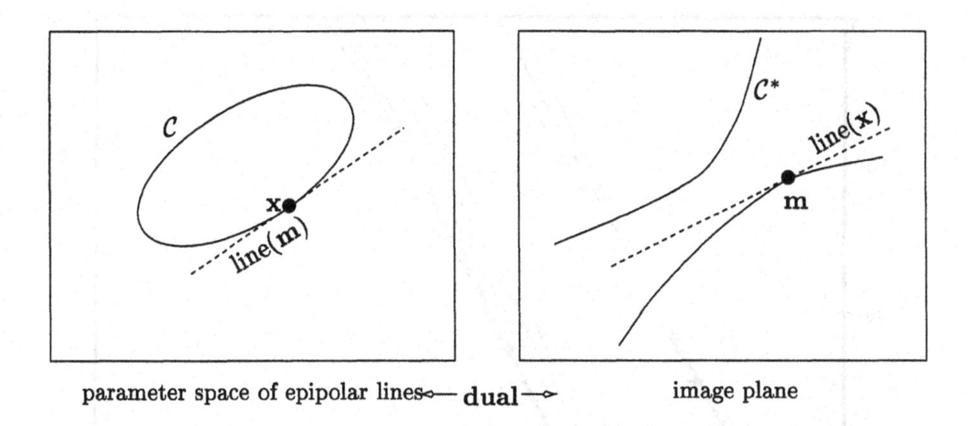

Figure 3.11 Duality between the image plane and the parameter space of the epipolar lines

image, its match should be searched in this region. Although, theoretically, the uncertainty conic defining the epipolar band could be an ellipse or parabola, it is always an hyperbola in practice (except when $\Lambda_{\mathbf{F}}$ is extremely huge).

We have estimated the uncertainty of the fundamental matrix for the image pair shown in Fig. 3.4. In Fig. 3.12, we show the epipolar bands of matches 1, 220, 0 and 183 in the second images, computed as described above. The displayed hyperbolas correspond to a probability of 70% ($k = 2.41$) with image point uncertainty of $\sigma_{uu} = \sigma_{vv} = 0.5^2$ and $\sigma_{uv} = 0$. We have also shown in Fig. 3.12 the epipolar lines drawn in dashed lines and the matched points indicated in +. An interesting thing is that the matched points are located in the area where the two sections of hyperbolas are closest to each other. This suggests that the covariance matrix of the fundamental matrix actually captures, to some extent, the matching information (disparity in stereo terminology). Such areas should be first examined in searching for point matches. This may, however, not be true if a significant depth discontinuity presents in the scene and if the point matches used in computing the fundamental matrix do not represent sufficiently enough the depth variation.

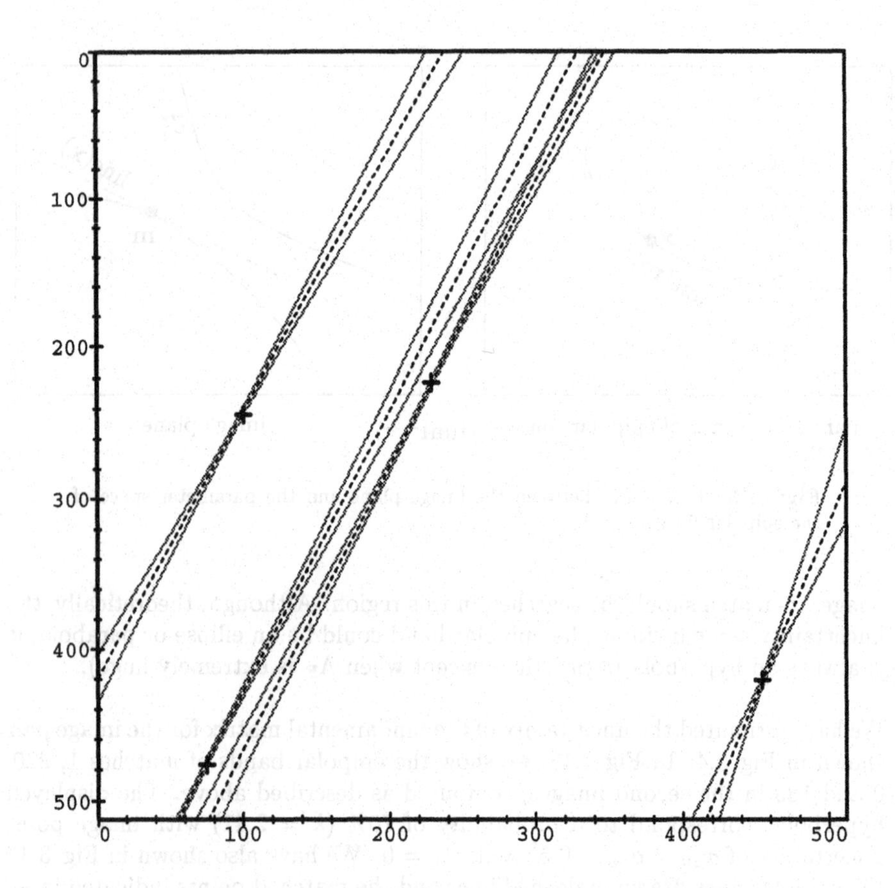

Figure 3.12 Epipolar bands for several point matches

3.2 DETERMINING FUNDAMENTAL MATRIX FOR AFFINE CAMERAS

In this section, we describe how to determine fundamental matrix for affine cameras from point matches in pixel coordinates, and how to determine the motion equations for the 2D affine motion as a degenerate case of the rigid motion. As in the last section, neither intrinsic nor extrinsic parameters are assumed to be known. The only assumption is that the points undergo a rigid transformation between the two camera coordinate systems.

Let the point match be $\mathbf{m}_i = [u_i, v_i]^T$ in the first image and $\mathbf{m}'_i = [u'_i, v'_i]^T$ in the second image. They must satisfy the epipolar equation $\tilde{\mathbf{m}}_i^T \mathbf{F}_A \tilde{\mathbf{m}}'_i = 0$. It is expanded as

$$f_{13}u_i + f_{23}v_i + f_{31}u_i' + f_{32}v_i' + f_{33} = 0 , \tag{3.36}$$

which is linear in image coordinates. If we define $\mathbf{f} = [f_{13}, f_{23}, f_{31}, f_{32}]^T$ and $\mathbf{u}_i = [u_i, v_i, u_i', v_i']^T$, then the above equation can be rewritten as

$$\mathbf{u}_i^T \mathbf{f} + f_{33} = 0 . \tag{3.37}$$

3.2.1 Exact Solution with 4 Point Matches

It is easy to see that there are 4 degrees of freedom in the epipolar equation. Thus in general, a minimum of 4 pairs of matched points are required to uniquely determine the affine fundamental matrix.

With 4 pairs of matched points, we can construct $\mathbf{U}_4$ as

$$\mathbf{U}_4 = \begin{bmatrix} \mathbf{u}_1^T \\ \mathbf{u}_2^T \\ \mathbf{u}_3^T \\ \mathbf{u}_4^T \end{bmatrix} .$$

If these 4 points are not from the same plane, and the motion is not a degenerate motion, or 2D affine motion, then the rank of $\mathbf{U}_4$ is 4. Thus we can determine the epipolar equation as

$$\mathbf{f} = -f_{33}\mathbf{U}_4^{-1}\mathbf{1} , \tag{3.38}$$

where $\mathbf{1} \equiv [1, 1, 1, 1]^T$. The solution is uniquely determined up to a scale factor.

3.2.2 Analytic Method with More than 4 Point Matches

If more than 4 point matches are available, we can use least-squares method to determine the epipolar equation more robustly.

Linear Least-Squares Technique

In the ideal case, the right hand side of (3.37) is zero for every point. But in practice due to discretization errors and noise, usually it is not true. Let us define the right hand side for point i as

$$\epsilon_i = \tilde{\mathbf{m}}_i^T \mathbf{F}_A \tilde{\mathbf{m}}_i' = \mathbf{u}_i^T \mathbf{f} + f_{33}\ , \quad i = 1, ..., n \ \ (n > 4)\ . \tag{3.39}$$

One way is to minimize the sum of all ϵ_i^2's. Let us define

$$C_1 = \sum_{i=1}^{n} \epsilon_i^2. \tag{3.40}$$

Minimizing C_1 is equivalent to

$$\min \sum_{i=1}^{n} \left(\tilde{\mathbf{m}}_i^T \mathbf{F}_A \tilde{\mathbf{m}}_i'\right)^2$$

and

$$\min \sum_{i=1}^{n} \left(\mathbf{u}_i^T \mathbf{f} + f_{33}\right)^2$$

Without loss of generality, we assume that f_{33} is not zero. (If f_{33} is indeed zero, then we can choose another unknown and solve the equation in a similar way.) Differentiating C_1 with respect to $\mathbf{f}$ and setting it to be zero vector, we have

$$2\mathbf{U}_n^T \mathbf{U}_n \mathbf{f} + 2 f_{33} \mathbf{U}_n \mathbf{1} = \mathbf{0}$$

This leads to the solution using pseudoinverse matrix of $\mathbf{U}_n$,

$$\mathbf{f} = -f_{33} \mathbf{U}_n^+ \mathbf{1} = -f_{33} (\mathbf{U}_n^T \mathbf{U}_n)^{-1} \mathbf{U}_n^T \mathbf{1}$$

The solution is unique up to a scale factor. Without loss of generality, we can let $f_{33} = 1$. So the solution is

$$\mathbf{f} = -\mathbf{U}_n^+ \mathbf{1} = -(\mathbf{U}_n^T \mathbf{U}_n)^{-1} \mathbf{U}_n^T \mathbf{1}\ , \tag{3.41}$$

where $\mathbf{U}_n^+$ is the pseudoinverse matrix of $\mathbf{U}_n$.

One problem with this method is that we do not know *a priori* which unknown is non-zero. We can actually try all possibilities by choosing each unknown to be non-zero, and retain the best estimation.

The other problem with this method is that the quantity which is minimized is not normalized with respect to $\mathbf{f}$, thus is dependent on $\mathbf{f}$. To tackle this problem, we can minimize another quantity which leads to the eigen analysis method.

Eigen Analysis

Now instead of minimizing C_1, we minimize

$$C_2 = \sum_{i=1}^{n} \frac{\epsilon_i^2}{\|\mathbf{f}\|^2} = \sum_{i=1}^{n} \frac{(\mathbf{u}_i^T\mathbf{f} + f_{33})^2}{\mathbf{f}^T\mathbf{f}} . \tag{3.42}$$

This formulation has a clear and intuitive interpretation. that is, here we are minimizing the sum of distances of each point $\mathbf{u}_i$ to the hyperplane $\mathbf{u}^T\mathbf{f}+f_{33} = 0$ in the 4-dimensional space. C_2 can be thought of as a projection of C_1 onto the normal direction of that hyperplane. As will be seen in next subsection, this formulation also has a nice physical interpretation.

Differentiating C_2 with respect to f_{33} and setting it to be zero, we have

$$\frac{dC_2}{df_{33}} = 2\sum_{i=1}^{n} \frac{\mathbf{u}_i{}^T\mathbf{f} + f_{33}}{\mathbf{f}^T\mathbf{f}} = 0 .$$

The solution of f_{33} is

$$f_{33} = -\frac{\sum_{i=1}^{n} \mathbf{u}_i^T\mathbf{f}}{n} = -\mathbf{u}_0^T\mathbf{f} , \tag{3.43}$$

where $\mathbf{u}_0 = [u_0, v_0, u_0', v_0']^T = \frac{1}{n}\sum_{i=1}^{n} \mathbf{u}_i$ is the average of all $\mathbf{u}_i$'s.

Substituting (3.43) for (3.42), C_2 becomes

$$C_2 = \sum_{i=1}^{n} \frac{(\mathbf{u}_i^T\mathbf{f} - \mathbf{u}_0^T\mathbf{f})^2}{\mathbf{f}^T\mathbf{f}} . \tag{3.44}$$

Let $\mathbf{v}_i = \mathbf{u}_i - \mathbf{u}_0$ and $\mathbf{W} = \sum_{i=1}^{n} \mathbf{v}_i\mathbf{v}_i^T$. Then we have

$$C_2 = \frac{\sum_{i=1}^{n} \mathbf{f}^T(\mathbf{u}_i - \mathbf{u}_0)(\mathbf{u}_i - \mathbf{u}_0)^T\mathbf{f}}{\mathbf{f}^T\mathbf{f}} = \frac{\mathbf{f}^T(\sum_{i=1}^{n} \mathbf{v}_i\mathbf{v}_i{}^T)\mathbf{f}}{\mathbf{f}^T\mathbf{f}} = \frac{\mathbf{f}^T\mathbf{W}\mathbf{f}}{\mathbf{f}^T\mathbf{f}} .$$

Now we can differentiate C_2 with respect to $\mathbf{f}$.

$$\frac{dC_2}{d\mathbf{f}} = \frac{2\mathbf{W}\mathbf{f}}{\mathbf{f}^T\mathbf{f}} - \frac{2\mathbf{f}^T\mathbf{W}\mathbf{f}\mathbf{f}}{(\mathbf{f}^T\mathbf{f})^2} = 2\frac{\mathbf{W}\mathbf{f} - C_2\mathbf{f}}{\mathbf{f}^T\mathbf{f}} .$$

Setting it to zero yields the eigen equation:

$$\mathbf{W}\mathbf{f} - C_2\mathbf{f} = \mathbf{0} . \tag{3.45}$$

As $\mathbf{W}$ is symmetric and positive semi-definite, it has n real non-negative eigenvalues and n associated eigenvectors. The solutions of C_2 and $\mathbf{f}$ in the above equation are the eigenvalues and their associated eigenvectors of $\mathbf{W}$, respectively. As we want to minimize C_2, we choose the minimal eigenvalue and its associated eigenvector of $\mathbf{W}$.

Note that this formulation is equivalent to minimizing C_1 subject to $\|\mathbf{f}\|^2 = 1$, as described in the last section.

Rank of W *and Relation with the 2D Affine Motion*

Let us look at the rank of $\mathbf{W}$. $\mathbf{W}$ can also be rewritten as

$$\mathbf{W} = \mathbf{V}_n^T \mathbf{V}_n \,, \tag{3.46}$$

where

$$\mathbf{V}_n = \begin{bmatrix} u_1 - u_0 & v_1 - v_0 & u'_1 - u'_0 & v'_1 - v'_0 \\ \cdot & \cdot & \cdot & \cdot \\ u_n - u_0 & v_n - v_0 & u'_n - u'_0 & v'_n - v'_0 \end{bmatrix} , \qquad (n \geq 4) \,.$$

We can rewrite (3.37) as

$$\mathbf{v}_i^T \mathbf{f} = 0 \,, \quad \text{for } i = 1, ..., n \quad (n \geq 4) \,, \tag{3.47}$$

because $f_{33} = -\mathbf{u}_0^T \mathbf{f}$. Note that it includes the case of $n = 4$. Putting the n equations into matrix form, we have

$$\mathbf{V}_n \mathbf{f} = \mathbf{0} \,. \tag{3.48}$$

where $\mathbf{V}_n$ is defined in (3.2.2), and its size is $n \times 4$. Thus, rank($\mathbf{V}_n$) is at most 4. In order for $\mathbf{V}_n \mathbf{f} = \mathbf{0}$ to have a unique solution (up to a scale factor), the rank of $\mathbf{V}_n$ (in turn, that of $\mathbf{W}$ as well) must be equal to 3.

If

$$\text{rank}(\mathbf{V}_n) = 2 \,,$$

then it is either because all the points lie on the same plane, or because $\beta = 0$, that is, the points undergo a 2D affine motion. Since it is not likely that all the

n ($n > 4$) points lie on the same plane, it is highly probable that the motion is a 2D affine motion, a degenerate case of the 3D rigid motion. Note that the rank of $\mathbf{V}_n$ cannot be 1 unless all the points in each image lie on a straight line.

Based on the above discussions, we now propose an algorithm that not only can find the effective rank of $\mathbf{V}_n$, but can also at the same time determine the epipolar equation.

By Singular Value Decomposition [46], we can write the left-hand side of (3.48) as the product of a $n \times 4$ matrix $\mathbf{U}$, a diagonal 4×4 matrix $\mathbf{S}$, a 4×4 matrix $\mathbf{V}^T$ and $\mathbf{f}$,

$$\mathbf{USV}^T\mathbf{f}$$

such that $\mathbf{U}^T\mathbf{U} = \mathbf{V}^T\mathbf{V} = \mathbf{VV}^T = \mathbf{I}$, where $\mathbf{I}$ is a 4×4 identity matrix. $\mathbf{S}$ is a diagonal matrix whose diagonal components are the singular values $\sigma_1 \geq ... \geq \sigma_4$ sorted in non-increasing order. The vectors in $\mathbf{U}$ and $\mathbf{V}$ are of unit length and orthogonal to each other respectively.

In the ideal case, as the rank of $\mathbf{V}_n$ is only 3, the last singular value should be 0. This should be always true in the case of $n = 4$, but not so in the case of $n > 4$ due to noise in the coordinates. Let us call the singular vector associated with the 4th singular value $\mathbf{s}_4$. Then the solution of $\mathbf{f}$ is

$$\mathbf{f} = \mathbf{s}_4 \ . \tag{3.49}$$

Substituting it for (3.48) yields,

$$\mathbf{USV}^T\mathbf{s}_4 = \mathbf{US}\begin{bmatrix}0\\0\\0\\1\end{bmatrix} = \mathbf{U}\begin{bmatrix}0\\0\\0\\\sigma_4\end{bmatrix} = \sigma_4\mathbf{l}_4 \ .$$

As the σ_4 is the smallest singular value, it is guaranteed that $\mathbf{f} = \mathbf{s}_4$ makes the left-hand side to be the shortest possible vector. It is worth mentioning that here matrix $\mathbf{V}$ is identical with the eigenvectors of matrix $\mathbf{W}$. We thus arrive at the same solution.

The effective rank of $\mathbf{V}_n$ can be determined from the singular values. If not only σ_4 is small, but σ_3 is small as well, then we know that the rank $\mathbf{V}_n$ is 2, and the motion is an 2D affine motion.

Also note that if σ_4 is not close to zero (this happens only when $n > 4$), then it means that the points are not undergoing an identical rigid motion. A small deviation from zero is understandable as there exist errors in the coordinates of points even if they do undergo an identical rigid motion.

3.2.3 Minimizing Distances of Points to Epipolar Lines

The above linear methods did not consider minimizing any physically meanful quantities. As in the last section, we can minimize the sum of squared Euclidean distances to the epipolar lines on which they are supposed to lie. For the first image, it is

$$d_1^2 = \sum_{i=1}^{n} \frac{\epsilon_i^2}{f_{13}^2 + f_{23}^2}, \tag{3.50}$$

and similarly, for the second image, it is

$$d_2^2 = \sum_{i=1}^{n} \frac{\epsilon_i^2}{f_{31}^2 + f_{32}^2}. \tag{3.51}$$

One might try to minimize only one of them. But since the two images play a symmetrical role, it is evident that some sort of summation of the two should perform better.

One criterion is to minimize the sum of half d_1^2 and half d_2^2,

$$C_3 = \frac{1}{2}d_1^2 + \frac{1}{2}d_2^2 = \frac{1}{2}\sum_{i=1}^{n} \epsilon_i^2 \left(\frac{1}{f_{13}^2 + f_{23}^2} + \frac{1}{f_{31}^2 + f_{32}^2}\right), \tag{3.52}$$

allowing the two to contribute equal to the final result.

However, as described in Sect. 2.5, the scales of the two images are generally different, that is, the spacings of the epipolar lines in the two images are generally different. It is reasonable to allow the two to make different contributions to the final result. Taking into account of this scale change ρ, the criterion can then be a weighted sum of the two with the weights representing the scale change:

$$C_4 = \frac{1}{1+\rho^2}d_1^2 + \frac{\rho^2}{1+\rho^2}d_2^2. \tag{3.53}$$

Substituting (3.50), (3.51) and (2.116) for d_1, d_2 and ρ, respectively, we have

$$C_4 = 2\sum_{i=1}^{n} \frac{\epsilon_i^2}{f_{13}^2 + f_{23}^2 + f_{31}^2 + f_{32}^2} = 2\sum_{i=1}^{n} \frac{(\mathbf{p}_i{}^T\mathbf{f} + f_{33})^2}{\mathbf{f}^T\mathbf{f}} . \tag{3.54}$$

When the scale is the same for the two images, $\rho = 1$. Then the two criteria C_3 and C_4 become identical. As the scale does not change very much between the two images, these two criteria should produce similar performances. This was confirmed by Shapiro *et al.* who did comparisons between the results using different criteria [137].

As a matter of fact, C_4 is twice of C_2. Thus minimizing C_2 and minimizing C_4 are equivalent to each other. That is, the eigen analysis method used to minimize C_2 can be used to minimize C_4. This means that minimizing C_4 also has an analytic solution which is described in the last subsection.

We will show some experimental results in Sect. 3.2.5, where results are compared between using C_1 and C_2. This is sufficient as C_2 is equivalent to C_4 and is close to C_3 if the scales of the two images are similar to each other.

3.2.4 Minimizing Distances Between Observation and Reprojection

Shapiro *et al.* also show that C_4 is also equivalent to minimizing the distances of observations and reprojections [137].

Suppose that the 3D points are $\mathtt{M}_i$. They are projected onto images as $[u_i, v_i]^T$ and $[u_i', v_i']^T$ by

$$\begin{bmatrix} u_i \\ v_i \end{bmatrix} = \mathbf{T}_A \mathtt{M}_i + \mathbf{t}_A , \tag{3.55}$$

$$\begin{bmatrix} u_i' \\ v_i' \end{bmatrix} = \mathbf{T}_A' \mathtt{M}_i + \mathbf{t}_A' , \tag{3.56}$$

where $\mathbf{T}_A = \frac{1}{P_{34}} \begin{bmatrix} \mathbf{p}_1^T \\ \mathbf{p}_2^T \end{bmatrix} = \frac{1}{P_{34}} \begin{bmatrix} P_{11} & P_{12} & P_{13} \\ P_{21} & P_{22} & P_{23} \end{bmatrix}$, $\mathbf{t}_A = \frac{1}{P_{34}} \begin{bmatrix} P_{14} \\ P_{24} \end{bmatrix}$, and $\mathbf{T}_A'$ and $\mathbf{t}_A'$ are defined similarly.

Since the centroid of the object $\mathtt{M}_0$ is projected onto the two images as centroids $[u_0, v_0]^T$ and $[u'_0, v'_0]^T$, we have

$$\mathbf{t}_A = \begin{bmatrix} u_0 \\ v_0 \end{bmatrix} - \mathbf{T}_A \mathtt{M}_0 , \tag{3.57}$$

$$\mathbf{t}'_A = \begin{bmatrix} u'_0 \\ v'_0 \end{bmatrix} - \mathbf{T}'_A \mathtt{M}_0 . \tag{3.58}$$

Now define C_5 as

$$C_5 = \sum_{i=1}^{n} \| \begin{bmatrix} u_i \\ v_i \end{bmatrix} - \mathbf{T}_A \mathtt{M}_i - \mathbf{t}_A \|^2 + \sum_{i=1}^{n} \| \begin{bmatrix} u'_i \\ v'_i \end{bmatrix} - \mathbf{T}'_A \mathtt{M}_i - \mathbf{t}'_A \|^2 . \tag{3.59}$$

Substituting (3.57) and (3.58) for (3.59), we have

$$C_5 = \sum_{i=1}^{n} \|\mathbf{v}_i - \mathbf{L}\bar{\mathtt{M}}_i\|^2 ,$$

where $\mathbf{v}_i = \begin{bmatrix} u_i - u_0 \\ v_i - v_0 \\ u'_i - u'_0 \\ v'_i - v'_0 \end{bmatrix}$, $\mathbf{L} = \frac{1}{P_{34}} \begin{bmatrix} \mathbf{T}_A \\ \mathbf{T}'_A \end{bmatrix} = \frac{1}{P_{34}} \begin{bmatrix} \mathbf{p}_1^T \\ \mathbf{p}_2^T \\ \mathbf{p}_1'^T \\ \mathbf{p}_2'^T \end{bmatrix}$ and $\bar{\mathtt{M}}_i = \mathtt{M}_i - \mathtt{M}_0$. Differentiating C_5 with respect to $\bar{\mathtt{M}}_i$ and setting it to be zero yields

$$\frac{dC_5}{d\bar{\mathtt{M}}_i} = -2\mathbf{L}^T \mathbf{v}_i + 2\mathbf{L}^T \mathbf{L} \bar{\mathtt{M}}_i = \mathbf{0} ,$$

that is,

$$\bar{\mathtt{M}}_i = \mathbf{L}^+ \mathbf{v}_i .$$

Substituting this back into C_5 gives

$$C_5 = \sum_{i=1}^{n} \| \left(\mathbf{I} - \mathbf{L}\mathbf{L}^+\right) \mathbf{v}_i \|^2 . \tag{3.60}$$

From (2.150) (i.e. $\mathbf{L}^T \mathbf{f} = \mathbf{0}$) and (2.179), we know

$$\mathbf{I} - \mathbf{L}\mathbf{L}^+ = \frac{\mathbf{f}\mathbf{f}^T}{\|\mathbf{f}\|^2} . \tag{3.61}$$

Substituting this for C_5 yields

$$C_5 = \sum_{i=1}^{n} \frac{\mathbf{f}^T \mathbf{v}_i \mathbf{v}_i^T \mathbf{f}}{\mathbf{f}^T \mathbf{f}} , \tag{3.62}$$

which is exactly the same as C_2.

Note that Reid et al. show that this is the cost function minimized by Tomasi and Kanade's factorization algorithm for shape and motion recovery [123, 155].

3.2.5 An Example of Affine Fundamental Matrix Estimation

A group of 20 3D points are randomly generated and two images are synthesized by applying a rigid transformation to the 3D points and applying a weak perspective projection. Then Gaussian noises of $\sigma = 1.0$ and $\sigma = 2.0$ are added to the image points.

Two methods are compared. The first is the pseudo-inverse matrix method (indicated by "A" in the table), and the other is the eigen analysis method (indicated by "B" in the table). Table 3.4 shows the results for 4 different sets of transformations. The first row shows the true data used for simulation. The 2nd, 3rd, 4th and 5th rows show the results of using method "A" to data with noise $\sigma = 1.0$, method "B" to data with noise $\sigma = 1.0$, using method "A" to data with noise $\sigma = 2.0$, method "B" to data with noise $\sigma = 2.0$, respectively. E_1 and E_2 are the sum of squared distances of the image points to the epipolar lines in the first and second images, respectively, and E_3 is twice the sum of squared distances of the 4D points to the 4D hyperplane determining the epipolar geometry. E_1, E_2 and E_3 are thus comparable.

It is easy to verify that the ratio of E_1 and E_2 is roughly the square of the scale change between the two images. And E_3 is always smaller than the average of E_1 and E_2. These agree with the theoretical analysis in the last subsections.

As for the performance of the two methods, the eigen analysis method is always superior to the pseudo-inverse method in terms of E_1, E_2 and E_3. But it is not always true in terms of accuracy of recovered motion parameters.

Table 3.4 Four examples for the estimation of affine fundamental matrix

	α	θ	ρ	λ	E_1	E_2	E_3
example 1							
true data ($\beta = 30.0$)	30.0	15.0	1.100	13.5			
result by A, $\sigma = 1.0$	29.7	14.6	1.101	13.5	43.2	52.4	47.4
result by B, $\sigma = 1.0$	29.7	14.8	1.102	13.2	42.0	51.0	46.1
result by A, $\sigma = 2.0$	28.7	11.7	1.099	14.8	70.3	84.9	76.9
result by B, $\sigma = 2.0$	28.7	12.0	1.100	14.3	67.7	81.8	76.0
example 2							
true data ($\beta = 10.0$)	30.0	15.0	0.950	11.6			
result by A, $\sigma = 1.0$	33.4	22.0	0.949	11.4	31.7	28.6	30.0
result by B, $\sigma = 1.0$	33.6	22.5	0.948	11.2	31.2	28.0	29.5
result by A, $\sigma = 2.0$	29.6	14.6	0.959	12.6	86.1	79.3	82.6
result by B, $\sigma = 2.0$	30.1	15.8	0.955	12.1	83.0	75.6	79.1
example 3							
true data ($\beta = 5.0$)	30.0	10.0	1.100	14.1			
result by A, $\sigma = 1.0$	25.2	0.1	1.098	15.0	42.8	51.6	46.8
result by B, $\sigma = 1.0$	27.0	3.8	1.097	14.5	41.7	50.2	45.6
result by A, $\sigma = 2.0$	29.4	9.2	1.082	14.4	84.0	98.4	90.6
result by B, $\sigma = 2.0$	34.6	20.0	1.081	13.3	77.9	91.0	84.0
example 4							
true data ($\beta = 3.0$)	30.0	10.0	0.950	12.2			
result by A, $\sigma = 1.0$	34.5	19.5	0.946	11.6	29.6	26.5	28.0
result by B, $\sigma = 1.0$	37.6	25.7	0.947	11.2	28.4	25.5	26.9
result by A, $\sigma = 2.0$	33.4	17.0	0.953	12.3	71.6	65.0	68.2
result by B, $\sigma = 2.0$	44.7	39.4	0.950	10.7	63.3	57.2	60.1

It is also noted that to such a larger number of points, adding noise of $\sigma = 1$ and adding noise of $\sigma = 2$ do not produce large difference in the recovered motion parameters as one might expect.

The increase of errors is mainly due to the decrease of rotation angle β, the rotation out of image plane. When it is large enough, adding noise does not significantly affect the quality of recovered motion parameters. But when it is as small as 5 degrees or 3 degrees, the errors in the recovered motion parameters become very large.

3.2.6 Charactering the Uncertainty of Affine Fundamental Matrix

In this subsection, we discuss the uncertainty of affine fundamental matrix computed from matched points which are always corrupted by noise. Similar to the full perspective projection, we use the covariance of $\mathbf{f}$ to characterize the uncertainty of the fundamental matrix. This analysis was originally given by Shapiro *et al* [137].

As seen in the last subsections, $\mathbf{f}$ is determined as the eigenvector $\mathbf{s}_4$ associated with the smallest eigenvalue of $\mathbf{W}$. We estimate the covariance matrix for $\mathbf{s}_4$ as the points are perturbed by independent isotropic additive Gaussian noise.

Let the Gaussian noise be $\delta\mathbf{u}_i$, and it is added to the real 4D point $\hat{\mathbf{u}}_i$, giving the measurement $\mathbf{u}_i = \hat{\mathbf{u}}_i + \delta\mathbf{u}_i$. In the following we use "hat" to indicate noise-free quantities. It is assumed that the noise has zero mean with variance σ^2. This leads to $E\{\delta\mathbf{u}_i\} = \mathbf{0}$, $E\{\mathbf{u}_i\} = \hat{\mathbf{u}}_i$, $Var\{\mathbf{u}_i\} = Var\{\delta\mathbf{u}_i\}$ and

$$\mathbf{\Gamma} = E\{\delta\mathbf{u}_i\delta\mathbf{u}_i^T\} = \sigma^2\mathbf{I}_4 .$$

Since the noise is independent from each other,

$$E\{\delta\mathbf{u}_i\delta\mathbf{u}_j^T\} = \delta_{ij}\mathbf{\Gamma} ,$$

where, $\delta_{ij} = \begin{cases} 1 & i = j \\ 0 & \text{otherwise} \end{cases}$ is the Kronecker delta product.

Now let us consider $\mathbf{v}_i = \mathbf{u}_i - \mathbf{u}_0$. The noise in $\mathbf{u}_i$ induces an error $\delta\mathbf{v}_i$ in $\mathbf{v}_i$. It is easy to derive

$$\delta\mathbf{u}_0 = \frac{1}{n}\sum_{i=1}^{n}\delta\mathbf{u}_i .$$

From it we get

$$E\{\delta\mathbf{v}_i\delta\mathbf{v}_i^T\} = E\{(\delta\mathbf{u}_i - \delta\mathbf{u}_0)(\delta\mathbf{u}_i - \delta\mathbf{u}_0)^T\} = (1 - \frac{1}{n})\mathbf{\Gamma} , \tag{3.63}$$

and for $i \neq j$,

$$E\{\delta\mathbf{v}_i\delta\mathbf{v}_j^T\} = E\{(\delta\mathbf{u}_i - \delta\mathbf{u}_0)(\delta\mathbf{u}_j - \delta\mathbf{u}_0)^T\} = -\frac{1}{n}\mathbf{\Gamma} . \tag{3.64}$$

When n is large, the covariance matrices for $\mathbf{v}_i$ tends towards those for $\mathbf{u}_i$.

Let us write $\mathbf{V}_n = [\mathbf{v}_1 \ \ \mathbf{v}_2 \ \ \ldots \ \ \mathbf{v}_n]$ and $\delta\mathbf{V}_n = [\delta\mathbf{v}_1 \ \ \delta\mathbf{v}_2 \ \ \ldots \ \ \delta\mathbf{v}_n]$. Since $\mathbf{W} = \mathbf{V}_n\mathbf{V}_n^T$, we get

$$\mathbf{W} = (\hat{\mathbf{V}}_n + \delta\hat{\mathbf{V}}_n)(\hat{\mathbf{V}}_n + \delta\hat{\mathbf{V}}_n)^T = \hat{\mathbf{V}}_n\hat{\mathbf{V}}_n^T + \hat{\mathbf{V}}_n\delta\hat{\mathbf{V}}_n^T + \delta\hat{\mathbf{V}}_n\hat{\mathbf{V}}_n^T + \delta\hat{\mathbf{V}}_n\delta\hat{\mathbf{V}}_n^T .$$

If we write $\mathbf{W} = \hat{\mathbf{W}} + \delta\mathbf{W}$, and ignore the second order term in the above expression, then we have

$$\delta\mathbf{W} \approx \hat{\mathbf{V}}_n\delta\hat{\mathbf{V}}_n^T + \delta\hat{\mathbf{V}}_n\hat{\mathbf{V}}_n^T .. \tag{3.65}$$

Now we can consider the eigenvector $\hat{\mathbf{s}}_4$. In the noise-free ideal case, $\hat{\mathbf{W}}\hat{\mathbf{s}}_4 = \lambda_4\hat{\mathbf{s}}_4 = \mathbf{0}$. Since $\delta\mathbf{W}$ is a real symmetric matrix, the first-order change in $\hat{\mathbf{s}}_4$ can be written as (see [46, 135])

$$\delta\mathbf{s}_4 = -\sum_{k=1}^{3} \frac{(\hat{\mathbf{s}}_k^T\delta\mathbf{W}\hat{\mathbf{s}}_4)\hat{\mathbf{s}}_k}{\hat{\lambda}_k} = \hat{\mathbf{J}}\delta\mathbf{W}\hat{\mathbf{s}}_4 ,$$

where

$$\hat{\mathbf{J}} = -\sum_{k=1}^{3} \frac{\hat{\mathbf{s}}_k\hat{\mathbf{s}}_k^T}{\hat{\lambda}_k} .$$

Substituting (3.65) for it, we have

$$\delta\mathbf{s}_4 = \hat{\mathbf{J}}(\hat{\mathbf{V}}_n\delta\hat{\mathbf{V}}_n^T + \delta\hat{\mathbf{V}}_n\hat{\mathbf{V}}_n^T)\hat{\mathbf{s}}_4 = \hat{\mathbf{J}}\sum_{i=1}^{n}\hat{\mathbf{v}}_i(\delta\mathbf{v}_i^T\hat{\mathbf{s}}_4) .$$

Finally the covariance matrix for $\mathbf{s}_4$ can be computed as

$$\begin{aligned} E\{\delta\mathbf{s}_4\delta\mathbf{s}_4^T\} &= E\{\hat{\mathbf{J}}\hat{\mathbf{V}}_n\delta\mathbf{V}_n^T\hat{\mathbf{s}}_4\hat{\mathbf{s}}_4^T\delta\mathbf{V}_n\hat{\mathbf{V}}_n^T\hat{\mathbf{J}}^T\} \\ &= \hat{\mathbf{J}}E\{\left(\sum_{i=1}^{n}\hat{\mathbf{v}}_i(\delta\mathbf{v}_i^T\hat{\mathbf{s}}_4)\right)\left(\sum_{j=1}^{n}\hat{\mathbf{v}}_j(\delta\mathbf{v}_j^T\hat{\mathbf{s}}_4)\right)\}\hat{\mathbf{J}}^T \\ &= \hat{\mathbf{J}}\{\sum_{i=1}^{n}\hat{\mathbf{v}}_i\left(\sum_{j=1}^{n}\hat{\mathbf{v}}_j^T\hat{\mathbf{s}}_4^T E\{\delta\mathbf{v}_i\delta\mathbf{v}_j^T\}\hat{\mathbf{s}}_4\right)\}\hat{\mathbf{J}}^T . \end{aligned}$$

From (3.63) and (3.64), and $\hat{\mathbf{s}}_4^T\hat{\mathbf{s}}_4 = 1$, we have

$$\hat{\mathbf{s}}_4^T E\{\delta\mathbf{v}_i\delta\mathbf{v}_j^T\}\hat{\mathbf{s}}_4 = \begin{cases} \sigma^2(1-\frac{1}{n}), & i = j \\ -\sigma^2\frac{1}{n}, & i \neq j \end{cases}$$

Since $\sum_{j=1}^{n} \hat{\mathbf{v}}_j = 0$, we get

$$\sum_{j=1}^{n} \hat{\mathbf{v}}_j^T \hat{\mathbf{s}}_4^T E\{\delta \mathbf{v}_i \delta \mathbf{v}_j^T\} \hat{\mathbf{s}}_4 = \sigma^2 \hat{\mathbf{v}}_i^T ,$$

and

$$\sum_{i=1}^{n} \hat{\mathbf{v}}_i \left(\sum_{j=1}^{n} \hat{\mathbf{v}}_j^T \hat{\mathbf{s}}_4^T E\{\delta \mathbf{v}_i \delta \mathbf{v}_j^T\} \hat{\mathbf{s}}_4 \right) = \sigma^2 \sum_{i=1}^{n} \hat{\mathbf{v}}_i \hat{\mathbf{v}}_i^T = \sigma^2 \hat{\mathbf{W}} .$$

Now the covariance becomes

$$\begin{aligned}
\sigma^2 \hat{\mathbf{J}} \hat{\mathbf{W}} \hat{\mathbf{J}}^T &= \sigma^2 \left(\sum_{k=1}^{3} \frac{\hat{\mathbf{s}}_k \hat{\mathbf{s}}_k^T}{\hat{\lambda}_k} \right) \left(\sum_{l=1}^{3} \frac{\hat{\mathbf{W}} \hat{\mathbf{s}}_l \hat{\mathbf{s}}_l^T}{\hat{\lambda}_l} \right) \\
&= \sigma^2 \left(\sum_{k=1}^{3} \frac{\hat{\mathbf{s}}_k \hat{\mathbf{s}}_k^T}{\hat{\lambda}_k} \right) \left(\sum_{l=1}^{3} \frac{\hat{\lambda}_l \hat{\mathbf{s}}_l \hat{\mathbf{s}}_l^T}{\hat{\lambda}_l} \right) \\
&= \sigma^2 \sum_{k=1}^{3} \frac{\hat{\mathbf{s}}_k \hat{\mathbf{s}}_k^T}{\hat{\lambda}_k} = -\sigma^2 \hat{\mathbf{J}} .
\end{aligned}$$

Most of the above equations require the noise-free quantities, which are not available in general. Weng *et al.* pointed out that if one writes $\hat{\mathbf{V}}_n = \mathbf{V}_n - \delta \mathbf{V}_n$ and substitute this in the relevant equations, the terms in $\delta \mathbf{V}_n$ disappear in the first-order approximations [174]. This allows us to replace the noise-free quantities by the directly measurable quantities. By doing so, we can represent the covariance of $\mathbf{f}$ as

$$\Lambda_{\mathbf{f}} = \sigma^2 \sum_{k=1}^{3} \frac{\mathbf{s}_k \mathbf{s}_k^T}{\lambda_k} . \tag{3.66}$$

It is easy to see that if $\lambda_k, k = 1, 2, 3$ are large, then the covariance is small. If λ_3 is also close to zero (remember that $\lambda_3 < \lambda_2 < \lambda_1$), that is, the rank of $\mathbf{W}$ is only 2, which corresponds to the 2D affine motion, then the covariance can be very large.

3.2.7 Determining Motion Equation in the 2D Affine Motion Case

In the 2D affine motion case, the motion can be completely and uniquely determined. Minimally, two pairs of points can determine the 2D affine motion equations. But since we have more than 2 pairs of points, again we can obtain the optimal solution using linear least-squares technique, or by minimizing the distances between observed points and reprojections.

Eq. (2.126) can be rewritten as

$$\begin{bmatrix} u_i \\ v_i \end{bmatrix} = \begin{bmatrix} u'_i & v'_i & 1 & 0 \\ v'_i & -u'_i & 0 & 1 \end{bmatrix} \begin{bmatrix} a \\ b \\ t_u \\ t_v \end{bmatrix} . \tag{3.67}$$

Using the pseudo-inverse matrix, we can determine the coefficients as

$$\begin{bmatrix} a \\ b \\ t_u \\ t_v \end{bmatrix} = \left(\mathbf{H}^T\mathbf{H}\right)^{-1} \mathbf{H}^T\mathbf{h} , \tag{3.68}$$

where

$$\mathbf{H} = \begin{bmatrix} u'_i & v'_i & 1 & 0 \\ v'_i & -u'_i & 0 & 1 \\ \vdots & \vdots & \vdots & \vdots \\ u'_n & v'_n & 1 & 0 \\ v'_n & -u'_n & 0 & 1 \end{bmatrix} ,$$

and

$$\mathbf{h} = \begin{bmatrix} u_i & v_i & \dots & u'_n & v'_n \end{bmatrix}^T .$$

We can also define a criterion as the sum of squared distances between the n observed corresponding points and their reprojections after the 2D affine motion,

$$C_6 = \sum_{i=1}^{n} \{(u_i - (au'_i + bv'_i + t_u))^2 + (v_i - (-bu'_i + av'_i + t_v))^2\} , \tag{3.69}$$

and minimize it with respect to a, b, t_u, and t_v.

Differentiating C_6 with respect to t_u and t_v yields

$$\begin{aligned} \sum_{i=1}^{n} \{u_i - (au'_i + bv'_i) - t_u\} &= 0 , \\ \sum_{i=1}^{n} \{v_i - (-bu'_i + av'_i) - t_v\} &= 0 . \end{aligned}$$

Thus, t_u and t_v can be obtained as

$$t_u = \frac{1}{n}\sum_{i=1}^{n} u_i - (a\frac{1}{n}\sum_{i=1}^{n} u'_i + b\frac{1}{n}\sum_{i=1}^{n} v'_i) = u_0 - (au'_0 + bv'_0) , \quad (3.70)$$

$$t_v = \frac{1}{n}\sum_{i=1}^{n} v_i - (-b\frac{1}{n}\sum_{i=1}^{n} u'_i + a\frac{1}{n}\sum_{i=1}^{n} v'_i) = v_0 - (-bu'_0 + av'_0) , \quad (3.71)$$

where

$$u_0 = \frac{1}{n}\sum_{i=1}^{n} u_i , \quad v_0 = \frac{1}{n}\sum_{i=1}^{n} v_i , \quad u'_0 = \frac{1}{n}\sum_{i=1}^{n} u'_i , \quad v'_0 = \frac{1}{n}\sum_{i=1}^{n} v'_i .$$

Defining

$$\begin{aligned} \bar{u}_i &= u_i - u_0 , \\ \bar{v}_i &= v_i - v_0 , \\ \bar{u}'_i &= u'_i - u'_0 , \\ \bar{v}'_i &= v'_i - v'_0 , \end{aligned}$$

and substituting these equations back for (3.69), we have

$$C_6 = \sum_{i=1}^{n} (\bar{u}_i - (a\bar{u}'_i + b\bar{v}'_i))^2 + \sum_{i=1}^{n} (\bar{v}_i - (-b\bar{u}'_i + a\bar{v}'_i))^2 . \quad (3.72)$$

Expanding it yields

$$C_6 = A - 2Ba - 2Cb + D(a^2 + b^2) , \quad (3.73)$$

where

$$\begin{aligned} A &= \sum_{i=1}^{n} (\bar{u}_i^2 + \bar{v}_i^2) , \\ B &= \sum_{i=1}^{n} (\bar{u}_i \bar{u}'_i + \bar{v}_i \bar{v}'_i) , \\ C &= \sum_{i=1}^{n} (-\bar{v}_i \bar{u}'_i + \bar{u}_i \bar{v}'_i) , \\ D &= \sum_{i=1}^{n} ((\bar{u}'_i)^2 + (\bar{v}'_i)^2) . \end{aligned}$$

Differentiating (3.73) with respect to a and b and setting them to zero, we have

$$a = \frac{B}{D}, \quad (3.74)$$

$$b = \frac{C}{D}. \quad (3.75)$$

3.3 RECOVERY OF MULTIPLE EPIPOLAR EQUATIONS BY CLUSTERING

In the last section we described how to estimate affine epipolar equations given a set of matched points in two images which belong to the same rigid object or motion. In this section, we describe the problem of finding multiple epipolar equations as a problem of unsupervised clustering in the parameter space.

Given two images and feature points in them, the task is to segment the feature points into groups that each represent a rigid motion. This can be understood as a combination of segmentation and outlier rejection. The only constraint we use is that *if the point matches belong to the same rigid motion, they must satisfy the same epipolar geometry.* The reverse, however, is not true. Take an instance. Two objects undergoing a pure translation in the same direction but with different magnitude satisfy the same epipolar geometry.

We employ the generate-and-clustering strategy. In the affine projection case, for each group of 4 neighboring point matches, we can determine one epipolar equation. One epipolar equation in turn is projected onto the parameter space as two points. The task then becomes one of finding clusters in that space, where each cluster represents an epipolar equation supported by those groups of corresponding points residing within that cluster.

Note that here we only find multiple epipolar equations for the affine camera whose optical axis is perpendicular to image plane, and whose horizontal and vertical scales are in the same unit. Under the general affine projection, the number of coefficients involved in the epipolar equation is only 4, while that under full perspective projection is 8. The general approach should work as well for the full perspective projection, though with more difficulty as the dimensionality increases [157, 156].

3.3.1 The Problem in Stereo, Motion and Object Recognition

In this subsection, we describe various cases of the same problem which arise in different situations in stereo, motion and object recognition.

In stereo, the whole scene is regarded as a single rigid object, there should be only one cluster in the parameter space. Generally, since the two views are uncalibrated, correpondences between feature points are not readily available. First, we can roughly match feature points, which are not unique in general, and divide them into groups, for each of them an epipolar equation is determined. If there is an error in the correspondence, the solution of the epipolar equation will be wrong. The estimated epipolar equation will be different from the majority. Thus, besides the points within the single cluster, random points scatter. The probility of random points is important because if the probility of random points is high, then the clusters would be difficult to find.

For motion images in which there are more than one rigid motions, the problem is more difficult than stereo as there are more than one clusters in the paramter space. Again we can roughly match feature points between the two images first, which is easier as the difference between consecutive motion images is relatively small. Then groups of correspondences are hypothesized and an epipolar equation is estimated for each group. Each cluster in the space corresponds to one epipolar equation.

For object recognition, generally we do not know the correspondences in advance as the model view and input view can be quite different in terms of viewing distance, viewing angle and lighting conditions. Though intensity patterns around each feature point do provide some constraints on the candidates of its correspondence, generally they are not strong enough to limit the number of candidates to be one. Thus what we can do is to hypothesize correspondences, and to estimate an epipolar equation for each local group of correspondences. Based on the same observation as above, there will be many random points in the parameter space, making the clusters less obvious. If the model and input images have only one identical object, then the cluster should be single, and obvious. If the input image includes not only the object in the model image, but also includes background objects, then there might be multiple clusters in the space, of which only one is true.

3.3.2 Definitions and Assumptions

Based on the above observations, we can model the points in the parameter space as a summation of probability processes representing the clusters, and a random process representing points resulted from wrong hypotheses of wrong correspondences. Since they are random, the probability of their forming clusters is small. Let the cluster centers be $\boldsymbol{\mu}_i, i = 1, ..., c$. The density function can be written as

$$p(\mathbf{x}|\boldsymbol{\mu}) = \sum_{i}^{c} P(\omega_i)p(\mathbf{x}|\omega_i, \boldsymbol{\mu}_i) + aP(\omega_0) , \tag{3.76}$$

subject to

$$\sum_{i=1}^{c} P(\omega_i) + P(\omega_0) = 1 , \tag{3.77}$$

where $p(\mathbf{x}|\omega_i, \boldsymbol{\mu}_i)$ is the probability distribution for cluster ω_i with center $\boldsymbol{\mu}_i$, $P(\omega_i)$ is the probability of belonging to i-th cluster, $P(\omega_0)$ is the probability of belonging to none of the clusters, and $aP(\omega_0)$ is the constant density function corresponding to the random process.

$p(\mathbf{x}|\omega_i, \boldsymbol{\mu}_i)$ has to be a function decreasing monotonically with distance from the center $\boldsymbol{\mu}_i$. Usually Gaussian function is used. It means that due to errors arising from discretization and calculations, there is a difference between the estimated epipolar equation and the true epipolar equation, and the difference obeys the Gaussian distribution. Suppose that the covariance matrix is identical for all clusters. Then the Gaussian distribution can be written as

$$p(\mathbf{x}|\omega_i, \boldsymbol{\mu}_i) = (2\pi)^{-\frac{d}{2}} \mid \boldsymbol{\Sigma} \mid^{-\frac{1}{2}} \exp\{-\frac{1}{2}(\mathbf{x} - \boldsymbol{\mu}_i)^T \boldsymbol{\Sigma}^{-1}(\mathbf{x} - \boldsymbol{\mu}_i)\} \tag{3.78}$$

$P(\omega_i), i = 0, ..., c$ are not known *a priori*. They can be estimated if the number of points for each cluster and the number of random points are known. Assuming that the total number of points in the space is n, the number of points for the i-th cluster is n_i, and the number of random points is n_0, subject to $\sum_{i=0}^{c} n_i = n$, then

$$P(\omega_i) = \frac{n_i}{n}, \quad i = 0, ..., c. \tag{3.79}$$

Needless to say, the ease of finding the clusters depends on how many random points exist. Thus it is vital to limit the number of random points, that is, $P(\omega_0)$.

3.3.3 Error Analysis of Motion Parameters

Errors in Motion Parameters

Let x, y, x', y' be the new coordinates after moving the image origins to the object centroids in the images such that λ is now zero. The third dimension z and z' are not our concerns but are needed to represent the 3D rotation. Let $\mathtt{X} = [x, y, z]^T$ be the 3D coordinates in the first view frame, and $\mathtt{X}' = [x', y', z']^T$ be the 3D coordinates in the second view frame. Now the relation between $\mathtt{X}$ and $\mathtt{X}'$ can be expressed as

$$\mathtt{X} = \rho \mathbf{R}_z(\alpha)\mathbf{R}_y(\beta)\mathbf{R}_z(-\alpha)\mathbf{R}_z(\theta)\mathtt{X}' , \tag{3.80}$$

where ρ represents the scale change due to translation in depth.

Now we can look at how the changes in image coordinates x, y, x', y' affect the change in those motion parameters. Differentiating (3.80) yields

$$d\rho = \frac{\mathbf{dx} - \rho \mathbf{R} \mathbf{dx}'}{\mathbf{R}\mathbf{x}'} , \tag{3.81}$$

$$d\theta = \frac{\mathbf{dx} - \rho \mathbf{R} \mathbf{dx}'}{\mathbf{R}_z(\alpha)\mathbf{R}_y(\beta)\mathbf{R}_z(-\alpha) \begin{bmatrix} -\sin\theta & \cos\theta & 0 \\ -\cos\theta & -\sin\theta & 0 \\ 0 & 0 & 0 \end{bmatrix} \mathbf{x}'} , \tag{3.82}$$

$$d\alpha = \frac{\mathbf{dx} - \rho \mathbf{R} \mathbf{dx}'}{\begin{bmatrix} 2\sin\alpha\cos\alpha(1-\cos\beta) & (1-2\sin^2\alpha)(1-\cos\beta) & \sin\alpha\sin\beta \\ (1-2\sin^2\alpha)(1-\cos\beta) & -2\sin\alpha\cos\alpha(1-\cos\beta) & \cos\alpha\sin\beta \\ \sin\alpha\sin\beta & \cos\alpha\sin\beta & \cos\beta \end{bmatrix} \mathbf{R}_z(\theta)\mathbf{x}'} . \tag{3.83}$$

Comparing the above equations, it is not difficult to see that the difference is only in the denominator. If any of the matrices in the denominators is close to zero, then the motion parameters are not stable.

In (3.81), only the rotation matrix is there, so there is no fear. In (3.82), the differential of the rotation matrix with respect to θ is

$$\begin{bmatrix} -\sin\theta & \cos\theta & 0 \\ -\cos\theta & -\sin\theta & 0 \\ 0 & 0 & 0 \end{bmatrix} ,$$

whose components have nothing to do with the depth but $\sin\theta$ and $\cos\theta$ cannot be zero simultaneously. In (3.83), the differential of the rotation matrix with respect to the α angle is

$$\begin{bmatrix} 2\sin\alpha\cos\alpha(1-\cos\beta) & (1-2\sin^2\alpha)(1-\cos\beta) & \sin\alpha\sin\beta \\ (1-2\sin^2\alpha)(1-\cos\beta) & -2\sin\alpha\cos\alpha(1-\cos\beta) & \cos\alpha\sin\beta \\ \sin\alpha\sin\beta & \cos\alpha\sin\beta & \cos\beta \end{bmatrix},$$

whose first two rows can all vanish if β is small. Therefore, it is easy to conclude that the scale change ρ is the most stable, the frontoparallel rotation angle θ is the next most stable, and the projection of 3D rotation axis is the least stable parameter. Note that the frontoparallel translation perpendicular to the epipolar direction, λ, is a function of the angle α (see 2.117), thus it has similar variance as α.

It is worth mentioning here that the above analysis is valid even if there are more than one pair of corresponding points.

Space of Motion Parameters vs Space of Equation Coefficients

There are two spaces that we can possibly use for clustering: one is the space specified by the coefficients of the epipolar equation, and the other is the space of motion parameters α, θ, ρ and λ, which can be computed from the epipolar equations.

As analyzed in Sect. 3.2.6, the coefficients have uniform variances, but they are not independent of each other. For instance, the squared sum of f_{31} and f_{32} and that of f_{13} and f_{23} have very high correlation. Also the ratio of f_{31} and f_{32} and that of f_{13} and f_{23} have high correlation.

While the motion parameters computed from those coefficients have different variances, they are generally independent of each other.

Based on the above property, we decided to use the space of motion parameters for clustering.

3.3.4 Estimating Covariance Matrix

Assuming that the distribution of points in the motion parameter space obeys Gaussian function, it is necessary to estimate its covariance matrix in order to use it for clustering. Though it is clear that the covariance matrix varies with different images and different motions, it is still considered feasible to assume that a few typical covariance matrices reflect in a large content the range within which any variance matrix resides. Thus the approach we take is to estimate a few variance matrices from examples, and use them for clustering.

Now we need to estimate the covariance matrix $\mathbf{\Sigma}$ in a general Gaussian distribution:

$$p(\mathbf{x}) = (2\pi)^{-\frac{d}{2}} \mid \mathbf{\Sigma} \mid^{-\frac{1}{2}} \exp[-\frac{1}{2}(\mathbf{x}-\boldsymbol{\mu})^T \mathbf{\Sigma}^{-1} (\mathbf{x}-\boldsymbol{\mu})] \, , \tag{3.84}$$

where $\mathbf{x}$ is a $d \times 1$ vector with mean vector $\boldsymbol{\mu}$ and covariance matrix $\mathbf{\Sigma}$. The symmetry of the $d \times d$ covariance matrix allows complete specification of this distribution. In our case, d is 4.

Now suppose we are given n points $\mathbf{x}_1, \mathbf{x}_2, ..., \mathbf{x}_k, ..., \mathbf{x}_n$. The maximum likelihood estimation tries to find a $\boldsymbol{\mu}$ and $\mathbf{\Sigma}$ so as to maximize the following likelihood function

$$P(\boldsymbol{\mu}, \mathbf{\Sigma}) = \prod_{k=1}^{n} p(\mathbf{x}_k) \, . \tag{3.85}$$

This is equivalent to maximizing the natural logarithm function of $P(\boldsymbol{\mu}, \mathbf{\Sigma})$,

$$l(\boldsymbol{\mu}, \mathbf{\Sigma}) = -\sum_{k=1}^{n} \frac{1}{2} \log(2\pi)^4 \mid \mathbf{\Sigma} \mid - \sum_{k=1}^{n} \frac{1}{2} (\mathbf{x}_k - \boldsymbol{\mu})^T \mathbf{\Sigma}^{-1} (\mathbf{x}_k - \boldsymbol{\mu}) \, . \tag{3.86}$$

Differentiating $l(\boldsymbol{\mu}, \mathbf{\Sigma})$ with respect to $\boldsymbol{\mu}$ and setting it to zero, we have

$$\frac{d}{d\boldsymbol{\mu}} l(\boldsymbol{\mu}, \mathbf{\Sigma}) = -\sum_{k=1}^{n} \mathbf{\Sigma}^{-1} (\mathbf{x}_k - \boldsymbol{\mu}) = \mathbf{0} \, .$$

The solution is

$$\boldsymbol{\mu} = \frac{1}{n} \sum_{k=1}^{n} \mathbf{x}_k \, , \tag{3.87}$$

which is exactly the mean vector of all the sample points.

In order to estimate the covariance matrix, we consider $\mathbf{\Sigma}$ as the variable. Differentiating $l(\boldsymbol{\mu}, \mathbf{\Sigma})$ with respect to $\mathbf{\Sigma}^{-1}$ and setting the differential to be a zero matrix, we have

$$\frac{d}{d\mathbf{\Sigma}^{-1}} l(\boldsymbol{\mu}, \mathbf{\Sigma}) = \sum_{k=1}^{n} \{\frac{1}{2}\mathbf{\Sigma} - \frac{1}{2}(\mathbf{x}_k - \boldsymbol{\mu})(\mathbf{x}_k - \boldsymbol{\mu})^T\} = \begin{bmatrix} 0 & \dots & 0 \\ \vdots & \vdots & \vdots \\ 0 & \dots & 0 \end{bmatrix} .$$

The solution is

$$\mathbf{\Sigma} = \frac{1}{n} \sum_{k=1}^{n} (\mathbf{x}_k - \boldsymbol{\mu})(\mathbf{x}_k - \boldsymbol{\mu})^T , \tag{3.88}$$

which is exactly the covariance matrix of the sample points. Thus we have shown that the defintions of the mean vector and covariance matrix are also the Maximum Likelihood Estimations based on Gaussian distribution from sample data.

Actually, (3.3.4) can also be understood of as minimizing the sum of squared Mahalanobis distances from each point to the mean vector

$$\sum_i (\mathbf{x}_k - \boldsymbol{\mu})^T \mathbf{\Sigma}^{-1} (\mathbf{x}_k - \boldsymbol{\mu})$$

Thus, the mean vector is also the optimal solution in the least squares sense.

The covariance matrix we use for clustering is

$$\begin{bmatrix} 1231.437134 & -32.256741 & -0.746871 & 440.431793 \\ -32.256741 & 8.477827 & 0.007116 & 22.509176 \\ -0.746871 & 0.007116 & 0.003669 & -0.882438 \\ 440.431793 & 22.509176 & -0.882438 & 539.617126 \end{bmatrix}$$

which is computed using (3.88) from local groups of known correspondences between two real images.

3.3.5 The Maximal Likelihood Approach

Assuming that the n samples $\mathbf{S} = \{\mathbf{x}_1, \mathbf{x}_2, ..., \mathbf{x}_n\}$ are independent of each other, the joint density is the mixture density,

$$p(\mathbf{S}|\boldsymbol{\mu}) = \prod_{k=1}^{n} p(\mathbf{x}_k|\boldsymbol{\mu}) , \tag{3.89}$$

where $\boldsymbol{\mu} = \{\boldsymbol{\mu}_1, \boldsymbol{\mu}_2, ..., \boldsymbol{\mu}_c\}$. The maximal likelihood approach tries to seek a $\boldsymbol{\mu}$ that maximizes $P(\mathbf{S}|\boldsymbol{\mu})$.

A $\boldsymbol{\mu}$ that maximizes $P(\mathbf{S}|\boldsymbol{\mu})$ also maximizes $\log P(\mathbf{S}|\boldsymbol{\mu})$. The constraint on $\boldsymbol{\mu}$ is

$$\nabla_{\boldsymbol{\mu}_i} \log P(\mathbf{S}|\boldsymbol{\mu}) = \mathbf{0} , \quad i = 1, ..., c. \tag{3.90}$$

From (3.89), we define $\mathbf{q}$ as

$$\mathbf{q} \equiv \nabla_{\boldsymbol{\mu}_i} \log P(\mathbf{S}|\boldsymbol{\mu}) = \nabla_{\boldsymbol{\mu}_i} \sum_{k=1}^{n} \log p(\mathbf{x}_k|\boldsymbol{\mu}) = \sum_{k=1}^{n} \nabla_{\boldsymbol{\mu}_i} \log p(\mathbf{x}_k|\boldsymbol{\mu}) , \quad i = 1, ..., c. \tag{3.91}$$

Substituting (3.76) for the above equation yields

$$\mathbf{q} = \sum_{k=1}^{n} \{\frac{1}{p(\mathbf{x}_k|\boldsymbol{\mu})} \nabla_{\boldsymbol{\mu}_i} \{\sum_{j=1}^{c} p(\mathbf{x}_k|\omega_j, \boldsymbol{\mu}_j) P(\omega_j) + aP(\omega_0)\}\} = \mathbf{0} , \quad i = 1, ..., c. \tag{3.92}$$

Assuming independence of $\boldsymbol{\mu}_i$ and $\boldsymbol{\mu}_j (i \neq j)$, we have

$$\nabla_{\boldsymbol{\mu}_i} p(\mathbf{x}_k|\boldsymbol{\mu}) = \nabla_{\boldsymbol{\mu}_i} \{p(\mathbf{x}_k|\omega_i, \boldsymbol{\mu}_i) P(\omega_i)\} = P(\omega_i) \nabla_{\boldsymbol{\mu}_i} p(\mathbf{x}_k|\omega_i, \boldsymbol{\mu}_i) , \quad i = 1, ..., c. \tag{3.93}$$

Thus,

$$\mathbf{q} = \nabla_{\boldsymbol{\mu}_i} \log P(\mathbf{S}|\boldsymbol{\mu}) = \sum_{k=1}^{n} \frac{P(\omega_i)}{p(\mathbf{x}_k|\boldsymbol{\mu})} \nabla_{\boldsymbol{\mu}_i} p(\mathbf{x}_k|\omega_i, \boldsymbol{\mu}_i) , \quad i = 1, ..., c. \tag{3.94}$$

Using the Bayesian rule, the above equation can be rewritten as

$$\mathbf{q} = \sum_{k=1}^{n} P(\omega_i|\mathbf{x}_k, \boldsymbol{\mu}) \nabla_{\boldsymbol{\mu}_i} \{\log[p(\mathbf{x}_k|\omega_i, \boldsymbol{\mu}_i)]\} , \quad i = 1, ..., c. \tag{3.95}$$

Working with Gaussian Distribution

Substituting (3.78) for the above equation and setting $\mathbf{q}$ to be zero yields

$$\mathbf{q} = \sum_{k=1}^{n} P(w_i|\mathbf{x}_k, \boldsymbol{\mu}) \boldsymbol{\Sigma}^{-1} (\mathbf{x} - \boldsymbol{\mu}_i) = \mathbf{0} , \quad i = 1, ..., c. \tag{3.96}$$

That is,

$$\boldsymbol{\mu}_i = \frac{\sum_{k=1}^{n} P(\omega_i|\mathbf{x}_k, \boldsymbol{\mu})\mathbf{x}_k}{\sum_{k=1}^{n} P(\omega_i|\mathbf{x}_k, \boldsymbol{\mu})}, \quad i = 1, ..., c. \tag{3.97}$$

From the above equation, $\boldsymbol{\mu}_i$ is formed as a weighted summation of the $\mathbf{x}_k$, where the weight for each sample is

$$\frac{P(w_i|\mathbf{x}_k, \boldsymbol{\mu})}{\sum_{k=1}^{n} P(\omega_i|\mathbf{x}_k, \boldsymbol{\mu})}.$$

For samples where $P(\omega_i|\mathbf{x}_k, \boldsymbol{\mu})$ is small, little is contributed to $\boldsymbol{\mu}_i$. This is intuitively appealing and suggests only using samples which are close to $\boldsymbol{\mu}_i$.

Equation (3.97) is difficult to apply directly, but it does suggest an iterative procedure. If we can obtain reasonable initial estimates for $\boldsymbol{\mu}_i(0), i = 1, ..., c$, they can be updated using

$$\boldsymbol{\mu}_i(t+1) = \frac{\sum_{k=1}^{n} P(\omega_i|\mathbf{x}_k, \boldsymbol{\mu}(t))\mathbf{x}_k}{\sum_{k=1}^{n} P(\omega_i|\mathbf{x}_k, \boldsymbol{\mu}(t))}, \quad i = 1, ..., c. \tag{3.98}$$

until no significant change is available. This procedure involves updating the class means by readjusting the weights on each sample at each iteration. It provides a theoretical basis for the c-means clustering algorithm [132].

From Bayesian rule, we get

$$P(\omega_i|\mathbf{x}_k, \boldsymbol{\mu}) = \frac{P(\omega_i)(2\pi)^{-\frac{d}{2}} \mid \boldsymbol{\Sigma} \mid^{-\frac{1}{2}} \exp\{-\frac{1}{2}\|\mathbf{x}_k - \boldsymbol{\mu}_i\|^2_{\boldsymbol{\Sigma}^{-1}}\}}{\sum_i P(\omega_i)(2\pi)^{-\frac{d}{2}} \mid \boldsymbol{\Sigma} \mid^{-\frac{1}{2}} \exp\{-\frac{1}{2}\|\mathbf{x}_k - \boldsymbol{\mu}_i\|^2_{\boldsymbol{\Sigma}^{-1}}\} + aP(\omega_0)}, \tag{3.99}$$

where $\|\mathbf{x}_k - \boldsymbol{\mu}_i\|^2_{\boldsymbol{\Sigma}^{-1}} = (\mathbf{x}_k - \boldsymbol{\mu}_i)^T\boldsymbol{\Sigma}^{-1}(\mathbf{x}_k - \boldsymbol{\mu}_i)$. It is clear from the above equation that the probability $P(\omega_i|\mathbf{x}_k, \boldsymbol{\mu})$ is large when $\mathbf{x}_k$ is close to $\boldsymbol{\mu}_i$. This suggests classifying $\mathbf{x}_k$ to class ω_i when $\|\mathbf{x}_k - \boldsymbol{\mu}_i\|^2_{\boldsymbol{\Sigma}^{-1}}$ is small.

If $\mathbf{x}_k$ is much closer to $\boldsymbol{\mu}_i$ than to any other $\boldsymbol{\mu}_j$, then

$$p(\mathbf{x}_k|\boldsymbol{\mu}) \approx P(\omega_i)(2\pi)^{-\frac{d}{2}} \mid \boldsymbol{\Sigma} \mid^{-\frac{1}{2}} \exp\{-\frac{1}{2}\|\mathbf{x}_k - \boldsymbol{\mu}_i\|^2_{\boldsymbol{\Sigma}^{-1}}\} + aP(\omega_0),$$

Further, if the number of random points is also small, i.e. $P(\omega_0)$ is small, then $P(\omega_i|\mathbf{x}_k, \boldsymbol{\mu}) \approx 1$. Substituting this for (3.98), it becomes a mere average, i.e.,

$$\boldsymbol{\mu}_i(t+1) = \frac{\sum_{\mathbf{x}_k \subset R} \mathbf{x}_k}{\sum_{\mathbf{x}_k \subset R} 1}, \tag{3.100}$$

where R denotes the range of $\mathbf{x}_k$ such that $\|\mathbf{x}_k - \boldsymbol{\mu}_i\|^2_{\boldsymbol{\Sigma}^{-1}} < T$. Here T is a suitable threshold determining the size of R.

If the number of random points is not small, then $P(\omega_i|\mathbf{x}_k, \boldsymbol{\mu})$ is not close to 1, which means that a mere average does not give correct answer.

Whether or not a class can be regarded as a cluster depends on how many points are in that class. For each class, we can count the number of points falling in that class and it has to be larger than a threshold for that class to be regarded as a cluster representing a true epipolar equation.

3.3.6 Robust Estimation Using Exponential of Gaussian Distribution

As analyzed in the last subsection, if there are no random points, the center can be determined as the mean vector of all the points falling within the range. However, if there are random points, merely taking the mean of all points is risky, because any random point within but near the boundary of the range brings a greater deviation from the true value. Thus, it is desirable to allow less contribution from points far from the center and large contribution from the points close to the center. (see Sect. 3.1.6 for a general discussion on robust estimation.)

To limit the influence of random points, we can use a function that decreases faster than Gaussian with the distance from the center. One such option is the exponential of Gaussian,

$$p(\mathbf{x}_k|\omega_i, \boldsymbol{\mu}_i) = c_w \exp\{\exp(-\frac{1}{2}\|\mathbf{x}_k - \boldsymbol{\mu}_i\|^2_{\boldsymbol{\Sigma}^{-1}})\} - c_w \,, \tag{3.101}$$

where c_w is a constant chosen such that the integral of the distribution over the whole space is equal to 1. Differentiating the function with respect to $\mathbf{x}_k$ yields

$$\exp\{\exp(-\frac{1}{2}\|\mathbf{x}_k - \boldsymbol{\mu}_i\|^2_{\boldsymbol{\Sigma}^{-1}})\} \exp(-\frac{1}{2}\|\mathbf{x}_k - \boldsymbol{\mu}_i\|^2_{\boldsymbol{\Sigma}^{-1}})\boldsymbol{\Sigma}^{-1}(\mathbf{x}_k - \boldsymbol{\mu}_i) \,.$$

Since $\exp\{\exp(-\frac{1}{2}\|\mathbf{x}_k - \boldsymbol{\mu}_i\|^2_{\boldsymbol{\Sigma}^{-1}})\}$ is always larger than 1, this function decreases faster than the Gaussian function itself. For its integral to be the same as that of Gaussian, c_w must be larger than $(2\pi)^{-\frac{d}{2}} \mid \boldsymbol{\Sigma} \mid^{-\frac{1}{2}}$, the coefficient of Gaussian.

Substituting it for (3.95) which does not depend on specific distribution functions, and setting $\mathbf{q}$ to be zero yields

$$\mathbf{q} = \sum_{k=1}^{n} P(\omega_i|\mathbf{x}_k, \boldsymbol{\mu}) c_w \exp(-\frac{1}{2}\|\mathbf{x}_k - \boldsymbol{\mu}_i\|^2_{\Sigma^{-1}}) \Sigma^{-1}(\mathbf{x}_k - \boldsymbol{\mu}_i) = 0 \ , \quad i = 1, ..., c.$$

That is,

$$\boldsymbol{\mu}_i = \frac{\sum_{k=1}^{n} P(\omega_i|\mathbf{x}_k, \boldsymbol{\mu}) \exp(-\frac{1}{2}\|\mathbf{x}_k - \boldsymbol{\mu}_i\|^2_{\Sigma^{-1}})\mathbf{x}_k}{\sum_{k=1}^{n} P(\omega_i|\mathbf{x}_k, \boldsymbol{\mu}) \exp(-\frac{1}{2}\|\mathbf{x}_k - \boldsymbol{\mu}_i\|^2_{\Sigma^{-1}})} \ , \quad i = 1, ..., c. \qquad (3.102)$$

This means that $\boldsymbol{\mu}_i$ is formed as a weighted summation of the $\mathbf{x}_k$, where the weight for each sample is

$$\frac{P(\omega_i|\mathbf{x}_k, \boldsymbol{\mu}) \exp(-\frac{1}{2}\|\mathbf{x}_k - \boldsymbol{\mu}_i\|^2_{\Sigma^{-1}})}{\sum_{k=1}^{n} P(\omega_i|\mathbf{x}_k, \boldsymbol{\mu}) \exp(-\frac{1}{2}\|\mathbf{x}_k - \boldsymbol{\mu}_i\|^2_{\Sigma^{-1}})} .$$

Again, Equation (3.102) is difficult to apply directly, but it suggests an iterative procedure. If we can obtain reasonable initial estimates for $\boldsymbol{\mu}_i(0), i = 1, ..., c$, they can be updated using

$$\boldsymbol{\mu}_i(t+1) = \frac{\sum_{k=1}^{n} P(\omega_i|\mathbf{x}_k, \boldsymbol{\mu}(t)) \exp(-\frac{1}{2}\|\mathbf{x}_k - \boldsymbol{\mu}_i(t)\|^2_{\Sigma^{-1}})\mathbf{x}_k}{\sum_{k=1}^{n} P(\omega_i|\mathbf{x}_k, \boldsymbol{\mu}(t)) \exp(-\frac{1}{2}\|\mathbf{x}_k - \boldsymbol{\mu}_i(t)\|^2_{\Sigma^{-1}})} \ , \quad i = 1, ..., c. \qquad (3.103)$$

until no significant change is available. This procedure involves updating the class means by readjusting the weights on each sample at each iteration.

From Bayesian rule,

$$P(\omega_i|\mathbf{x}_k, \boldsymbol{\mu}) = \frac{P(\omega_i) c_w \exp\{\exp(-\frac{1}{2}\|\mathbf{x}_k - \boldsymbol{\mu}_i\|^2_{\Sigma^{-1}})\}}{\sum_i P(\omega_i) c_w \exp\{\exp(-\frac{1}{2}\|\mathbf{x}_k - \boldsymbol{\mu}_i\|^2_{\Sigma^{-1}})\} + aP(\omega_0)} . \qquad (3.104)$$

It is clear from the above equation that the probability $P(\omega_i|\mathbf{x}_k, \boldsymbol{\mu})$ is large when $\mathbf{x}_k$ is close to $\boldsymbol{\mu}_i$. This suggests classifying $\mathbf{x}_k$ to class ω_i when $\|\mathbf{x}_k - \boldsymbol{\mu}_i\|^2_{\Sigma^{-1}}$ is small. Also, when $\|\mathbf{x}_k - \boldsymbol{\mu}_i\|^2_{\Sigma^{-1}}$ is large, $P(\omega_i|\mathbf{x}_k, \boldsymbol{\mu})$ is small, and little is contributed to $\boldsymbol{\mu}_i$. Compared with using Gaussian distribution, this property is even more enhanced. It suggests only using samples close to $\boldsymbol{\mu}_i$.

Now if $\mathbf{x}_k$ is much closer to $\boldsymbol{\mu}_i$ than any other $\boldsymbol{\mu}_j$, then

$$p(\mathbf{x}_k|\boldsymbol{\mu}) \approx P(\omega_i) c_w \exp\{\exp(-\frac{1}{2}\|\mathbf{x}_k - \boldsymbol{\mu}_i\|^2_{\Sigma^{-1}})\} + aP(\omega_0) \ ,$$

Compared with using Gaussian distribution, for points close to $\boldsymbol{\mu}_i$, $P(\omega_i|\mathbf{x}_k, \boldsymbol{\mu})$ is even closer to 1, because c_w is larger than $(2\pi)^{-\frac{d}{2}} \mid \boldsymbol{\Sigma} \mid^{-\frac{1}{2}}$. This confirms that using exponential of Gaussian does produce more robust result than using Gaussian itself.

Substituting this for (3.103), it becomes a weighted average, i.e.,

$$\boldsymbol{\mu}_i(t+1) = \frac{\sum_{\mathbf{x}_k \subset R} w(\mathbf{x}_k, \boldsymbol{\mu}_i(t))\mathbf{x}_k}{\sum_{\mathbf{x}_k \subset R} w(\mathbf{x}_k, \boldsymbol{\mu}_i(t))}\,, \tag{3.105}$$

where $w(\mathbf{x}_k, \boldsymbol{\mu}_i(t)) = \exp\{-\frac{1}{2}(\mathbf{x}_k - \boldsymbol{\mu}_i(t))^T \boldsymbol{\Sigma}(\mathbf{x}_k - \boldsymbol{\mu}_i(t))\}$ and R is defined as the range within which points are classified as belonging to class ω_i. This equation means that the center is weighted mean of the points, with the points close to the center having larger weights, while points less closer to the center having smaller weights. It agrees with our intuition. It can be proven that Eq. (3.105) is actually an extension to a larger dimension of the 1D Welsch M-estimator described in Sect. 3.1.6.

The denominator can actually been used as a measure of concentration of points. It does not merely count the number of points, but also takes the distribution into account. The more concentrated the points are, the higher the value. Renaming the denominator as C, we have

$$C = \sum_{\mathbf{x}_k \subset R} \exp(-\frac{1}{2}(\mathbf{x}_k - \boldsymbol{\mu}_i)^T \boldsymbol{\Sigma}^{-1}(\mathbf{x}_k - \boldsymbol{\mu}_i))\,. \tag{3.106}$$

Only those classes whose concentrations are higher than a threshold are regarded as indicating true epipolar equations.

3.3.7 A Clustering Algorithm

Suppose that we are given a set of n unlabeled points in a space $\mathbf{S} = \{\mathbf{x}_1, \ldots, \mathbf{x}_k, \ldots, \mathbf{x}_n\}$. These points can either form clusters or not. The number of clusters can be single or multiple.

The procedure goes like this. Let each point be an initial position $\boldsymbol{\mu}_k(0) = \mathbf{x}_k, k = 1, ..., n$. And for each point and each iteration, we compute

$$C_i(t+1) = \sum_{\|\mathbf{x}_k - \boldsymbol{\mu}_i(t)\|^2_{\boldsymbol{\Sigma}^{-1}} < T} \exp\{-\tfrac{1}{2}\|\mathbf{x}_k - \boldsymbol{\mu}_i(t)\|^2_{\boldsymbol{\Sigma}^{-1}}\}\,, \tag{3.107}$$

$$X_i(t+1) = \sum_{\|\mathbf{x}_k - \boldsymbol{\mu}_i(t)\|^2_{\boldsymbol{\Sigma}^{-1}} < T} \exp\{-\tfrac{1}{2}\|\mathbf{x}_k - \boldsymbol{\mu}_i(t)\|^2_{\boldsymbol{\Sigma}^{-1}}\}\mathbf{x}_k\,, \tag{3.108}$$

$$\boldsymbol{\mu}_i(t+1) = \frac{X_i(t+1)}{C_i(t+1)}\,, \tag{3.109}$$

till $\|\mu_i(t+1) - \mu_i(t)\| = 0$.

As a result of running this process, points converge to a smaller number of class centers. And for each class, we have a measure of concentration. Then the task is only to choose those classes with concentration larger than a threshold. They are regarded as valid clusters representing epipolar equations. Epipolar equations are then computed for each cluster from the corresponding points associated with each cluster.

3.3.8 An Example of Clustering

Figure 3.13 shows an example of clustering. Since we cannot visualize a 4D space, it is shown in two 2D figures. For each epipolar geometry, we have two clusters, for which the two α's are separated by 180 degrees, the two λ's are opposite in sign, while the ρ's and θ's are the same.

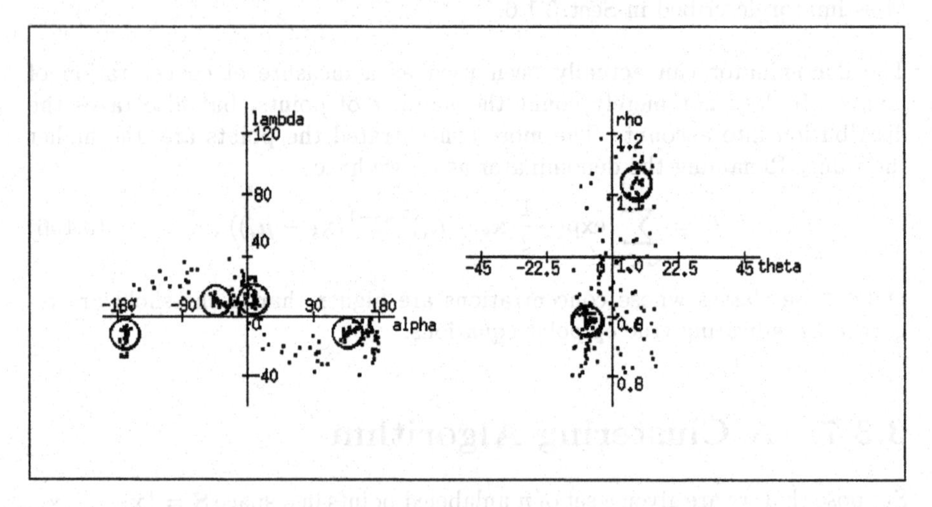

Figure 3.13 An example of clustering

The four found clusters are marked by four circles, whose centers are the centers of the found clusters. For the first epipolar geometry, $\alpha_1 = 137.460464$, $\alpha_2 = -42.655895$, $\theta_1 = -9.003800$, $\theta_2 = -8.993267$, $\rho_1 = 0.894926$, $\rho_2 = 0.895087$, $\lambda_1 = -11.565264$, and $\lambda_2 = 11.624498$.

For the second epipolar geometry, $\alpha_1 = 9.602901$, $\alpha_2 = -170.052139$, $\theta_1 = 8.066992$, $\theta_2 = 8.072144$, $\rho_1 = 1.115633$, $\rho_2 = 1.115886$,$\lambda_1 = 11.836720$ and $\lambda_2 = -11.770472$

3.4 PROJECTIVE RECONSTRUCTION

We show in this section how to estimate the position of a point in space, given its projections in two images whose epipolar geometry is known. The problem is known as *3D reconstruction* in general, and *triangulation* in particular. In the calibrated case, the relative position (i.e. the rotation and translation) of the two cameras is known, and 3D reconstruction has already been extensively studied in stereo [5]. In the uncalibrated case, like the one considered here, we assume that the fundamental matrix between the two images is known (e.g. computed with the methods described in Sect. 3.1), and we say that they are *weakly calibrated*.

3.4.1 Projective Structure from Two Uncalibrated Images

In the calibrated case, a 3D structure can be recovered from two images only up to a rigid transformation and an unknown scale factor (this transformation is also known as a *similarity*), because we can choose an arbitrary coordinate system as a world coordinate system (although one usually chooses it to coincide with one of the camera coordinate systems). Similarly, in the uncalibrated case, a 3D structure can only be performed up to a projective transformation of the 3D space [34, 60, 98, 35].

At this point, we have to introduce a few notations from Projective Geometry (a good introduction can be found in [35]). For a 3D point $\mathtt{M} = [X, Y, Z]^T$, its homogeneous coordinates are $\tilde{\mathbf{x}} = [U, V, W, S]^T = \lambda\widetilde{\mathtt{M}}$ where λ is any nonzero scalar and $\widetilde{\mathtt{M}} = [X, Y, Z, 1]^T$. This implies: $U/S = X$, $V/S = Y$, $W/S = Z$. If we include the possibility that $S = 0$, then $\tilde{\mathbf{x}} = [U, V, W, S]^T$ are called the *projective coordinates* of the 3D point M, which are not all equal to zero and defined up to a scale factor. Therefore, $\tilde{\mathbf{x}}$ and $\lambda\tilde{\mathbf{x}}$ ($\lambda \neq 0$) represent the same projective point. When $S \neq 0$, $\tilde{\mathbf{x}} = S\widetilde{\mathtt{M}}$. When $S = 0$, we say that the point is at infinity. A 4×4 nonsingular matrix $\mathbf{H}$ defines a linear transformation from one projective point to another, and is called the *projective transformation*. The matrix $\mathbf{H}$, of course, is also defined up to a nonzero scale factor, and we write

$$\rho\widetilde{\mathbf{y}} = \mathbf{H}\widetilde{\mathbf{x}}\,, \tag{3.110}$$

if $\widetilde{\mathbf{x}}$ is mapped to $\widetilde{\mathbf{y}}$ by $\mathbf{H}$. Here ρ is a nonzero scale factor.

Proposition 3.2. *Given two (perspective) images with unknown intrinsic parameters of a scene, the 3D structure of the scene can be reconstructed up to an unknown projective transformation as soon as the epipolar geometry (i.e. the fundamental matrix) between the two images is known.*

Assume that the true camera projection matrices are $\mathbf{P}$ and $\mathbf{P}'$. From (2.89), we have the following relation

$$\mathbf{F} = [\mathbf{P}\mathbf{p}'^{\perp}]_{\times}\mathbf{P}\mathbf{P}'^{+}\,,$$

where $\mathbf{F}$ is the known fundamental matrix. The 3D structure thus reconstructed is $\mathtt{M}$. The proposition says that the 3D structure $\mathbf{H}^{-1}\widetilde{\mathtt{M}}$, where $\mathbf{H}$ is any projective transformation of the 3D space, is still consistent with the observed image points and the fundamental matrix. Following the pinhole model, the camera projection matrices corresponding to the new structure $\mathbf{H}^{-1}\widetilde{\mathtt{M}}$ are

$$\widehat{\mathbf{P}} = \mathbf{P}\mathbf{H} \quad \text{and} \quad \widehat{\mathbf{P}}' = \mathbf{P}'\mathbf{H}\,,$$

respectively. In order to show the above proposition, we only need to prove

$$[\widehat{\mathbf{P}}\widehat{\mathbf{p}}'^{\perp}]_{\times}\widehat{\mathbf{P}}\widehat{\mathbf{P}}'^{+} = \lambda\mathbf{F} \equiv \lambda[\mathbf{P}\mathbf{p}'^{\perp}]_{\times}\mathbf{P}\mathbf{P}'^{+}\,, \tag{3.111}$$

where $\widehat{\mathbf{p}}'^{\perp} = (\mathbf{I} - \widehat{\mathbf{P}}'^{+}\widehat{\mathbf{P}}')\widehat{\omega}$ with $\widehat{\omega}$ any 4-vector, and λ is a scalar since $\mathbf{F}$ is defined up to a scale factor.

Proof. After some simple algebra, we have

$$\widehat{\mathbf{P}}\widehat{\mathbf{p}}'^{\perp} = \mathbf{P}\mathbf{x}\,,$$

where $\mathbf{x}$ is a 4-vector given by

$$\mathbf{x} = (\mathbf{I} - \mathbf{H}\mathbf{H}^T\mathbf{P}'^T(\mathbf{P}'\mathbf{H}\mathbf{H}^T\mathbf{P}'^T)^{-1}\mathbf{P}')\,\mathbf{H}\widehat{\omega}\,.$$

Multiplying $\mathbf{x}$ by $\mathbf{P}'$ from the left yields $\mathbf{P}'\mathbf{x} = \mathbf{0}$, which implies that $\mathbf{x}$ is a null vector of $\mathbf{P}'$. Since we assume that $\text{rank}(\mathbf{P}') = 3$ (i.e. the three row vectors are independent to each other, which is usually the case for both perspective projections and affine cameras), there is a unique null vector of $\mathbf{P}'$, defined up to a scale factor; we thus have $\mathbf{x} = \lambda\mathbf{p}'^{\perp}$ where $\mathbf{p}'^{\perp}$ is given by (2.81) and λ is a scale factor. Therefore, we have

$$\widehat{\mathbf{P}}\widehat{\mathbf{p}}'^{\perp} = \lambda\mathbf{P}\mathbf{p}'^{\perp} . \tag{3.112}$$

Next, let us examine $\widehat{\mathbf{P}}\widehat{\mathbf{P}}'^{+}$, which is equal to

$$\widehat{\mathbf{P}}\widehat{\mathbf{P}}'^{+} = \mathbf{P}\mathbf{H}\mathbf{H}^T\mathbf{P}'^T(\mathbf{P}'\mathbf{H}\mathbf{H}^T\mathbf{P}'^T)^{-1} \equiv \mathbf{P}\mathbf{N} ,$$

where $\mathbf{N} = \mathbf{H}\mathbf{H}^T\mathbf{P}'^T(\mathbf{P}'\mathbf{H}\mathbf{H}^T\mathbf{P}'^T)^{-1}$ is a 4×3 matrix. It is easy to verify that $\mathbf{P}'\mathbf{N} = \mathbf{I}$, so we must have

$$\mathbf{N} = \mathbf{P}'^{+} + \mathbf{Q} ,$$

where $\mathbf{Q}$ is a 4×3 matrix such that $\mathbf{P}'\mathbf{Q} = \mathbf{0}$, i.e. each column vector $\mathbf{q}_i$ ($i = 1, 2, 3$) of $\mathbf{Q}$ must be the null vector of $\mathbf{P}'$. Since the null vector is unique (up to a scale factor), we have $\mathbf{q}_i = \alpha_i\mathbf{p}'^{\perp}$, where α_i is some scalar. Therefore, we have

$$\widehat{\mathbf{P}}\widehat{\mathbf{P}}'^{+} = \mathbf{P}\mathbf{P}'^{+} + \mathbf{P}\mathbf{Q} . \tag{3.113}$$

Combining (3.112) and (3.113) gives

$$[\widehat{\mathbf{P}}\widehat{\mathbf{p}}'^{\perp}]_\times\widehat{\mathbf{P}}\widehat{\mathbf{P}}'^{+} = \lambda[\mathbf{P}\mathbf{p}'^{\perp}]_\times\mathbf{P}\mathbf{P}'^{+} + \lambda[\mathbf{P}\mathbf{p}'^{\perp}]_\times\mathbf{P}\mathbf{Q} .$$

Because of the structure of matrix $\mathbf{Q}$ and the operator $[\cdot]_\times$, the second term of the right side of the above equation is a zero matrix, i.e. $[\mathbf{P}\mathbf{p}'^{\perp}]_\times\mathbf{P}\mathbf{Q} = \mathbf{0}$. The above equation is finally reduced to (3.111). □

Although the above result has been known for several years, the proof through pure linear algebra like the one described here is much simpler than what was known in the literature before.

3.4.2 Computing Camera Projection Matrices

The projective reconstruction is very similar to the 3D reconstruction when cameras are calibrated. First, we need to compute the camera projection matrices from the fundamental matrix $\mathbf{F}$ with respect to a projective basis, which can be arbitrary because of Proposition 3.2.

Factorization Method

Let $\mathbf{F}$ be the fundamental matrix for the two cameras. There are an infinite number of projective bases which all satisfy the epipolar geometry. One possibility is to factor $\mathbf{F}$ as a product of a skew matrix $[\mathbf{e}]_\times$ ($\mathbf{e}$ is in fact the epipole in the first image) and a matrix $\mathbf{M}$, i.e., $\mathbf{F} = [\mathbf{e}]_\times \mathbf{M}$. A canonical representation can then be used:

$$\mathbf{P} = [\mathbf{M}\ \mathbf{e}] \quad \text{and} \quad \mathbf{P}' = [\mathbf{I}\ \mathbf{0}] .$$

It is easy to verify that the above $\mathbf{P}$ and $\mathbf{P}'$ do induce the fundamental matrix.

The factorization of $\mathbf{F}$ into $[\mathbf{e}]_\times \mathbf{M}$ is in general not unique, because if $\mathbf{M}$ is a solution then $\mathbf{M} + \mathbf{e}\mathbf{v}^T$ is also a solution for any vector $\mathbf{v}$ (indeed, we have always $[\mathbf{e}]_\times \mathbf{e}\mathbf{v}^T = \mathbf{0}$). One way to do the factorization is as follow [93]. Since $\mathbf{F}^T\mathbf{e} = \mathbf{0}$, the epipole in the first image is given by the eigenvector of matrix $\mathbf{F}\mathbf{F}^T$ associated to the smallest eigenvalue. Using the relation

$$\|\mathbf{v}\|^2 \mathbf{I}_3 = \mathbf{v}\mathbf{v}^T - [\mathbf{v}]_\times^2 ,$$

we can obtain the $\mathbf{M}$ matrix as

$$\mathbf{M} = -\frac{1}{\|\mathbf{e}\|^2}[\mathbf{e}]_\times \mathbf{F} .$$

This decomposition is used in [11]. Numerically, better results of 3D reconstruction are obtained when the epipole $\mathbf{e}$ is normalized such that $\|\mathbf{e}\| = 1$.

Choosing a Projective Basis

Another possibility is to choose effectively five pairs of points, each of four points not being coplanar, between the two cameras as a projective basis. We can of course choose five corresponding points we have identified. However, the precision of the final projective reconstruction will depend heavily upon the precision of the pairs of points. In order to overcome this problem, we have chosen in [189] the following solution. We first choose five arbitrary points in the first image, noted by $\mathbf{m}_i$ ($i = 1, \ldots, 5$). Although they could be chosen arbitrarily, they are chosen such that they are well distributed in the image to have a good numerical stability. For each point $\mathbf{m}_i$, its corresponding epipolar line in the second image is given by $\mathbf{l}'_i = \mathbf{F}^T\mathbf{m}_i$. We can now choose an arbitrary point on $\mathbf{l}'_i$ as $\mathbf{m}'_i$, the corresponding point of $\mathbf{m}_i$. Finally, we should verify that none of four points is coplanar, which can be easily done using the fundamental matrix [34, credited to Roger Mohr]. The advantage of this method is that the five pairs of points satisfy exactly the epipolar constraint.

Once we have five pairs of points $(\mathbf{m}_i, \mathbf{m}'_i)$, $(i = 1, \ldots, 5)$, we can compute the camera projection matrices as described in [34]. Assigning the projective coordinates (somewhat arbitrarily) to the five reference points, we have five image points and space points in correspondence, which provides 10 constraints on each camera projection matrix, leaving only one unknown parameter. This unknown can then be solved using the known fundamental matrix.

3.4.3 Reconstruction Techniques

Now that the camera projection matrices of the stereo with respect to a projective basis are available, we can reconstruct 3-D structures *with respect to that projective basis* from point matches.

Linear Methods

Given a pair of points in correspondence: $\mathbf{m} = [u, v]^T$ and $\mathbf{m}' = [u', v']^T$. Let $\tilde{\mathbf{x}} = [x, y, z, t]^T$ be the corresponding 3-D point in space with respect to the projective basis chosen before. Following the pinhole model, we have:

$$s\,[u,\ v,\ 1]^T = \mathbf{P}\,[x,\ y,\ z,\ t]^T\ , \tag{3.114}$$

$$s'\,[u',\ v',\ 1] = \mathbf{P}'\,[x,\ y,\ z,\ t]^T\ , \tag{3.115}$$

where s and s' are two arbitrary scalars. Denote $\mathbf{p}_i$ be the vector corresponding to the ith row of $\mathbf{P}$, and $\mathbf{p}'_i$ be the vector corresponding to the ith row of $\mathbf{P}'$. The two scalars can then be computed as:

$$s = \mathbf{p}_3^T \tilde{\mathbf{x}}\ , \quad s' = \mathbf{p}_3'^T \tilde{\mathbf{x}}\ .$$

Eliminating s and s' from (3.114) and (3.115) yields the following equation:

$$\mathbf{A}\tilde{\mathbf{x}} = \mathbf{0}\ , \tag{3.116}$$

where $\mathbf{A}$ is a 4×4 matrix given by

$$\mathbf{A} = [\mathbf{p}_1 - u\mathbf{p}_3,\ \mathbf{p}_2 - v\mathbf{p}_3,\ \mathbf{p}'_1 - u'\mathbf{p}'_3,\ \mathbf{p}'_2 - v'\mathbf{p}'_3]^T\ .$$

As the projective coordinates $\tilde{\mathbf{x}}$ are defined up to a scale factor, we can impose $\|\tilde{\mathbf{x}}\| = 1$, then the solution to (3.116) is well known (see also the description on page 83) to be the eigenvector of the matrix $\mathbf{A}^T\mathbf{A}$ associated to the smallest eigenvalue.

If we assume that no point is at infinity, then we can impose $t = 1$, and the projective reconstruction can be done exactly in the same way as for the Euclidean reconstruction. The set of homogeneous equations, $\mathbf{A}\tilde{\mathbf{x}} = \mathbf{0}$, is reduced to a set of 4 non-homogeneous equations in 3 unknowns (x, y, z). A linear least-squares technique can be used to solve this problem.

Iterative Linear Methods

The previous approach has the advantage of providing a closed-form solution, but it has the disadvantage that the criterion that is minimized does not have a good physical interpretation. Let us consider the first of the equations (3.116). In general, the point $\tilde{\mathbf{x}}$ found will not satisfy this equation exactly; rather, there will be an error $\epsilon_1 = \mathbf{p}_1^T\tilde{\mathbf{x}} - u\mathbf{p}_3^T\tilde{\mathbf{x}}$. What we really want to minimize is the difference between the measured image coordinate u and the projection of $\tilde{\mathbf{x}}$, which is given by $\mathbf{p}_1^T\tilde{\mathbf{x}}/\mathbf{p}_3^T\tilde{\mathbf{x}}$. That is, we want to minimize

$$\epsilon_1' = \mathbf{p}_1^T\tilde{\mathbf{x}}/\mathbf{p}_3^T\tilde{\mathbf{x}} - u = \epsilon_1/\mathbf{p}_3^T\tilde{\mathbf{x}} \,.$$

This means that if the equation had been weighted by the factor $1/w_1$ where $w_1 = \mathbf{p}_3^T\tilde{\mathbf{x}}$, then the resulting error would have been precisely what we wanted to minimize. Similarly, the weight for the second equation of (3.116) would be $1/w_2 = 1/w_1$, while the weight for the third and fourth equation would be $1/w_3 = 1/w_4 = 1/\mathbf{p}_3'^T\tilde{\mathbf{x}}$. Finally, the solution could be found by applying exactly the same method described in the last subsection (either eigenvector computation or linear least-squares).

Like the method for estimating the fundamental matrix described in Sect. 3.1.4, the problem is that the weights w_i depends themselves on the solution $\tilde{\mathbf{x}}$. To overcome this difficulty, we apply an iterative linear method. We first assume that all $w_i = 1$ and run a linear algorithm to obtain an initial estimation of $\tilde{\mathbf{x}}$. The weights w_i are then computed from this initial solution. The weighted linear least-squares is then run for an improved solution. This procedure can be repeated several times until convergence (either the solution or the weight does not change between successive iterations). Two iterations are usually sufficient.

Nonlinear Methods

As said in the last paragraph, the quantity we want to minimize is the error measured in the image plane between the observation and the projection of the reconstruction, that is

$$(u - \frac{\mathbf{p}_1^T\tilde{\mathbf{x}}}{\mathbf{p}_3^T\tilde{\mathbf{x}}})^2 + (v - \frac{\mathbf{p}_2^T\tilde{\mathbf{x}}}{\mathbf{p}_3^T\tilde{\mathbf{x}}})^2 + (u' - \frac{\mathbf{p}_1'^T\tilde{\mathbf{x}}}{\mathbf{p}_3'^T\tilde{\mathbf{x}}})^2 + (v' - \frac{\mathbf{p}_2'^T\tilde{\mathbf{x}}}{\mathbf{p}_3'^T\tilde{\mathbf{x}}})^2 .$$

However, there does not exist any closed-form solution, and we must use any standard iterative minimization technique, such as the Levenberg-Marquardt. The initial estimate of $\tilde{\mathbf{x}}$ can be obtained by using any linear technique described before.

Hartley in [61] reformulates the above criterion in terms of the distance between a point and its corresponding epipolar line defined by the ideal space point being sought. By parameterizing the pencil of epipolar lines in one image by a parameter t (which defines also the corresponding epipolar line in the other image by using the fundamental matrix), he is able to transform the minimization problem to the resolution of a polynomial of degree 6 in t. There may exist up to 6 real roots, and the global minimum can be found by evaluating the minimization function for each real root.

More projective reconstruction techniques can be found in [61, 129], but it seems to us that the techniques presented here, especially the one based on the image errors, are the best that one can recommend.

3.4.4 Use of Projective Structure

The reader may wonder the usefulness of such a projective structure defined up to a 4×4 matrix transformation. We cannot obtain any metric information from a projective structure: measurements of lengths and angles do not make sense. However, a projective structure still contains rich information, such as coplanarity, collinearity, and cross ratios (ratio of ratios of distances), which is sometimes sufficient for artificial systems, such as robots, to perform tasks such as navigation and object recognition [139, 181, 11].

In many applications such as the reconstruction of the environment from a sequence of video images where the parameters of the video lens is submitted to continuous modification, camera calibration in the classical sense is not possible. We cannot exact any metric information, but a projective structure is still possible if the camera can be considered as a pinhole. Furthermore, if we can introduce some knowledge of the scene into the projective structure, we can obtain more specific structure of the scene. For example, by specifying a plane at infinity (in practice, we need only to specify a plane sufficiently far away), an affine structure can be computed, which preserves parallelism and ratios of distances [120, 35]. Hartley et al [61] first reconstruct a projective structure, and then use 8 ground reference points to obtain the Euclidean structure and the camera parameters. Mohr et al. [103] embed constraints such as location of points, parallelism and vertical planes (e.g. walls) directly into a minimization procedure to determine a Euclidean structure. Robert and Faugeras [126] show that the 3D convex hull of an object can be computed from a pair of images whose epipolar geometry is known.

If we assume that the camera parameters do not change between successive views, the projective invariants can even be used to calibrate the cameras in the classical sense without using any calibration apparatus (known as *self-calibration*) [99, 38, 92, 191].

3.5 AFFINE RECONSTRUCTION

This section deals with two images taken by an affine camera at two different instants or by two different affine cameras. We show that the structure can only be recovered up to an affine transformation in 3D space, and a new method for affine reconstruction is developed.

3.5.1 Affine Structure from Two Uncalibrated Affine Views

Given a sufficient number of point matches (at least 4) between two images, the affine fundamental matrix $\mathbf{F}_A$ can be estimated (see Sect. 3.2). We are now interested in recovering $\mathbf{P}_A$ and $\mathbf{P}'_A$ from $\mathbf{F}_A$, and once they are recovered, the structure can be redressed in 3D space.

In Sect. 2.5.3, we have already studied the constraints on the affine projection matrices $\mathbf{P}_A$ and $\mathbf{P}'_A$ if the affine fundamental matrix $\mathbf{F}_A$ is given. The constraints (2.150) and (2.151) can be rewritten in matrix form as

$$\begin{bmatrix} a_{13} \\ a_{23} \\ a_{31} \\ a_{32} \end{bmatrix}^T \begin{bmatrix} P_{11} & P_{12} & P_{13} & P_{14} \\ P_{21} & P_{22} & P_{23} & P_{24} \\ P'_{11} & P'_{12} & P'_{13} & P'_{14} \\ P'_{21} & P'_{22} & P'_{23} & P'_{24} \end{bmatrix} = \begin{bmatrix} 0 \\ 0 \\ 0 \\ -a_{33} \end{bmatrix}^T . \tag{3.117}$$

We now show the following proposition.

Proposition 3.3. *Given two images of a scene taken by an affine camera, the 3D structure of the scene can be reconstructed up to an unknown affine transformation as soon as the epipolar geometry (i.e. the affine fundamental matrix) between the two images is known.*

Let the 3D structure corresponding to the true camera projection matrices $\mathbf{P}_A$ and $\mathbf{P}'_A$ be $\mathtt{M}$. We need to show that the new structure $\widetilde{\widehat{\mathtt{M}}} = \mathbf{H}_A^{-1}\widetilde{\mathtt{M}}$ is still consistent with the same sets of image points (i.e. with the affine fundamental matrix $\mathbf{F}_A$), where

$$\mathbf{H}_A = \begin{bmatrix} \mathbf{A} & \mathbf{t} \\ \mathbf{0}^T & 1 \end{bmatrix}$$

is an affine transformation of the 3D space, $\mathbf{A}$ is a 3×3 matrix, and $\mathbf{t}$ is a 3-vector. It follows that $\widehat{\mathtt{M}} = \mathbf{A}\mathtt{M} + \mathbf{t}$.

Proof. The camera projection matrices corresponding to the new structure $\widetilde{\widehat{\mathtt{M}}}$ are:

$$\hat{\mathbf{P}}_A = \mathbf{P}_A\mathbf{H}_A \quad \text{and} \quad \hat{\mathbf{P}}'_A = \mathbf{P}'_A\mathbf{H}_A \,.$$

We only need to show that the new affine projection matrices $\hat{\mathbf{P}}_A$ and $\hat{\mathbf{P}}'_A$ satisfy the same relation as (3.117), where P_{ij} and P'_{ij} should be replaced by $\hat{P}_{ij}$ and $\hat{P}'_{ij}$. Indeed, multiplying both sides of (3.117) by $\mathbf{H}_A$ from the right, i.e.

$$\begin{bmatrix} a_{13} \\ a_{23} \\ a_{31} \\ a_{32} \end{bmatrix}^T \begin{bmatrix} P_{11} & P_{12} & P_{13} & P_{14} \\ P_{21} & P_{22} & P_{23} & P_{24} \\ P'_{11} & P'_{12} & P'_{13} & P'_{14} \\ P'_{21} & P'_{22} & P'_{23} & P'_{24} \end{bmatrix} \mathbf{H}_A = \begin{bmatrix} 0 \\ 0 \\ 0 \\ -a_{33} \end{bmatrix}^T \mathbf{H}_A$$

yields

$$\begin{bmatrix} a_{13} \\ a_{23} \\ a_{31} \\ a_{32} \end{bmatrix}^T \begin{bmatrix} \hat{P}_{11} & \hat{P}_{12} & \hat{P}_{13} & \hat{P}_{14} \\ \hat{P}_{21} & \hat{P}_{22} & \hat{P}_{23} & \hat{P}_{24} \\ \hat{P}'_{11} & \hat{P}'_{12} & \hat{P}'_{13} & \hat{P}'_{14} \\ \hat{P}'_{21} & \hat{P}'_{22} & \hat{P}'_{23} & \hat{P}'_{24} \end{bmatrix} = \begin{bmatrix} 0 \\ 0 \\ 0 \\ -a_{33} \end{bmatrix}^T .$$

This completes the proof. □

Because of the above result, there is no unique determination of $\mathbf{P}_A$ and $\mathbf{P}'_A$ from $\mathbf{F}_A$ based on (3.117). One simply way is the following:

$$\mathbf{P}_A = \begin{bmatrix} a_{31} & 0 & 0 & 0 \\ 0 & a_{32} & 0 & 0 \\ 0 & 0 & 0 & 1 \end{bmatrix}$$

$$\mathbf{P}'_A = \begin{bmatrix} -a_{13} & 0 & a_{32} & -a_{33}/a_{31} \\ 0 & -a_{23} & -a_{31} & 0 \\ 0 & 0 & 0 & 1 \end{bmatrix}.$$

Once $\mathbf{P}_A$ and $\mathbf{P}'_A$ are determined from $\mathbf{F}_A$, the 3D structure can be uniquely recovered. Let $\mathbf{m} = [u, v]^T$ and $\mathbf{m}' = [u', v']^T$ be the observed image points which have been matched between the two images. Let $\mathtt{M} = [X, Y, Z]^T$ be the corresponding space point to be estimated, which projects on to the two cameras $\mathbf{P}_A$ and $\mathbf{P}'_A$ as

$$\widehat{\mathbf{m}} = \begin{bmatrix} \hat{u} \\ \hat{v} \end{bmatrix} = \begin{bmatrix} a_{31}X \\ a_{32}Y \end{bmatrix}$$

$$\widehat{\mathbf{m}}' = \begin{bmatrix} \hat{u}' \\ \hat{v}' \end{bmatrix} = \begin{bmatrix} -a_{13}X + a_{32}Z - a_{33}/a_{31} \\ -a_{23}Y - a_{31}Z \end{bmatrix},$$

Because the observations are made in image plane and the noise level can be reasonably assumed to be the same for each extracted image point, a physically meaningful criterion is to minimize, over the structure parameter $\mathtt{M}$, the point-to-point distances between the observed locations ($\mathbf{m}$ and $\mathbf{m}'$) and the reprojections of the estimated scene structure ($\widehat{\mathbf{m}}$ and $\widehat{\mathbf{m}}'$):

$$\mathcal{F}(\mathtt{M}) = \|\mathbf{m} - \widehat{\mathbf{m}}\|^2 + \|\mathbf{m}' - \widehat{\mathbf{m}}'\|^2 .$$

The solution is obtained by setting the derivative of $\mathcal{F}(\mathtt{M})$ with respect to $\mathtt{M}$ to zero, i.e. $\partial\mathcal{F}(\mathtt{M})/\partial\mathtt{M} = \mathbf{0}$. This yields a vector equation

$$\mathbf{B}\mathtt{M} = \mathbf{b},$$

where

$$\mathbf{B} = \begin{bmatrix} a_{31}^2 + a_{13}^2 & 0 & -a_{13}a_{32} \\ 0 & a_{32}^2 + a_{23}^2 & a_{31}a_{23} \\ -a_{13}a_{32} & a_{31}a_{23} & a_{31}^2 + a_{32}^2 \end{bmatrix}$$

$$\mathbf{b} = \begin{bmatrix} a_{31}u - a_{13}u' - a_{13}a_{33}/a_{31} \\ a_{32}v - a_{23}v' \\ a_{32}u' - a_{31}v' + a_{32}a_{33}/a_{31} \end{bmatrix}.$$

The 3D reconstructed point is then given by $\mathtt{M} = \mathbf{B}^{-1}\mathbf{b}$.

3.5.2 Relation to Previous Work

There already exist a number of algorithms for the recovery of affine structure from two affine images. They can be divided into two categories. The first relies on use of a local coordinate frame by choosing four non-coplanar points to form the affine basis [80, 24, 121, 171]. One drawback is that the error in the basis points directly affects the precision of the entire solution. The second category is characterized by the work of Shapiro [134]. Inspired by the work of Tomasi and Kanade [155] for a long image sequence under orthography, Shapiro uses the singular value decomposition technique (SVD) to determine the affine cameras and the scene structure simultaneously with the whole set of points. Our work uses also the whole set of points, but we first recover the affine epipolar geometry and then determine the scene structure. Instead of conducting a SVD of a $4 \times n$ matrix as in [134] where n is the number of point matches, we solve now two smaller problems:

- determination of the affine epipolar geometry, which involves the computation of the eigenvector of a 4×4 symmetric matrix associated with the smallest eigenvalue (see Sect. 3.2);
- 3D reconstruction, which involves an inverse of a 3×3 symmetric matrix, which is the same for all points, and a multiplication of a 3×3 matrix with a 3-vector for each point.

The new technique is thus more efficient.

3.5.3 Experimental Results

We have tested the proposed technique with computer simulated data under affine projection, and very good results have been obtained. In this subsection, we show the result with data obtained *under full perspective projection* but treated as if it were obtained under affine projection.

The parameters of the camera set-up are taken from a real stereovision system. The two cameras are separated by an almost pure translation (the rotation angle is only 6 degrees). The baseline is about 350 mm (millimeters). An object of size $400 \times 250 \times 300$ mm^3 is placed in front of the cameras at a distance of about 2500 mm. Two images of this object under full perspective projection are generated as shown in Fig. 3.14. Line segments are drawn only for visual effect, and only the endpoints (12 points) are used in our experiment.

The image resolution is 512×512 pixels^2, and the projection of the object occupies a surface of about 130×120 pixels^2. Because the object only occupies a small portion of the image, the affine projection is a good approximation.

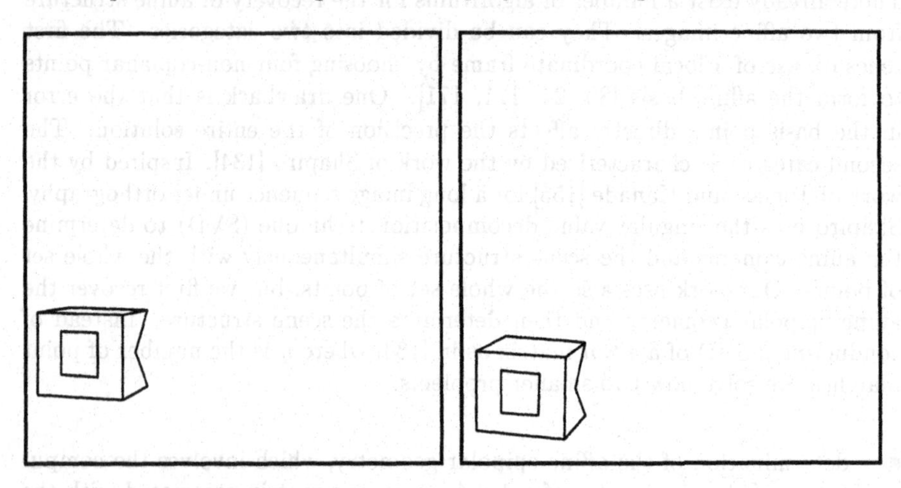

Figure 3.14 Two perspective images of a synthetic object

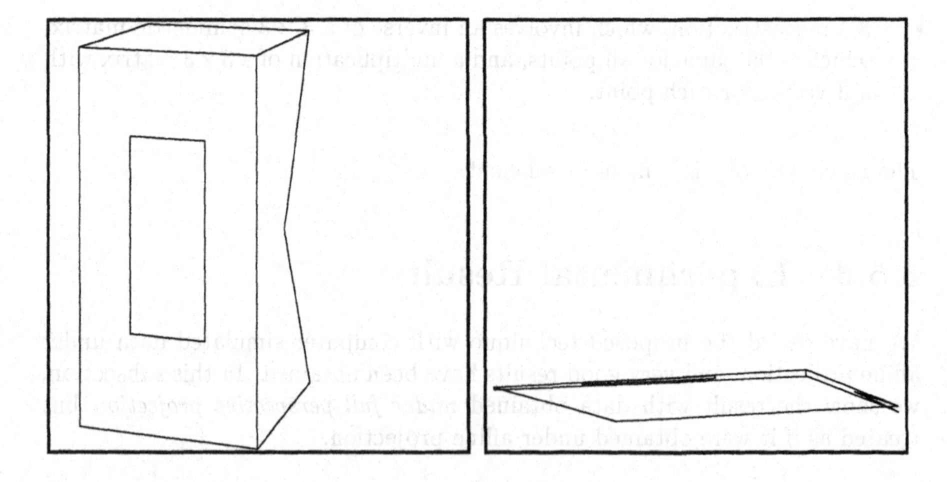

Figure 3.15 Two orthographic views of the affine reconstruction

The method described in [138] (see also Sect. 3.2) is used to compute the affine epipolar geometry, and the root of the mean point-to-point distance is 0.065 pixels. This implies that even the images are perspective, their relation can be quite reasonably described by the affine epipolar geometry. The affine reconstruction result obtained with the technique described in this paper is shown in Fig. 3.15.

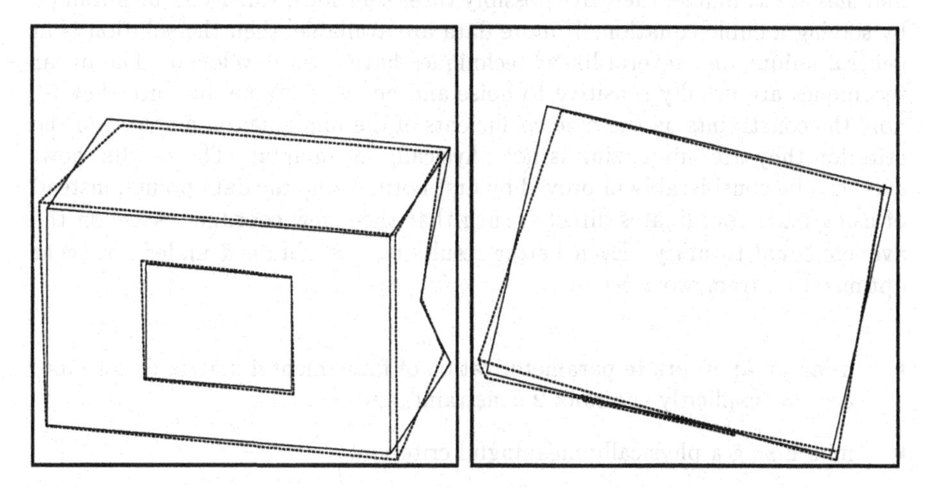

Figure 3.16 Two orthographic views of the superposition of the original 3D data (in solid lines) and the transformed affine reconstruction (in dashed lines)

In order to have a quantitative measure of the reconstruction quality, we estimate, in a least-squares sense, the affine transformation which brings the set of affinely reconstructed points to the original set of 3D points. The reader is referred to the appendix of this chapter (Sect. 3.A.2) for details on how to estimate the affine transformation between two sets of 3D points. The root of the mean of the squared distances between the corresponding points is 10.4 mm, thus the error is less than 5%. The superposition of the two sets of data is shown in Fig. 3.16. It is interesting to observe that the reconstruction of the near part is larger than the real size while that of the distant part is smaller. This is because the assumption of an affine camera ignores the perspective distortion in the image.

3.6 SUMMARY

In this chapter, we have described a number of techniques for estimating the epipolar geometry between two images and recovering the 3D structure from point matches, and this has been done for both full perspective and affine cameras. Point matches are assumed to be given, but some of them may have been incorrectly paired. How to establish point matches is the topic of Chap. 6.

For two uncalibrated images under full perspective projection, at least 7 point matches are necessary to determine the epipolar geometry. When only 7 matches are available, there are possibly three solutions, which can be obtained by solving a cubic equation. If more data are available, then the solution is in general unique and several linear techniques have been developed. The linear techniques are usually sensitive to noise and not very stable, because they ignore the constraints on the nine coefficients of the fundamental matrix and the criterion they are minimizing is not physically meaningful. The results, however, can be considerably improved by first normalizing the data points, instead of using pixel coordinates directly, such that their new coordinates are on the average equal to unity. Even better results can be obtained under nonlinear optimization framework by

- using an appropriate parameterization of fundamental matrix to take into account explicitly the rank-2 constraint, and
- minimizing a physically meaningful criterion.

Two choices are available for the latter: the distances between points and their corresponding epipolar lines, and the distances between points and the reprojections of their corresponding points reconstructed in space. Experiments show that the two give essentially the same results, but the second is much more time consuming. Therefore, the nonlinear optimization technique based on the distances between points and their correspopnding epipolar lines is recommended.

However, point matches are obtained by using some heuristic techniques such as correlation and relaxation, and they usually contain false matches. Also, due to the limited performance of a corner detector or poor contrast of an image, a few points are possibly incorrectly localized. These outliers (sometimes even one) will severely affect the precision of the fundamental matrix if we directly apply the methods described above, which are all least-squares techniques. We have thus presented in detail two commonly used robust techniques: M-Estimators and Least Median of Squares (LMedS). M-estimators try to reduce the effect of outliers by replacing the squared residuals by another function of the residuals which is less increasing than square. They can be implemented as an iterated reweighted least-squares. Experiments show that they are robust to outliers due to bad localization, but not robust to false matches. This is because they depend tightly on the initial estimation of the fundamental matrix. The LMedS method solves a nonlinear minimization problem which yields the smallest value for the median of squared residuals computed for the entire data set. It turns out that this method is very robust to false matches as well as outliers due to bad localization. Unfortunately, there is no straightforward formula for the LMedS estimator. It must be solved by a search in the space of possible estimates generated from the data. Since this space is too large, only a randomly chosen subset of data can be analyzed. We have proposed a regularly random selection method to improve the efficiency.

Almost the same thing can be said for the determination of the fundamental matrix between two images taken by affine cameras. The problem, however, is simpler because the linear relation between 3D coordinates and image data. We need at least 4 point matches to uniquely determine the affine fundamental matrix.

Since the data points are always corrupted by noise, one should model the uncertainty of the estimated fundamental matrix in order to exploit its underlying geometric information correctly and effectively. We have modelled the fundamental matrix as a random vector in its parameterization space and described methods to estimate the covariance matrix of this vector under the first order approximation. This has been done for the general fundamental matrix as well as the affine version. In [23], we show how the uncertainty can be used to define the epipolar band for matching, to compute the uncertainty of the projective reconstruction, and to improve the self-calibration based on Kruppa equations.

Usually, in a dynamic scene, the observer moves and at the same time there exist other moving objects, or the observer is fixed but there exist more than one moving objects (surveillance applications, for example). This is known as the multiple-object-motion problem. We tackle it by finding multiple fundamental matrices through unsupervised clustering in the parameter space. This is based on the fact that if the point matches belong to the same rigid motion, they must satisfy the same epipolar geometry. Note that the reverse is not always true. For example, two objects undergoing a pure translation in the same direction but with different magnitude satisfy the same epipolar geometry. Thus, object segmentation based on the epipolar constraint is necessary, but not sufficient.

Regarding the structure of the perceived environment, we have shown, in an easily understandable way, the now well-known result: The 3D scene structure can only be reconstructed up to an unknown projective (resp. affine) transformation from two perspective (resp. affine) images with unknown intrinsic parameters. Techniques for projective/affine reconstruction have also been detailed. Although we cannot obtain any metric information from a projective/affine structure (measurements of lengths and angles do not make sense), it still contains rich information, such as coplanarity, collinearity, and ratios, which is sometimes sufficient for artificial systems, such as robots, to perform tasks such as navigation and object recognition.

In order to achieve a 3D, either Euclidean or projective, reconstruction with high precision, one has to consider lens distortion. As said in Sect. 2.6, lens distortion has usually been corrected off-line. A preliminary investigation has been conducted [187], which considers lens distortion as an integral part of a camera. In this case, for a point in one image, its corresponding point does not lie on a line anymore. As a matter of fact, it lies on the so-called *epipolar curve*. Preliminary results show that the distortion can be corrected on-line if cameras have a strong lens distortion. More work still needs to be done to understand better the epipolar geometry with lens distortion.

3.A APPENDIX

In this appendix, we give details of how to obtain a rank-2 matrix from a general 3×3 matrix, and how to estimate the affine transformation between two sets of 3D points.

3.A.1 Approximate Estimation of Fundamental Matrix from General Matrix

We first introduce the Frobenius norm of a matrix $\mathbf{A} = [a_{ij}]$ $(i = 1, \ldots, m;\ j = 1, \ldots, n)$, which is defined by

$$\|\mathbf{A}\| = \sqrt{\sum_{i=1}^{m} \sum_{j=1}^{n} a_{ij}^2} \,. \tag{3.118}$$

It is easy to show that for all orthogonal matrices $\mathbf{U}$ and $\mathbf{V}$ of appropriate dimensions, we have

$$\|\mathbf{U}\mathbf{A}\mathbf{V}^T\| = \|\mathbf{A}\| \,.$$

Proposition 3.4. *We are given a 3×3 matrix $\mathbf{F}$, whose singular value decomposition (SVD) is*

$$\mathbf{F} = \mathbf{U}\mathbf{S}\mathbf{V}^T \,,$$

where $\mathbf{S} = diag(\sigma_1, \sigma_2, \sigma_3)$ *and* σ_i $(i = 1, 2, 3)$ *are singular values satisfying* $\sigma_1 \geq \sigma_2 \geq \sigma_3 \geq 0$. *Let* $\hat{\mathbf{S}} = diag(\sigma_1, \sigma_2, 0)$, *then*

$$\hat{\mathbf{F}} = \mathbf{U}\hat{\mathbf{S}}\mathbf{V}^T$$

is the closest matrix to $\mathbf{F}$ *that has rank 2. Here, "closest" is quantified by the Frobenius norm of* $\mathbf{F} - \hat{\mathbf{F}}$, *i.e.* $\|\mathbf{F} - \hat{\mathbf{F}}\|$.

Proof. We show it in two parts. First, the Frobenius norm of $\mathbf{F} - \hat{\mathbf{F}}$ is given by

$$\|\mathbf{F} - \hat{\mathbf{F}}\| = \|\mathbf{U}^T(\mathbf{F} - \hat{\mathbf{F}})\mathbf{V}\| = \|\mathrm{diag}(0, 0, \sigma_3)\| = \sigma_3 \, .$$

Second, for some 3×3 matrix $\mathbf{G}$ of *rank 2*, we can always find an orthogonal vector $\mathbf{z}$ such that $\mathbf{Gz} = \mathbf{0}$, i.e. $\mathbf{z}$ is the null vector of matrix $\mathbf{G}$. Since

$$\mathbf{Fz} = \sum_{i=1}^{3} \sigma_i (\mathbf{v}_i^T \mathbf{z}) \mathbf{u}_i \, ,$$

where $\mathbf{u}_i$ and $\mathbf{v}_i$ are the ith column vectors of $\mathbf{U}$ and $\mathbf{V}$, we have

$$\|\mathbf{F} - \mathbf{G}\|^2 \geq \|(\mathbf{F} - \mathbf{G})\mathbf{z}\|^2 = \|\mathbf{Fz}\|^2 = \sum_{i=1}^{3} \sigma_i^2 (\mathbf{v}_i^T \mathbf{z})^2 \geq \sigma_3^2 \, .$$

This implies that $\hat{\mathbf{F}}$ is indeed the closest to $\mathbf{F}$, which completes the proof. □

In the above derivation, we have used the following inequality which relates the Frobenius norm to the vector norm:

$$\|\mathbf{A}\| \geq \max_{\|\mathbf{z}\|=1} \|\mathbf{Az}\| \geq \|\mathbf{Az}\| \text{ with } \|\mathbf{z}\| = 1 .$$

The reader is referred to [46] for more details.

3.A.2 Estimation of Affine Transformation

In this section, we present a technique which computes the affine transformation from a set of affinely reconstructed 3D points, denoted here by $\mathbf{x}_i = [x_i, y_i, z_i]^T$, to a set of 3D reference points, denoted here by $\mathbf{x}'_i = [x'_i, y'_i, z'_i]^T$. Let n be the number of points. Let $\mathbf{A}$ and $\mathbf{t}$ be the 3×3 matrix and 3-vector representing the affine transformation. For each pair of points, we then have

$$\mathbf{x}'_i = \mathbf{A}\mathbf{x}_i + \mathbf{t} \, .$$

The estimation of the affine transformation can be formulated as a least-squares by minimizing the following cost function:

$$\mathcal{F}(\mathbf{A},\mathbf{t})=\sum_{i=1}^{n}(\mathbf{A}\mathbf{x}_i+\mathbf{t}-\mathbf{x}'_i)^T(\mathbf{A}\mathbf{x}_i+\mathbf{t}-\mathbf{x}'_i)\ .$$

The solution of $\mathbf{t}$ is obtained by setting the first derivative of $\mathcal{F}(\mathbf{A},\mathbf{t})$ with respect to zero:

$$\frac{\partial\mathcal{F}(\mathbf{A},\mathbf{t})}{\partial\mathbf{t}}=2\sum_{i=1}^{n}(\mathbf{A}\mathbf{x}_i+\mathbf{t}-\mathbf{x}'_i)=\mathbf{0}\ ,$$

which leads to

$$\mathbf{t}=\bar{\mathbf{x}}'-\mathbf{A}\bar{\mathbf{x}}\ ,$$

where $\bar{\mathbf{x}}=\frac{1}{n}\sum_i\mathbf{x}_i$ and $\bar{\mathbf{x}}'=\frac{1}{n}\sum_i\mathbf{x}'_i$ are the centroids of the two point sets. The optimum solution $\mathbf{A}$ thus passes through the data centroids $\bar{\mathbf{x}}$ and $\bar{\mathbf{x}}'$.

Substituting $\mathbf{t}$ into $\mathcal{F}(\mathbf{A},\mathbf{t})$ and denoting the centered points by $\mathbf{y}_i=\mathbf{x}_i-\bar{\mathbf{x}}$ and $\mathbf{y}'_i=\mathbf{x}'_i-\bar{\mathbf{x}}'$, we get

$$\mathcal{F}'(\mathbf{A})=\sum_{i=1}^{n}(\mathbf{A}\mathbf{y}_i-\mathbf{y}'_i)^T(\mathbf{A}\mathbf{y}_i-\mathbf{y}'_i)\ .$$

Now let us define the derivative of a scalar λ with respect to a matrix $\mathbf{A}$:

$$\mathcal{D}(f,\mathbf{A})=\begin{bmatrix}\frac{\partial f}{\partial a_{11}} & \frac{\partial f}{\partial a_{12}} & \frac{\partial f}{\partial a_{13}}\\ \frac{\partial f}{\partial a_{21}} & \frac{\partial f}{\partial a_{22}} & \frac{\partial f}{\partial a_{23}}\\ \frac{\partial f}{\partial a_{31}} & \frac{\partial f}{\partial a_{32}} & \frac{\partial f}{\partial a_{33}}\end{bmatrix}\ .$$

The solution for $\mathbf{A}$ is then given by setting $\mathcal{D}(\mathcal{F}'(\mathbf{A}),\mathbf{A})=\mathbf{O}$, where $\mathbf{O}$ is the 3×3 zero matrix. If we consider only one term in $\mathcal{F}'(\mathbf{A})$, it can be easily verified that

$$\mathcal{D}((\mathbf{A}\mathbf{y}_i-\mathbf{y}'_i)^T(\mathbf{A}\mathbf{y}_i-\mathbf{y}'_i),\mathbf{A})=2\mathbf{A}\mathbf{y}_i\mathbf{y}_i^T-2\mathbf{y}'_i\mathbf{y}_i^T\ .$$

We have thus

$$\mathcal{D}(\mathcal{F}'(\mathbf{A}),\mathbf{A})=2\mathbf{A}\mathbf{Y}\mathbf{Y}^T-2\mathbf{Y}'\mathbf{Y}^T\ ,$$

where $\mathbf{Y}$ and $\mathbf{Y}'$ are $3\times n$ matrices given by

$$\mathbf{Y}=[\mathbf{y}_1,\ldots,\mathbf{y}_n]\ ,\qquad \mathbf{Y}'=[\mathbf{y}'_1,\ldots,\mathbf{y}'_n]\ .$$

The solution of $\mathbf{A}$ is then given by

$$\mathbf{A}=\mathbf{Y}'\mathbf{Y}^T(\mathbf{Y}\mathbf{Y}^T)^{-1}\ .$$

4

RECOVERY OF EPIPOLAR GEOMETRY FROM LINE SEGMENTS OR STRAIGHT LINES

Essentially, two types of geometric primitives have been used in solving motion and structure problem, namely points and straight lines. When points are used, two perspective views are sufficient to recover the motion and structure of the scene, as was described in the last chapter. When straight lines are used, three perspective views are necessary. Closed-form solutions are available either for point correspondences [90, 162] (and see the last chapter for uncalibrated images) or for line correspondences [147, 89, 58]. Algorithms using both points and lines are also available [146, 59]. However, another important type of geometric primitives, namely that of *line segments*, has been since long ignored in motion and structure from motion[1], although the importance of line segments in computer vision has never been underestimated (as a matter of a fact, straight lines are merely the geometric abstraction of line segments by ignoring their endpoints). The overlook of line segments in the domain of motion and structure from motion is probably due to the lack of mathematical elegance in representing line segments.

In this chapter, we first describe a technique for determining motion and structure from correspondences of line segments. Unlike the case of straight lines, we show that two views are generally enough to recover the motion and structure of the scene. The only assumption we use is that two matched line segments contain the projection of a *common part* of the corresponding line segment in space (and we say that the two 2D line segments *overlap*). Indeed, this assumption is minimal, and is what we already use in matching line segments between different views. The analysis is done for calibrated images, but it ap-

[1]3D line segments, reconstructed by a stereo system, have been used in motion analysis by Zhang and Faugeras [190], but the problem there is different from the one addressed here. We are dealing with 2D images.

plies equally well to uncalibrated images. To our knowledge, this is the first investigation in computer vision on use of line segments for motion recovery from two perspective views.

If we ignore the endpoints of line segments, that is, if we use infinite support lines, then three views are required to recover the motion and structure. In the second part of this chapter, we will address the geometric constraints between three views (the so-called *trifocal constraint*, and finally present a unified algorithm for determining the epipolar geometry between three views from either points alone, or straight lines alone, or combination of points and lines. Since this topic has been studied extensively in the calibrated image case [146, 147, 172, 41], we will focus our discussion on the case of three uncalibrated images.

4.1 LINE SEGMENTS OR STRAIGHT LINES

An image line segment **l** is represented by its starting point **s** and its endpoint **e**[2]. We assume that line segments are *orientated*, which can be obtained from the intensity contrast information. For example, we can define an orientation signing convention such that the intensity changes from a low value to a high one when we cross the line segment from left to right. As we observe later, the orientation information is dispensable in the motion and structure problem. However, in general this information is available, and has usually already been used in the matching process.

The endpoints **s** and **e** are not reliable, mainly for three reasons:

- The first is purely algorithmic: because of noise in the images and because sometimes we approximate contours which are significantly curved with line segments, the polygonal approximation may vary from frame to frame, inducing a variation in the segments endpoints.
- The second is physical: because of partial occlusion in the scene, a segment can be considerably shortened or lengthened, and the occluded part may change over time.

[2]Please note that **e** in this chapter denotes the endpoint of a line segment, but not an epipole. This does not introduce any confusion since the epipoles will not be used explicitly in this chapter.

- The third is photometric: because lighting and surface reflection often change when the view point changes, the segments endpoints may vary from frame to frame.

However, the location and orientation of a line segment can generally be reliably determined by fitting a line to a set of linked edge points [45].

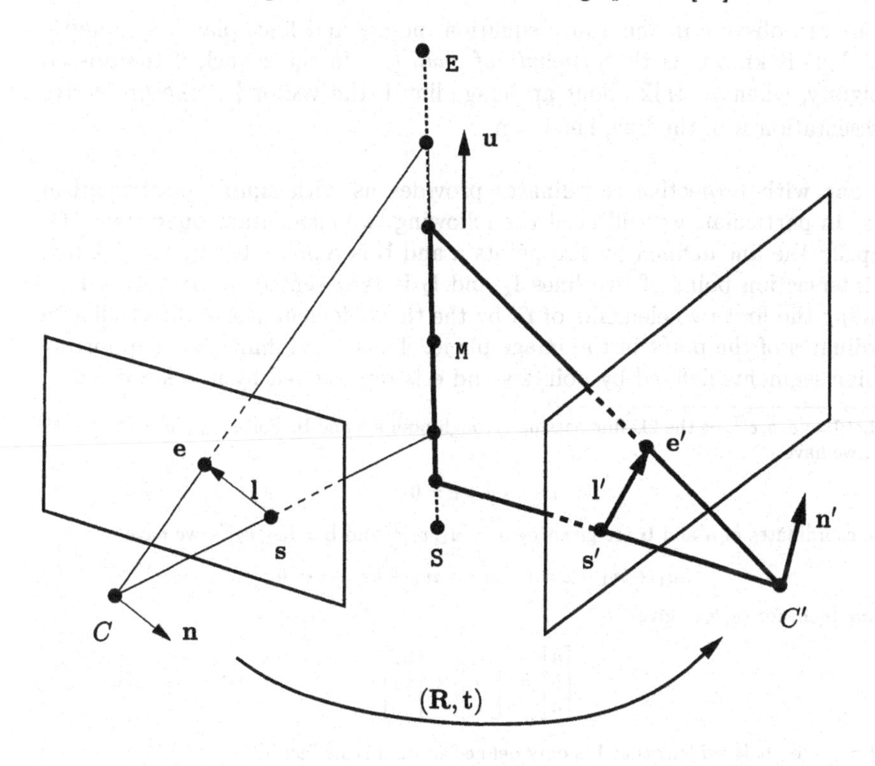

Figure 4.1 The geometry of the motion of line segments

Let $\mathbf{n}$ be the normal vector of the plane which passes through the line segment $\mathbf{l}$ and the optical center C (which is sometimes called the *projection plane* of the line segment, see Fig. 4.1). The vector $\mathbf{n}$ actually defines the infinite line supporting the line segment. More precisely, $\mathbf{n}$ is the *projective representation* of the line $\mathbf{l}$ in image plane (without ambiguity, we use $\mathbf{l}$ to denote both the line segment and its supporting line). As the image plane is parallel to y_1y_2 plane in the coordinate frame associated to the first camera, the coordinate frame attached to the image plane is Oy_1y_2. For a point $\mathbf{a} = [u, v]^T$ in the image plane, its homogeneous coordinates are $\tilde{\mathbf{a}} = [u, v, 1]^T$ corresponding to the same point in $Cy_1y_2y_3$ coordinate frame, and its projective coordinates will be $\lambda\tilde{\mathbf{a}}$ for any nonzero scalar λ. For any point $\mathbf{m} = [u, v]^T$ on the infinite supporting line $\mathbf{l}$, we have the following relation:

$$\mathbf{n}^T \tilde{\mathbf{m}} = 0 .$$

As one can observe in the above equation, points and lines play a symmetric role. This is known as the *principle of duality.* In the sequel, if there is no ambiguity, when we talk about an image line $\mathbf{l}$, the vector $\mathbf{l}$ is the projective representation $\mathbf{n}$ of the line, i.e. $\mathbf{l} = \mathbf{n}$.

Working with projective coordinates provides us with simple mathematical tools. In particular, we will need the following two elementary operations [41, Chap. 2]: the line defined by two points $\mathbf{a}$ and $\mathbf{b}$ is represented by $\mathbf{l} = \tilde{\mathbf{a}} \times \tilde{\mathbf{b}}$ [3]; the intersection point of two lines $\mathbf{l}_1$ and $\mathbf{l}_2$ is represented by $\tilde{\mathbf{m}} = \mathbf{l}_1 \times \mathbf{l}_2$ [4]. Dividing the first two elements of $\tilde{\mathbf{m}}$ by the third element yields the Euclidean coordinates of the point in the image plane. Thus, the infinite line supporting the line segments defined by points $\mathbf{s}$ and $\mathbf{e}$ is represented by $\mathbf{n} = \tilde{\mathbf{s}} \times \tilde{\mathbf{e}}$.

---

[3] Let $\mathbf{l} = [a, b, c]^T$ be the 2D line passing through points $\mathbf{a}$ and $\mathbf{b}$. For any point $\mathbf{m} = [u, v]^T$ on it, we have

$$au + bv + c = 0 .$$

If the coordinates of $\mathbf{a}$ and $\mathbf{b}$ are given by $\mathbf{a} = [u_1, v_1]^T$ and $\mathbf{b} = [u_2, v_2]^T$, we have

$$au_1 + bv_1 + c = 0 \quad \text{and} \quad au_2 + bv_2 + c = 0 .$$

Solving them for (a, b, c) gives

$$\begin{bmatrix} a \\ b \\ c \end{bmatrix} = \begin{bmatrix} u_1 \\ v_1 \\ 1 \end{bmatrix} \times \begin{bmatrix} u_2 \\ v_2 \\ 1 \end{bmatrix}$$

i.e. $\mathbf{l} = \widetilde{\mathbf{a}} \times \widetilde{\mathbf{b}}$. It is evident that $\mathbf{l}$ is only defined up to a scale factor.

[4] Let $\mathbf{m} = [u, v]^T$ be the intersection of the two lines $\mathbf{l}_1 = [a_1, b_1, c_1]^T$ and $\mathbf{l}_2 = [a_2, b_2, c_2]^T$. We have

$$a_1u + b_1v + c_1 = 0 \quad \text{and} \quad a_2u + b_2v + c_2 = 0 .$$

Thus

$$\begin{bmatrix} u \\ v \end{bmatrix} = \frac{1}{a_1b_2 - a_2b_1} \begin{bmatrix} c_2b1 - c1b2 \\ c_1a_2 - c_2a_1 \end{bmatrix} .$$

Let $s = a_1b_2 - a_2b_1$, then

$$s\tilde{\mathbf{m}} = \mathbf{l}_1 \times \mathbf{l}_2 ,$$

i.e. $\mathbf{l}_1 \times \mathbf{l}_2$ is the projective coordinates of the intersection point $\mathbf{m}$.

We are given two line segments, $\mathbf{l}$ in camera 1 and $\mathbf{l}'$ in camera 2, in correspondence. The basic assumption we will use is that they are projections of two portions of a line segment $\mathtt{SE}$ in space and that the two portions *share a common part* (i.e. they overlap). We do not assume that the starting points ($\mathbf{s}$ and $\mathbf{s}'$) and the endpoints ($\mathbf{e}$ and $\mathbf{e}'$) are in correspondence. Since points $\mathbf{s}$ and $\mathbf{e}$ are not reliable, one way is to consider only the infinite support lines. It has been well known that motion cannot be determined from two views of straight lines [89, 41]. Assume the world coordinate frame coincides with the coordinate frame of the first camera. Given two line segments $\mathbf{l}$ and $\mathbf{l}'$ in correspondence, let the corresponding 3D line segment be represented by its direction vector $\mathbf{u}$ and a point $\mathtt{M}$ on it. Since we ignore the endpoints, point $\mathtt{M}$ can be anywhere on the infinite support line. Clearly, we have $\mathbf{n} = \mathtt{M} \times \mathbf{u}$ for the first camera. Working in the coordinate frame of the second camera, we have

$$\mathbf{n}' = (\mathbf{R}\mathtt{M} + \mathbf{t}) \times (\mathbf{R}\mathbf{u}) = \mathbf{R}(\mathtt{M} \times \mathbf{u}) + \mathbf{t} \times (\mathbf{R}\mathbf{u}) ,$$

i.e.,

$$\mathbf{n}' = \mathbf{R}\mathbf{n} + \mathbf{t} \times (\mathbf{R}\mathbf{u}) .$$

Multiplying the above equation by $\mathbf{t}$ to eliminate the unknown $\mathbf{u}$, we have

$$\mathbf{t}^T(\mathbf{n}' - \mathbf{R}\mathbf{n}) = 0 .$$

This equation is useless, however, because during its derivation, $\mathbf{n}$ and $\mathbf{n}'$ were related to the unknown 3D line, which implies that the magnitude of $\mathbf{n}$ and $\mathbf{n}'$ cannot be anything. We have only a projective representation of the line, and so $\mathbf{n}$ and $\mathbf{n}'$ can be multiplied by an arbitrary scalar which is not equal to zero.

Geometrically, it is obvious: let us fix the position and orientation of the first camera. Now we move the second camera to another position and orientation. For each image line, its corresponding 3D line must lie on the plane (called the *projection plane* of the line) passing through the optical center and the image line. For each pair of lines in correspondence, we have a pair of projection planes, whose intersection determines the 3D line in space. The structure of the scene can so be determined. However, any two planes define a line. We can move the second camera to an arbitrary position and orientation, and we still obtain a 3D structure consistent with the two images. In other words, two sets of lines do not constrain the motion of the camera. If a third image is available, the motion and structure can in general be uniquely determined because three planes generally do not define a line (they define in general three lines).

When line segments are considered, the motion of the second camera can no longer be arbitrary [184]. Indeed, each line segment defines a *generalized triangle* in space with the first side passing through the optical center C and the

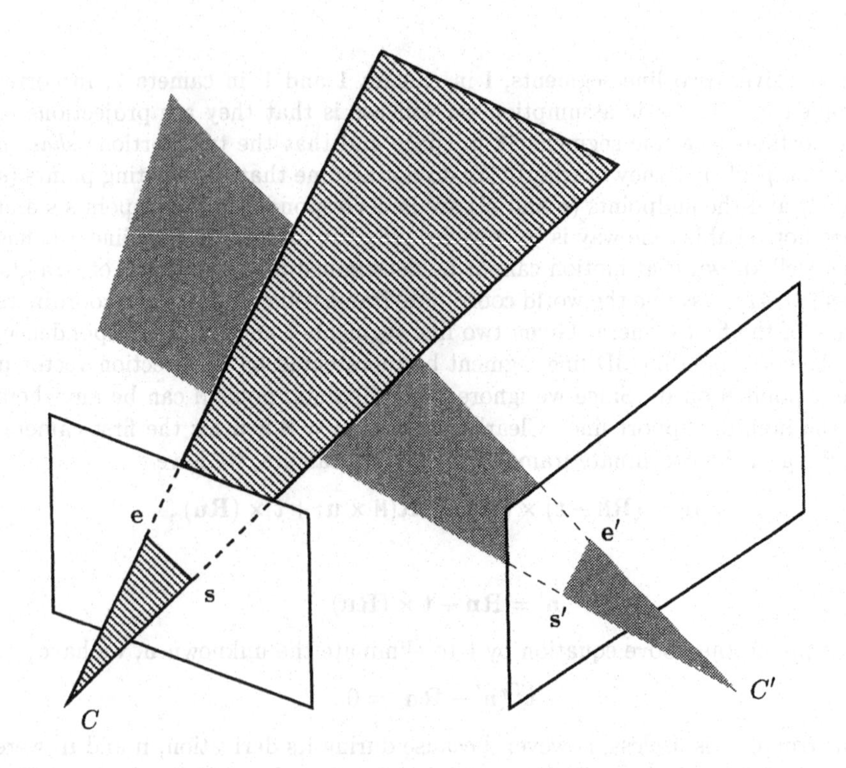

Figure 4.2 Motion and structure from line segments

starting point **s** of the line segment, the second side passing through C and the endpoint **e**, and the third side at infinity (see Fig. 4.2). Two such triangles generally do not intersect. By requiring a pair of matched line segments to overlap in space, we add a constraint on the family of feasible motions. The set of all constraints for all correspondences of line segments define an open set in motion parameter space. The larger the number of correspondences is, the smaller the extent of the open set is, and the more the motion is constrained. If we have only a few correspondences of line segments, the motion might not be well constrained and the corresponding reconstruction of the scene geometry will vary widely.

4.2 SOLVING MOTION USING LINE SEGMENTS BETWEEN TWO VIEWS

From the above discussion, we see that the motion between two views is constrained if line segments are used. In this section, we present the algorithm for solving the motion problem by maximizing the overlap of line segments. The intrinsic parameters of each image are assumed to be known, so we can use the normalized image coordinates. The motion between the two images is $(\mathbf{R}, \mathbf{t})$. The essential matrix is $\mathbf{E} = [\mathbf{t}]_\times \mathbf{R}$, as defined in (2.35).

4.2.1 Overlap of Two Corresponding Line Segments

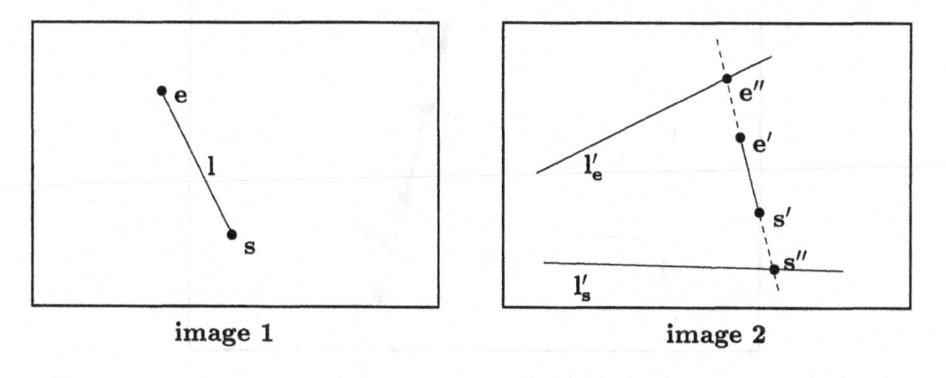

Figure 4.3 Overlap of two line segments in correspondence

Let us consider the situation illustrated in Fig. 4.3. We are given a pair of line segments $(\mathbf{l}, \mathbf{l}')$ in correspondence. The line $\mathbf{l}'_s$ in the second image is the epipolar line of $\mathbf{s}$, i.e. $\mathbf{l}'_s = \mathbf{E}^T\tilde{\mathbf{s}}$; the line $\mathbf{l}'_e$ is the epipolar line of $\mathbf{e}$, i.e. $\mathbf{l}'_e = \mathbf{E}^T\tilde{\mathbf{e}}$. We denote the intersection of $\mathbf{l}'_s$ with line $\mathbf{l}'$ by $\tilde{\mathbf{s}}'' = \mathbf{l}' \times \mathbf{l}'_s$, and the intersection of $\mathbf{l}'_e$ with line $\mathbf{l}'$ by $\tilde{\mathbf{e}}'' = \mathbf{l}' \times \mathbf{l}'_e$. (Please refer to Sect. 2.3.2 for the exposition of the epipolar geometry with calibrated images.)

Provided that the epipolar geometry (i.e. matrix $\mathbf{E}$, or the motion $(\mathbf{R}, \mathbf{t})$) between two images is correct, then $\mathbf{s}$ and $\mathbf{s}''$ correspond to a single point in space; so do $\mathbf{e}$ and $\mathbf{e}''$. Thus, the statement that two line segments $\mathbf{l}$ and $\mathbf{l}'$ share a common part of a 3D line segment is equivalent to saying that line segment $\mathbf{s''e''}$ and line segment $\mathbf{s'e'}$ (i.e. $\mathbf{l'}$) overlap. In order for $\mathbf{s'e'}$ and $\mathbf{s''e''}$ to overlap, one of the following two conditions must be satisfied:

1. Either $\mathbf{s''}$ or $\mathbf{e''}$ or both are between $\mathbf{s'}$ and $\mathbf{e'}$.

2. $\mathbf{s'}$ and $\mathbf{e'}$ are both between $\mathbf{s''}$ and $\mathbf{e''}$.

This implies that only when (here $\|$ stands for the or logic)

$$\begin{aligned}(\mathbf{s''}-\mathbf{s'})\cdot(\mathbf{e'}-\mathbf{s''})>0 \quad &\| \quad (\mathbf{e''}-\mathbf{s'})\cdot(\mathbf{e'}-\mathbf{e''})>0 \\ \| \quad (\mathbf{s'}-\mathbf{s''})\cdot(\mathbf{e''}-\mathbf{s'})>0 \quad &\| \quad (\mathbf{e'}-\mathbf{s''})\cdot(\mathbf{e''}-\mathbf{e'})>0\,,\end{aligned} \tag{4.1}$$

the two line segments $\mathbf{s'e'}$ and $\mathbf{s''e''}$ overlap.

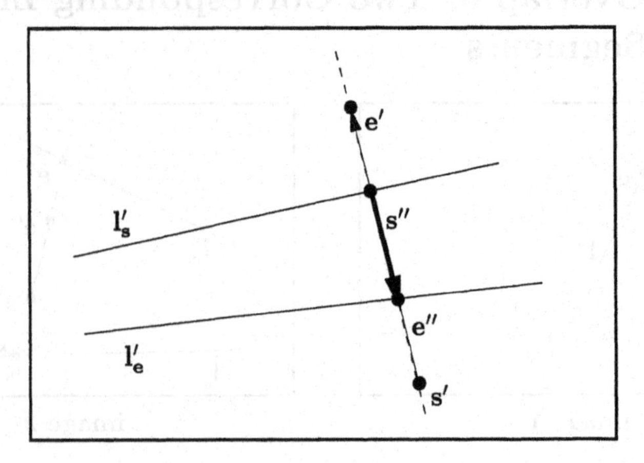

Figure 4.4 Two incongruent line segments in orientation

The above constraint does not use the fact that the line segments are oriented. The configuration in Fig. 4.4 satisfies the above constraint, but does not satisfy the orientation congruence. If the rotation between two images is not very big, then the orientation of the projected line segment in image cannot change abruptly. In order to assure the orientation congruence, we must impose another constraint:

3. Line segments $\mathbf{s'e'}$ and $\mathbf{s''e''}$ should be oriented in the same way. This implies:

$$(\mathbf{e'}-\mathbf{s'})\cdot(\mathbf{e''}-\mathbf{s''})>0\,.$$

Note that it is here that the orientation information of a line segment is used. Remove this constraint, and the proposed algorithm will work for line segments which are not oriented.

Thus, the problem of motion and structure from correspondences of line segments can be solved by nonlinear programming such that the above two constraints are satisfied for each correspondence. However, we can only obtain a feasible *region* in the motion space, and we rather need a unique solution. In the next, we solve the problem by maximizing the overlap of line segments.

4.2.2 Estimating Motion by Maximizing Overlap

We first define a measure of overlap, which we will call the *overlap length*, for two line segments in correspondence. The overlap length is positive if two line segments overlap; otherwise, it is negative.

If the two constraints described in the last subsection are satisfied, the two line segments overlap, and we can easily see that there exist only four configurations of overlap as illustrated in Fig. 4.5. The overlap length, denoted by $\mathcal{L}'$, is defined

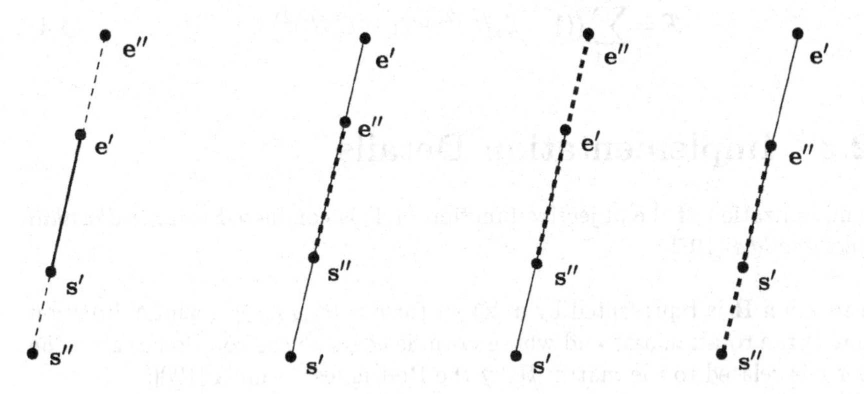

Figure 4.5 Four configurations of two line segments with overlap

as the length of the common segment, which is given by

$$\mathcal{L}' = \min(\|\mathbf{e}' - \mathbf{s}'\|, \|\mathbf{e}'' - \mathbf{s}'\|, \|\mathbf{e}' - \mathbf{s}''\|, \|\mathbf{e}'' - \mathbf{s}''\|) . \tag{4.2}$$

If two segments do not overlap (i.e. the inequalities in Sect. 4.2.1 are not satisfied), we define the overlap length as

$$\mathcal{L}' = -\min(\|\mathbf{e}' - \mathbf{s}''\|, \|\mathbf{e}'' - \mathbf{s}'\|) , \tag{4.3}$$

which corresponds to the gap between the two segments[5]. The reader is referred to [182] for more details.

The above overlap measure of a given pair of line segments is defined in the second image. We have no reason for one image to prevail over another. In order for the two images to play a symmetric role, we compute the overlap length in the first image, denoted by $\mathcal{L}$, exactly in the same way.

Since a small overlap length for a short line segment is as important as a large overlap length for a long line segment, it is more reasonable to use the *relative* overlap length, and thus we should use $\mathcal{L}_i/l_i$ and $\mathcal{L}'_i/l'_i$ to measure the overlap of a pair of line segments $(\mathbf{l}_i, \mathbf{l}'_i)$, where l_i and l'_i are the length of the line segments $\mathbf{l}_i$ and $\mathbf{l}'_i$, respectively. The relative overlap length takes a value between 0 and 1 when two segments overlap; otherwise it will be negative. Now we can formulate the motion problem as follows: Given n correspondences of line segments, $\{(\mathbf{l}_i, \mathbf{l}'_i) \mid i = 1, \ldots, n\}$, estimate the camera motion parameters $(\mathbf{R}, \mathbf{t})$ by minimizing the following objective function

$$\mathcal{F} = \sum_{i=1}^{n} \left((1 - \mathcal{L}_i/l_i)^2 + (1 - \mathcal{L}'_i/l'_i)^2\right) . \tag{4.4}$$

4.2.3 Implementation Details

The minimization of the objective function (4.4) is conducted using a *downhill simplex method* [107].

The rotation $\mathbf{R}$ is represented by a 3D vector $\mathbf{r} = [r_1, r_2, r_3]^T$, whose direction is that of the rotation axis and whose norm is equal to the rotation angle. The vector $\mathbf{r}$ is related to the matrix $\mathbf{R}$ by the Rodrigues' formula [190]:

$$\mathbf{R} = \mathbf{I}_3 + \frac{\sin\vartheta}{\vartheta}[\mathbf{r}]_\times + \frac{1 - \cos\vartheta}{\vartheta^2}[\mathbf{r}]_\times^2 ,$$

[5] Another possible definition is

$$\mathcal{L}' = -\min(\|\mathbf{s}' - \mathbf{s}''\|, \|\mathbf{e}' - \mathbf{e}''\|) ,$$

which corresponds to the minimum distance between the endpoints. It gives in general a larger value than that of (4.3), thus providing a higher penality for non-overlapping segments.

where $\mathbf{I}_3$ is the 3×3 identity matrix, and $\vartheta = \|\mathbf{r}\|$.

Because the magnitude of $\mathbf{t}$ is inherently unrecoverable, the translation $\mathbf{t}$ may be assumed to be of unit length, and hence is represented by a point on the unit sphere. The spherical coordinates (ϕ, θ) is used to represent $\mathbf{t}$.

As the problem is nonlinear, an initial guess of the motion is required. We have tried to estimate the motion by assuming the correspondences of endpoints or midpoints, but the results are useless. For the solution that works best, we choose to sample the parameter space to obtain a global minimum. The space of rotation can be thought of as a solid ball of radius π. Assume that the motion between two successive views is small, we sample the range $[-\frac{\pi}{4}, \frac{\pi}{4}]$ with step equal to $\frac{\pi}{8}$ in each direction. The maximum rotation angle is $\frac{\sqrt{3}}{4}\pi$ (i.e. 78°). This range is sufficient for most applications of motion analysis. It is rare that the rotation angle goes beyond 60° between successive views. We have thus $5^3 = 125$ samples of rotation[6].

The samples of translation are obtained through a uniform partition of a Gauss sphere based on the icosahedron [9]. The icosahedron has 12 vertices, 20 faces and 30 edges. Basically, we obtain 20 samples of 3D directions. Adding its dual (vertices) yields in total 32 samples. To obtain more samples, we further divide each icosahedral edge into n equal lengths and construct n^2 congruent equilateral triangles on each face, pushing them out to the radius of the sphere for their final position. In particular, for $n = 2$, we have 80 samples; for $n = 3$, we have 180 samples. From our experience, we found that 80 samples are sufficient for solving the problem in hand.

As a matter of fact, we do not need to use all 80 samples because of the following proposition. We only need half of them, i.e. the samples from a hemisphere.

Proposition 4.1. *We consider two given sets of points* $\{(\mathbf{m}_i, \mathbf{m}'_i)\}$ *(the same for line segments) in correspondence. If* $(\mathbf{R}, \mathbf{t}, \{\mathtt{M}_i\})$ *is a solution of the motion and structure, then* $(\mathbf{R}, -\mathbf{t}, \{-\mathtt{M}_i\})$ *is also a solution.*

Proof. Under the pinhole model, we have

$$\begin{cases} s_i\tilde{\mathbf{m}}_i = & [\mathbf{R}\ \mathbf{t}]\tilde{\mathtt{M}}_i & = \mathbf{R}\mathtt{M}_i + \mathbf{t} & \text{(for the first image)} \\ s'_i\tilde{\mathbf{m}}'_i = & [\mathbf{I}\ \mathbf{0}]\tilde{\mathtt{M}}_i & = \mathtt{M}_i & \text{(for the second image)} \end{cases}$$

[6]This sampling is not uniform. A better way may be the following. The solid ball of radius π is first sampled by a set of spheres of radius from 0 to π with step equal to, say, $\frac{\pi}{8}$. Each sphere can then be quasi-uniformly sampled as we will do for translation.

where s_i and s'_i are arbitrary scalars. It is evident that if $(\mathbf{R}, \mathbf{t}, \{\mathtt{M}_i\})$ is a solution to the motion and structure problem, then $(\mathbf{R}, -\mathbf{t}, \{-\mathtt{M}_i\})$ is also a solution. This is because if s_i and s'_i are the scale factors for the first solution, we obtain the second solution with scale factors $-s_i$ and $-s'_i$. Both solutions are compatible with the observed data. □

It is thus inherently impossible to determine geometrically the sign of the translation vector from two perspective images. So we only need to sample a hemisphere for the translation. The ambiguity can be resolved by imposing some physical constraint, e.g. the reconstructed points should be in front of the cameras (i.e. they have positive depth). If their depths are negative, it is sufficient, from the above proposition, to multiply $\mathbf{t}$ and $\{\mathtt{M}_i\}$ by -1 to obtain the physical solution.

In passing, if we do not impose that matrix $\mathbf{R}$ is a rotation matrix, then $(-\mathbf{R}, \mathbf{t}, \{-\mathtt{M}_i\})$ and $(-\mathbf{R}, -\mathbf{t}, \{\mathtt{M}_i\})$ are two other solutions. However, if the original solution $\mathbf{R}$ is a rotation, i.e. $\det \mathbf{R} = 1$, then these two solutions correspond to a reflection of the camera coordinate frame because $\det(-\mathbf{R}) = -1$. They are thus excluded on physical grounds.

To summarize, we have $125 \times 40 = 5000$ sample points in the motion space. We evaluate the objective function for each sample, and retain 10 samples which yield the smallest values of the objective function. All 10 samples are used as the initial guess to carry out the minimization procedure independently. At the end, the one which produces the smallest value of the objective function is considered as the solution of the motion. To give an idea of the time complexity, it takes about 4.3 seconds on a SPARC 10 station to perform a complete run of the algorithm for 35 line segments correspondences.

It is well-known that the relative position of two cameras is determined up to a π radians twist about the line joining the two optical centers [98]. The reader may wonder why we do not consider this ambiguity in the above sampling technique. The reason is that we have assumed a small rotation (less than $\pi/2$) between two cameras. In that case the twisted solution never appears in our sampling range because its corresponding rotation is always bigger than $\pi/2$, and it is not a physical solution. More precisely, we have

Proposition 4.2. *If* $(\mathbf{R}, \mathbf{t})$ *is a solution of the motion between two sets of points, and the rotation angle of* $\mathbf{R}$ *is less than* $\pi/2$, *then the twisted solution* $(\mathbf{S}, \mathbf{t})$ *has a rotation angle larger than* $\pi/2$.

Proof. Let σ be a rotation through π radians with axis $\mathbf{t}$. By definition, we have $\mathbf{S} = \mathbf{R}\sigma$. A rotation can be represented by a unit quaternion [41, 190]. If $\mathbf{R}$ is a rotation through θ radians with axis $\mathbf{u}$, its quaternion is $(\cos(\theta/2), \sin(\theta/2)\mathbf{u})$. The quaternion corresponding to σ is simply $(0, \mathbf{t})$. Following the multiplication rule of quaternion, the quaternion corresponding to $\mathbf{S}$ is given by

$$\begin{aligned}\mathbf{q} &= (\cos(\theta/2), \sin(\theta/2)\mathbf{u}) \star (0, \mathbf{t}) \\ &= (-\sin(\theta/2)\mathbf{u} \cdot \mathbf{t}, \cos(\theta/2)\mathbf{t} + \sin(\theta/2)\mathbf{u} \times \mathbf{t}) ,\end{aligned}$$

where $\star$ denotes the quaternion multiplication, and $\cdot$ and $\times$ are, respectively, the dot and cross product of vectors. Let ϕ be the rotation angle of S, and consider only the scalar part of $\mathbf{q}$, we have

$$\cos(\phi/2) = -\sin(\theta/2)\mathbf{u} \cdot \mathbf{t} .$$

Thus, $\phi = \pi + \theta$ if $\mathbf{u} \cdot \mathbf{t} = 1$; $\phi = \pi$ if $\mathbf{u} \cdot \mathbf{t} = 0$; $\phi = \pi - \theta$ if $\mathbf{u} \cdot \mathbf{t} = -1$. Following the property of cosine and sine, we have $\frac{\pi}{2} \leq \phi \leq \frac{3\pi}{2}$ if $|\theta| < \frac{\pi}{2}$. □

If we do not restrict the rotation angle between the two cameras, then we need to consider the twisted pair ambiguity to find the physical solution. The property of the twisted pair can be used to cut the number of rotations by half, thus halving the search space.

4.2.4 Reconstructing 3D Line Segments

Once we have an estimate of the motion between two images, we can reconstruct the 3D line segment for each pair of image line segments $(\mathbf{l}, \mathbf{l}')$ in correspondence.

We first compute the infinite 3D line, which is the intersection of the two projection planes. The line can be represented by its direction vector $\mathbf{u}$ and a point $\mathbf{x}$ on it, say, the point which is the closest to the origin of the coordinate frame of the first camera. For the direction vector $\mathbf{u}$, we have

$$\begin{cases} \mathbf{n}^T\mathbf{u} = 0 & (\mathbf{u} \text{ is in the first projection plane}) \\ \mathbf{n}'^T(\mathbf{R}^T\mathbf{u}) = 0 & (\mathbf{u} \text{ is in the second projection plane}) \end{cases}$$

which gives

$$\mathbf{u} = \mathbf{n} \times (\mathbf{R}\mathbf{n}') .$$

For the point $\mathbf{x}$, we have

$$\begin{cases} \mathbf{n}^T\mathbf{x} = 0 & (\mathbf{x} \text{ is in the first projection plane}) \\ \mathbf{n}'^T(\mathbf{R}^T\mathbf{x} - \mathbf{R}^T\mathbf{t}) = 0 & (\mathbf{x} \text{ is in the second projection plane}) \\ \mathbf{u}^T\mathbf{x} = 0 & (\mathbf{x} \text{ is the point on the line closest to } C) \end{cases}$$

The solution is:

$$\mathbf{x} = [\mathbf{n} \quad \mathbf{R}\mathbf{n}' \quad \mathbf{n} \times (\mathbf{R}\mathbf{n}')]^{-T} \begin{bmatrix} 0 \\ \mathbf{t}^T\mathbf{R}\mathbf{n}' \\ 0 \end{bmatrix} .$$

It is then trivial to recover the point on the 3D line corresponding to each endpoint of the image line segments, and we have two 3D line segments. It remains the choice of the appropriate 3D line segments. Due to the reasons described in Sect. 4.1, a 2D line segments is only an observation of a portion of the real line segment in space. Two segments are considered to be matched if they have a common part. Their corresponding segment in space can be expected not to be shorter than either of the two segments. That is, the *union* of the two segments can be reasonably considered as a better estimate of the corresponding segment in space. In passing, our trinocular stereo algorithm [6] uses the *intersection* strategy, that is, only the part of line segment which is perceived by all of the three cameras is reconstructed in space. Thus, it usually reconstructs line segments much shorter than the real segments in space.

4.2.5 Experimental Results

The proposed algorithm has been tested with both synthetic and real data. The reader is referred to [182] for the results with synthetic data.

For comparison reason, we have also tried to apply a point-based method [41] to the endpoints or midpoints of line segments, but the results are useless. For example, for the **Modig** scene described below (Figs. 4.6-4.10), the motion estimated when applying the point-based method to the endpoints is:

$$\mathbf{r} = [-2.244e{-2}, -2.284e{-2}, 1.224e{-4}]^T \ , \ \mathbf{t} = [3.577e{-1}, 8.368e{-1}, 4.146e{-1}]^T \ ,$$

which is completely different from that obtained through stereo calibration:

$$\mathbf{r} = [6.250e{-2}, -7.196e{-2}, 2.109e{-2}]^T \ , \ \mathbf{t} = [4.757e{-1}, 8.661e{-1}, 1.536e{-1}]^T \ .$$

The difference in the rotation angle is 5.164°, but the angle between the rotation axes is as high as 89.25°, and the angle between the translation vectors is 18.55°.

The 3D reconstruction provided by the point-based method is meaningless, and is thus not shown here. This is why we resort to the sampling technique described in Sect. 4.2.3 for searching an initial estimate of motion.

We have tested our algorithm with success on more than ten real image pairs, except for one which contains many line segments *aligned with the epipolar lines* (the latter case is well known to cause problems for binocular stereo). This is because our algorithm computes the overlap from the intersections of line segments with their corresponding epipolar lines, and the intersections will be unstable when they are almost aligned. One solution to this would be to compute the angles between line segments and their corresponding epipolar lines, and we could just discard those line segments that form a small angle with the epipolar lines.

In the following, we describe four sets of real data which were extracted from a trinocular stereo system [6]. We have chosen the stereo data because the stereo system has been calibrated which serves as a ground truth [40].

The first set of real data is an image pair of a scene named **Modig** because it contains a painting by the Italian painter Modigliani (see Fig. 4.6). There are 121 line segments matched by the trinocular stereo (see Fig. 4.7), among which there exist a few false matches. One can also notice several multiple matches: several segments on the painting are fragmented in one view, and the fragments are matched to a single segment in the other view.

Figure 4.6 Image pair of a **Modig** scene

Figure 4.7 Matched line segments of the **Modig** image pair

We searched for the initial motion estimate by sampling as described in Sect. 4.2.3. The ten best samples all converge to the good solution. The best solution is the one which minimizes (4.4). The final motion estimation is:

$$\mathbf{r} = [8.481e{-}2, -8.545e{-}2, 2.064e{-}2]^T \ , \ \mathbf{t} = [4.847e{-}1, 8.665e{-}1, 1.199e{-}1]^T \ ,$$

which should be compared with the estimation through stereo calibration already given at the beginning of this section. The difference in the rotation angle is 1.406°; the angle between the rotation axes is 4.642°; the angle between the translation vectors is 2.002°.

To better understand how the proposed algorithm works, we have extracted the intermediate results for one selected hypothesis of motion. Recall that our algorithm tries to maximize the overlap of two sets of line segments. Figure 4.8 shows how the overlap of the two sets of line segments of the **Modig** scene evolves during the optimization process. The four pictures correspond to the results obtained with the initial motion estimate and those obtained after five, ten and 24 iterations. The first set of line segments are projected onto the second image using the motion estimate as described in Sect. 4.2.1, and are shown in Fig. 4.8 as solid lines. The second set of line segments are shown as dashed lines. The matched line segments are shown to be collinear because of the way of the projection performed. As can be observed, the overlap by the initial motion estimate is very bad. Significant improvement is achieved after five iterations. Very good result is already obtained after ten iterations. Later

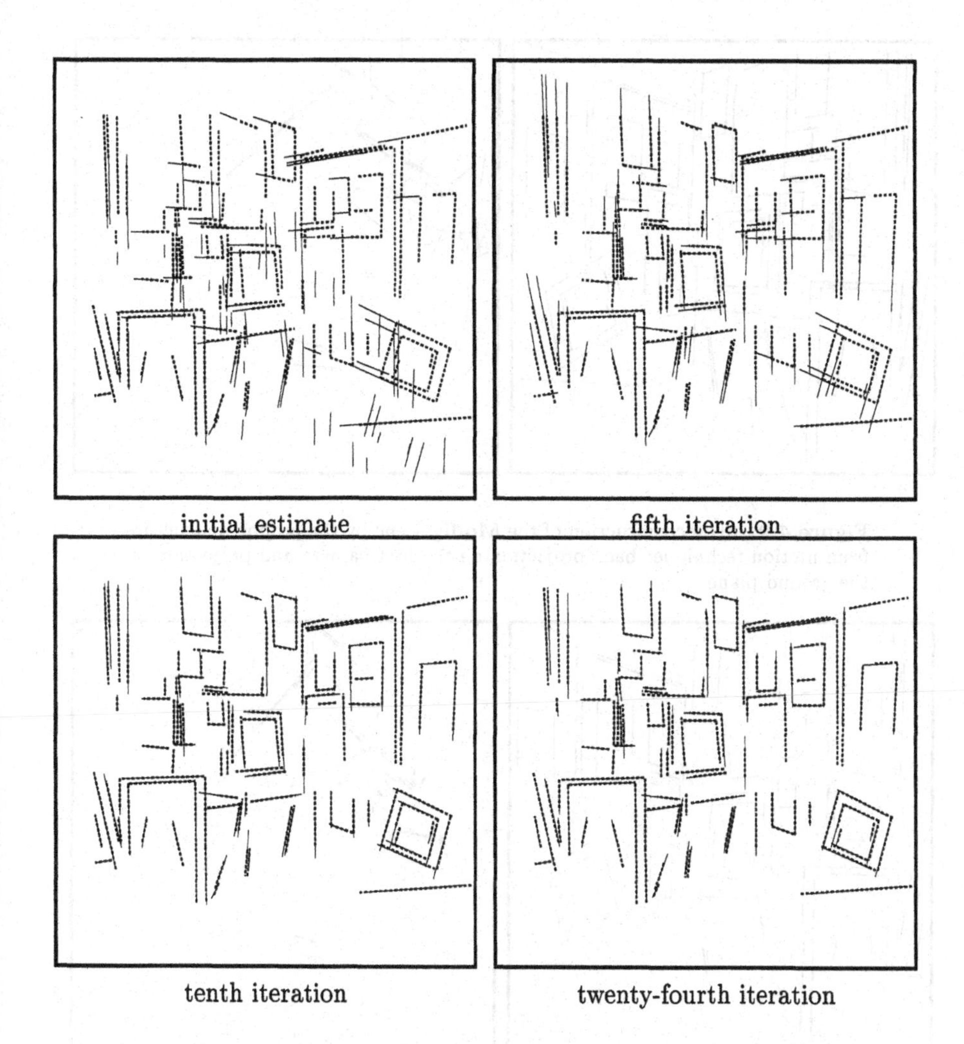

Figure 4.8 Evolution of the overlap of the line segments of the **Modig** scene during the optimization process: They correspond to the results of the initial, fifth, tenth and 24th iteration

on, the improvement is small, as can be asserted from the comparison of the two pictures of the lower row in Fig. 4.8.

The 3D reconstruction produced by the algorithm described in Sect. 4.2.4 is shown in Fig. 4.9, where the picture on the left is the perspective view from the first camera and the one on the right is a top view. This result should be com-

Figure 4.9 3D reconstruction of the **Modig** scene by the proposed structure from motion technique: back projection on the first camera and projection on the ground plane

Figure 4.10 3D reconstruction of the **Modig** scene by a classical trinocular stereo: back projection on the first camera and projection on the ground plane

pared with that reconstructed by our trinocular stereo which uses *three* images (note that only two images are used for our algorithm) and whose geometry has been previously calibrated (see Fig. 4.10). The two results are comparable. Because of use of the *union* strategy described in Sect. 4.2.4, the 3D reconstruction shown in Fig. 4.9 appears more complete than that shown in Fig. 4.10.

The second set of real data is an image pair of a **RobotLab** scene (see Fig. 4.11). 45 line segments have been matched by our trinocular stereo system[7], as shown in Fig. 4.12. One can notice several false matches.

Figure 4.11 Image pair of a **RobotLab** scene

Figure 4.12 Matched line segments of the **RobotLab** image pair

Through searching for the initial motion estimation by sampling as described in Sect. 4.2.3, six among the ten best samples converge to the good solution.

[7]The stereo can in fact match more line segments, but we have imposed a tighter epipolar constraint to limit the number of matches for this example.

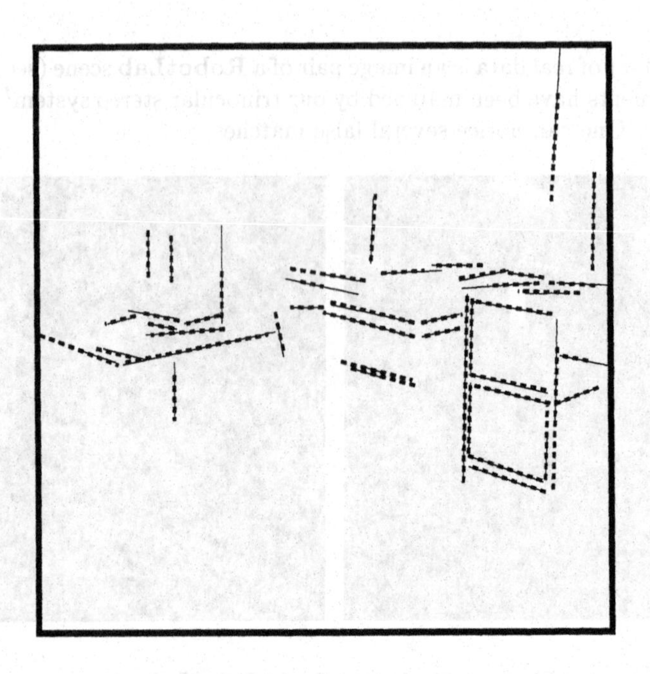

Figure 4.13 Overlap of the two sets of line segments given by the final motion estimate

The motion estimation given by the proposed algorithm is:

$$\mathbf{r} = [1.859e{-}1, 1.218e{-}1, 3.707e{-}2]^T \ , \ \mathbf{t} = [-5.939e{-}1, 6.341e{-}1, -4.951e{-}1]^T \ ,$$

while the estimation through stereo calibration is:

$$\mathbf{r} = [1.927e{-}1, 1.195e{-}1, 3.407e{-}2]^T \ , \ \mathbf{t} = [-5.927e{-}1, 7.325e{-}1, -3.349e{-}1]^T \ .$$

The difference in the rotation angle is 0.228°; the angle between the rotation axes is 1.682°; the angle between the translation vectors is 10.788°. The translation is not very well estimated. The overlap of the two sets of line segments given by the final motion estimate is shown in Fig. 4.13.

The projection on the ground plane of the 3D reconstruction based on the technique described in Sect. 4.2.4 is displayed on the left in Fig. 4.14. This result should be compared with that reconstructed by our trinocular stereo which uses *three* images and whose geometry has been previously calibrated (see the picture on the right in Fig. 4.14). Essentially, almost the same result can be observed. The two very long, almost vertical line segments in the left

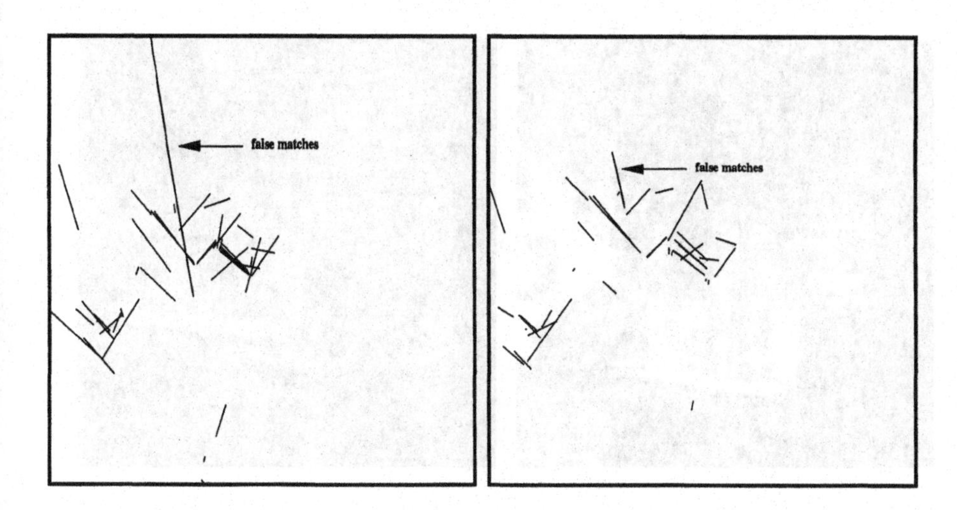

Figure 4.14 Comparison of the 3D reconstruction of the **RobotLab** scene by the proposed structure from motion technique (left) and that by a classical trinocular stereo (right): projection on the ground plane

picture of Fig. 4.14 correspond to two false matches on the floor. As we use the *union* strategy in 3D reconstruction, the reconstruction of the false matches becomes outstanding.

The third set of real data is an image pair of a scene named **Room** (see Fig. 4.15). 90 line segments have been matched by our trinocular stereo system, as shown in Fig. 4.16. One can easily notice two false matches located at the lower right corner near the border of the table.

The same algorithm has been applied to this set of data. Probably because of the gross error made in matching, only six of the ten best samples converge to the good solution. The final motion estimation is:

$$\mathbf{r} = [1.124e{-1}, 1.807e{-1}, 1.850e{-2}]^T \ , \ \mathbf{t} = [-7.939e{-1}, 5.904e{-1}, -1.451e{-1}]^T \ ,$$

while the estimation through stereo calibration is:

$$\mathbf{r} = [9.965e{-2}, 1.584e{-1}, 2.226e{-2}]^T \ , \ \mathbf{t} = [-7.768e{-1}, 6.182e{-1}, -1.198e{-1}]^T \ .$$

The difference in the rotation angle is 1.445°; the angle between the rotation axes is 1.839°; the angle between the translation vectors is 2.366°. The overlap of the two sets of line segments given by the final motion estimate is shown in Fig. 4.17.

Figure 4.15 Image pair of a **Room** scene

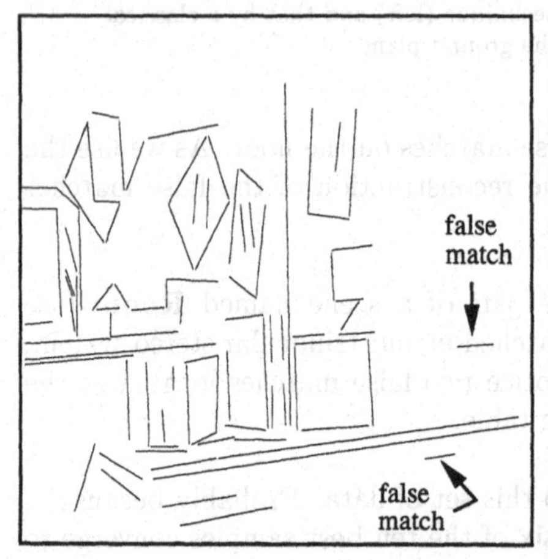

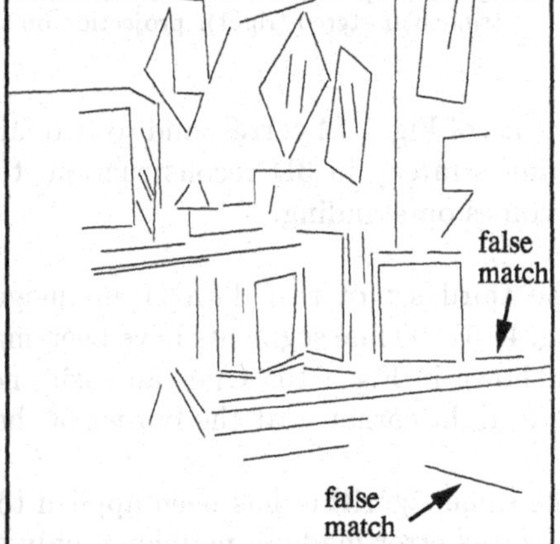

Figure 4.16 Matched line segments of the **Room** image pair

The projection on the ground plane of the 3D line segments reconstructed by our algorithm are shown on the left in Fig. 4.18, while those reconstructed by the trinocular stereo are shown on the right in Fig. 4.18. The reconstruction corresponding to the two false matches is easily identified to be the isolated line segments near the bottom edge of the picture.

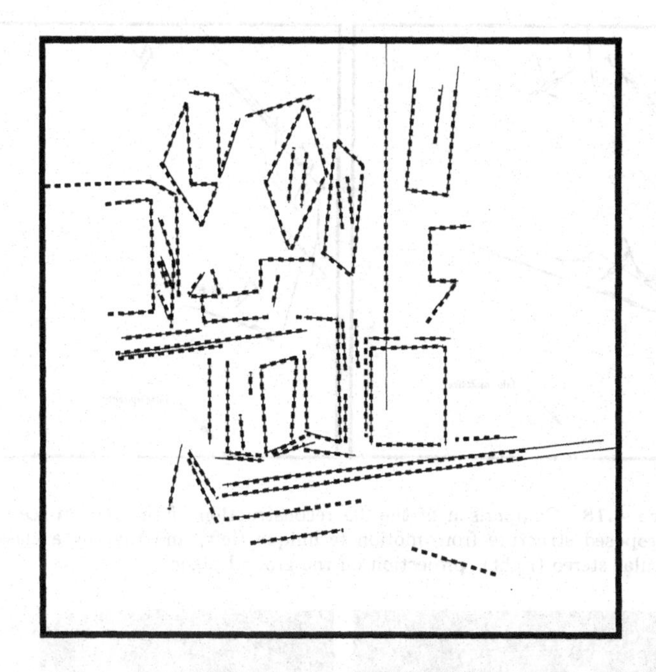

Figure 4.17 Overlap of the two sets of line segments given by the final motion estimate

The fourth set of real data is an image pair of a scene named **Table** because it contains a turntable (see Fig. 4.19). There are 128 line segments matched by the trinocular stereo (see Fig. 4.20). As usual, there exist a few false matches. We also notice that the border of the turntable is differently segmented because of curvature.

Again, we searched for the initial motion estimate by sampling. For each sample, we evaluated the cost function (4.4), and ten best samples were retained for further optimization. Five among these ten samples converge to the good solution. The final motion estimation is:

$$\mathbf{r} = [-8.924e{-3}, 2.194e{-2}, 2.224e{-2}]^T \ , \ \mathbf{t} = [-5.517e{-1}, -7.523e{-1}, -3.601e{-1}]^T$$

while the estimation through stereo calibration is:

$$\mathbf{r} = [-8.520e{-3}, 2.660e{-2}, 2.140e{-2}]^T \ , \ \mathbf{t} = [-5.784e{-1}, -7.355e{-1}, -3.527e{-1}]^T$$

The difference in the rotation angle is 0.155°; the angle between the rotation axes is 6.643°; the angle between the translation vectors is 1.856°. The overlap

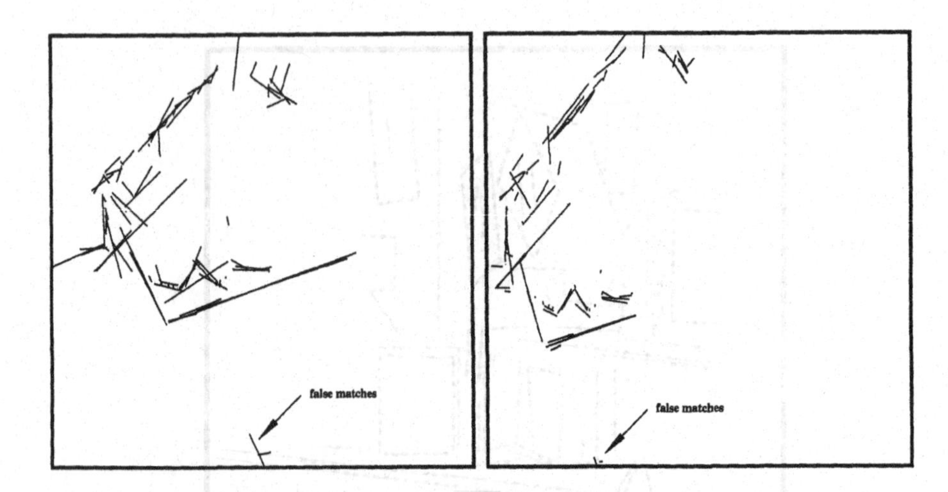

Figure 4.18 Comparison of the 3D reconstruction of the **Room** scene by the proposed structure from motion technique (left) and that by a classical trinocular stereo (right): projection on the ground plane

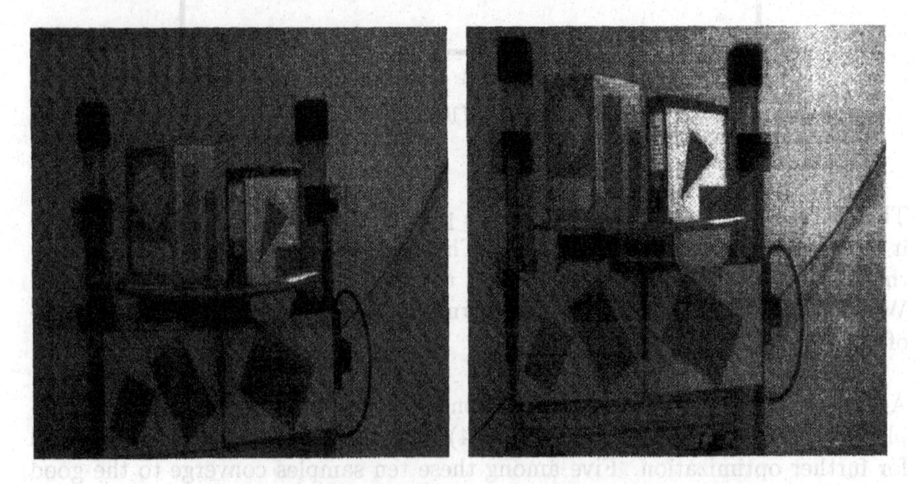

Figure 4.19 Image pair of a **Table** scene

of the two sets of line segments given by the final motion estimate is shown in Fig. 4.21.

The projection on the ground plane of the 3D reconstruction produced by our algorithm is shown on the left in Fig. 4.22, while that produced by the trinocular

Figure 4.20 Matched line segments of the **Table** image pair

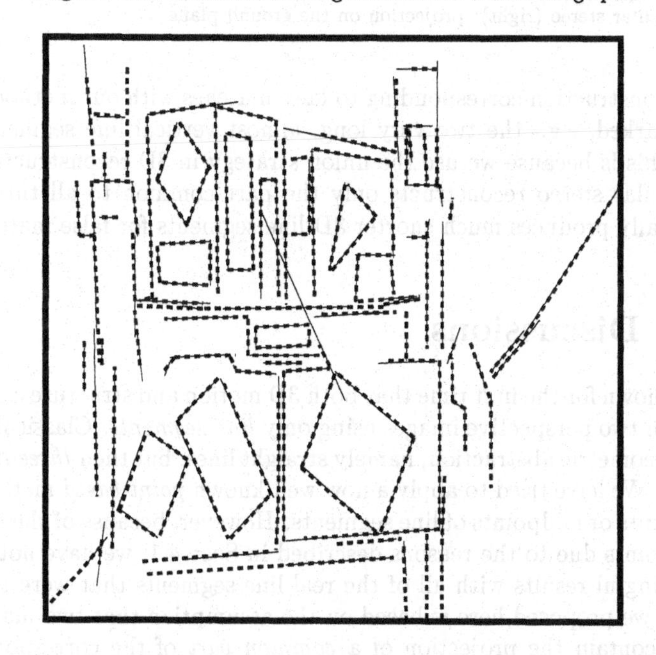

Figure 4.21 Overlap of the two sets of line segments given by the final motion estimate

stereo is shown on the right in Fig. 4.22. The two results are comparable, but

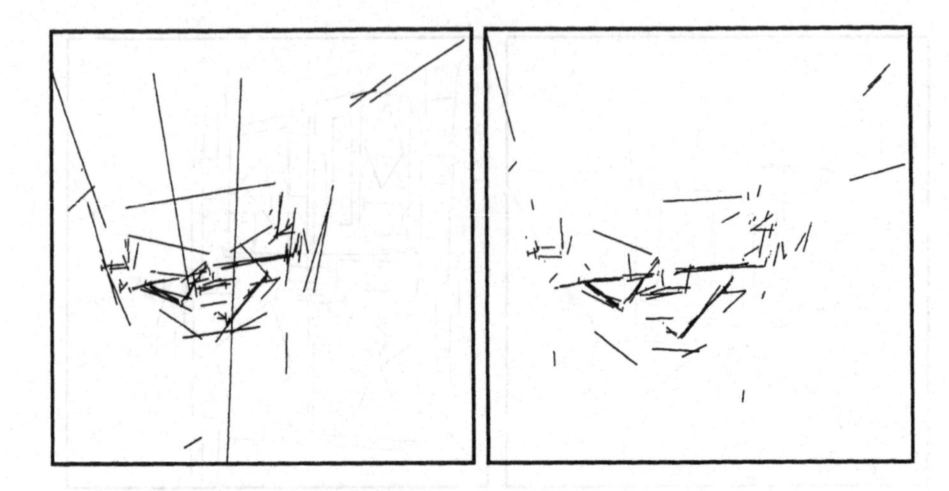

Figure 4.22 Comparison of the 3D reconstruction of the **Table** scene by the proposed structure from motion technique (left) and that by a classical trinocular stereo (right): projection on the ground plane

the 3D reconstruction corresponding to false matches with our method is easier to be remarked, e.g. the two very long, almost vertical line segments in the middle. This is because we use the union strategy in 3D reconstruction, while the trinocular stereo reconstructs only the part common to all three images which usually produces much shorter 3D line segments for false matches.

4.2.6 Discussions

We have shown for the first time that both 3D motion and structure can be computed from two perspective images using only *line segments*. Classical methods use their geometric abstraction, namely straight lines, but then *three images* are necessary. We have tried to apply a now well-known *point-based* method [41] to the endpoints or midpoints of line segments. However, because of the instability of these points due to the reasons described in Sect. 4.1, we have not obtained any meaningful results with all of the real line segments that were tried. The algorithm we proposed here is based on the assumption that two matched line segments contain the projection of a *common part* of the corresponding line segment in space. Indeed, this is what we use to match line segments between different views. This assumption has been implemented through the use of the epipolar geometry, which is of course unknown. Because a closed-form solution is not available, we have proposed a solution which samples the motion space (which is five-dimensional). Both synthetic and real data have been used to test the proposed algorithm, and excellent results have been obtained with real data containing about one hundred line segments. The results are comparable with those obtained with stereo calibration.

As described in Sect. 4.1, by requiring a pair of matched line segments to overlap in space, we add a constraint on the family of feasible motions. It is this constraint that allows the computation of motion from correspondences of two sets of line segments. If we have only a small set of matches, the feasible motion may not be well constrained, and the result will be poor. From our experience, we observe that in order to obtain a usable result, we need a relatively large set of correspondences of line segments (say 50). To alleviate this requirement, we can use points and line segments in combination, and the investigation of this issue is currently under way.

The proposed algorithm tries to find the motion by maximizing the overlap of line segments. This is a heuristic and a biased estimate may be obtained if the variation of the line segments is not random. When a large set of line segments are used, this problem is not severe because the motion is well constrained, as we have seen from the results with real data.

Another problem we encountered with the proposed algorithm, as already mentioned in Sect. 4.2.5, arises when line segments are aligned with the epipolar lines. This is because the overlap is computed from the intersection of line segments with their epipolar lines, and the intersections are instable when they are almost parallel. This case is also well-known to cause problems for binocular stereo. One solution to this would be just to discard the line segments that form a small angle with the epipolar lines.

The proposed algorithm is based on the assumption that the intrinsic parameters of the images are known, but it can easily be extended to estimate the epipolar geometry between two uncalibrated images using line segments. The only problem is that we need to sample a higher parameter space (seven dimensions now) to find an initial estimate.

4.3 DETERMINING EPIPOLAR GEOMETRY OF THREE VIEWS

Traditionally in computer vision, the endpoints of line segments are completely ignored because of their instability, and instead, the straight lines supporting them are considered. Then, at least three views are necessary to recover the motion and structure. We address in this section the epipolar geometry of three views. Unlike the case described in the last section, the intrinsic parameters of the images are assumed not to be known. The reader is referred to [146, 147, 172] for the work on three calibrated images. The reader is referred to [93, 167, 84] for the geometry of multiple uncalibrated images.

4.3.1 Trifocal Constraints for Point Matches

We are given three images, and we assume that no two of the images are taken from the same point in space. Consider first the case where a set of point matches are available, denoted by $\{\mathbf{m}_i, \mathbf{m}'_i, \mathbf{m}''_i\}$ $(i = 1, \ldots, n)$. To derive the constraints on the images between three views, we follow the same idea as that presented in [146] for calibrated images.

Consider one point match $(\mathbf{m}, \mathbf{m}', \mathbf{m}'')$ (we omit here the subscript to simplify the notation). Let the corresponding structure in 3D space be $\mathtt{M}$. In general (full perspective or affine projection), we have

$$s\widetilde{\mathbf{m}} = \mathbf{P}\widetilde{\mathtt{M}}, \tag{4.5}$$

$$s'\widetilde{\mathbf{m}}' = \mathbf{P}'\widetilde{\mathtt{M}}, \tag{4.6}$$

$$s''\widetilde{\mathbf{m}}'' = \mathbf{P}''\widetilde{\mathtt{M}}. \tag{4.7}$$

Using pseudo-inverse matrices, we can get

$$\widetilde{\mathtt{M}} = s\mathbf{P}^{+}\widetilde{\mathbf{m}} + \mathbf{p}^{\perp}. \tag{4.8}$$

Substituting this for (4.6) and (4.7) yields

$$s'\widetilde{\mathbf{m}}' = s\mathbf{P}'\mathbf{P}^{+}\widetilde{\mathbf{m}} + \mathbf{P}'\mathbf{p}^{\perp},$$

$$s''\widetilde{\mathbf{m}}'' = s\mathbf{P}''\mathbf{P}^{+}\widetilde{\mathbf{m}} + \mathbf{P}''\mathbf{p}^{\perp}.$$

Define $\mathbf{B}' \equiv \mathbf{P}'\mathbf{P}^{+}$, $\mathbf{B}'' \equiv \mathbf{P}''\mathbf{P}^{+}$, $\mathbf{b}' \equiv \mathbf{P}'\mathbf{p}^{\perp}$ and $\mathbf{b}'' \equiv \mathbf{P}''\mathbf{p}^{\perp}$. Then we have

$$s'\widetilde{\mathbf{m}}' = s\mathbf{B}'\widetilde{\mathbf{m}} + \mathbf{b}',$$

$$s''\widetilde{\mathbf{m}}'' = s\mathbf{B}''\widetilde{\mathbf{m}} + \mathbf{b}''.$$

To eliminate the unknown structure parameters s' and s'', we take the cross product of the above two equations with $\tilde{\mathbf{m}}'$ and $\tilde{\mathbf{m}}''$, respectively, which leads to

$$s\tilde{\mathbf{m}}' \times (\mathbf{B}'\tilde{\mathbf{m}}) + \tilde{\mathbf{m}}' \times \mathbf{b}' = \mathbf{0} ,$$
$$s\tilde{\mathbf{m}}'' \times (\mathbf{B}''\tilde{\mathbf{m}}) + \tilde{\mathbf{m}}'' \times \mathbf{b}'' = \mathbf{0} .$$

There is still one unknown s. We rearrange the terms of these equations as

$$s[\tilde{\mathbf{m}}']_\times \mathbf{B}'\tilde{\mathbf{m}} = -[\tilde{\mathbf{m}}']_\times \mathbf{b}' ,$$
$$-[\tilde{\mathbf{m}}'']_\times \mathbf{b}'' = s[\tilde{\mathbf{m}}'']_\times \mathbf{B}''\tilde{\mathbf{m}} .$$

Remember that $[\cdot]_\times$ denotes an antisymmetric matrix defined by a vector. To eliminate s, we take the outer product[8] of both sides of the above two equations, which gives

$$[\tilde{\mathbf{m}}']_\times \mathbf{B}'\tilde{\mathbf{m}}\mathbf{b}''^T[\tilde{\mathbf{m}}'']_\times = [\tilde{\mathbf{m}}']_\times \mathbf{b}'\tilde{\mathbf{m}}^T\mathbf{B}''^T[\tilde{\mathbf{m}}'']_\times ,$$

or

$$[\tilde{\mathbf{m}}']_\times \mathbf{G}(\tilde{\mathbf{m}})[\tilde{\mathbf{m}}'']_\times = \mathbf{O}_3 ,$$

where $\mathbf{O}_3$ is a 3×3 zero matrix, and

$$\mathbf{G}(\tilde{\mathbf{m}}) \equiv \mathbf{b}'\tilde{\mathbf{m}}^T\mathbf{B}''^T - \mathbf{B}'\tilde{\mathbf{m}}\mathbf{b}''^T .$$

If $\tilde{\mathbf{m}} = [u, v, t]^T$ (in general $t = 1$), then $\mathbf{G}(\tilde{\mathbf{m}})$ can be expressed as the sum of three matrices:

$$\mathbf{G}(\tilde{\mathbf{m}}) = u\mathbf{K} + v\mathbf{L} + t\mathbf{M}$$

with

$$\mathbf{K} = \mathbf{b}'\mathbf{b}_1''^T - \mathbf{b}_1'\mathbf{b}''^T , \tag{4.9}$$
$$\mathbf{L} = \mathbf{b}'\mathbf{b}_2''^T - \mathbf{b}_2'\mathbf{b}''^T , \tag{4.10}$$
$$\mathbf{M} = \mathbf{b}'\mathbf{b}_3''^T - \mathbf{b}_3'\mathbf{b}''^T . \tag{4.11}$$

Here, $\mathbf{b}_i'$ is the ith column vector of $\mathbf{B}'$ (i.e. $[\mathbf{b}_1', \mathbf{b}_2', \mathbf{b}_3'] \equiv \mathbf{B}'$]); similarly for $\mathbf{b}_i''$ (i.e. $[\mathbf{b}_1'', \mathbf{b}_2'', \mathbf{b}_3''] \equiv \mathbf{B}''$]). It is easy to see that the three matrices $\mathbf{K}$, $\mathbf{L}$, and $\mathbf{M}$ are all singular. By defining the following operation:

$$(\mathbf{K}, \mathbf{L}, \mathbf{M}) * \tilde{\mathbf{m}} = u\mathbf{K} + v\mathbf{L} + t\mathbf{M} , \tag{4.12}$$

[8] The outer product of two vectors $\mathbf{x}$ and $\mathbf{y}$ is defined as $\mathbf{x}\mathbf{y}^T$.

we finally obtain the following matrix equation for points between three views

$$[\widetilde{\mathbf{m}}']_\times[(\mathbf{K},\mathbf{L},\mathbf{M}) * \widetilde{\mathbf{m}}][\widetilde{\mathbf{m}}'']_\times = \mathbf{O}_3 \,. \tag{4.13}$$

We have therefore a set of nine equations, of which not all are independent.

Indeed, there are only four independent equations. Let $\widetilde{\mathbf{m}} = [u,v,t]^T$, $\widetilde{\mathbf{m}}' = [u',v',t']^T$, $\widetilde{\mathbf{m}}'' = [u'',v'',t'']^T$. Let $\boldsymbol{\tau}_{ij} = [K_{ij}, L_{ij}, M_{ij}]^T$ where K_{ij}, L_{ij} and M_{ij} are respectively the (i,j) element of matrices $\mathbf{K}$, $\mathbf{L}$ and $\mathbf{M}$. Then, (4.13) can be expanded as the following nine scalar equations:

$$-t't''\widetilde{\mathbf{m}}^T\boldsymbol{\tau}_{22} + t'v''\widetilde{\mathbf{m}}^T\boldsymbol{\tau}_{23} + v't''\widetilde{\mathbf{m}}^T\boldsymbol{\tau}_{32} - v'v''\widetilde{\mathbf{m}}^T\boldsymbol{\tau}_{33} = 0 \tag{4.14}$$

$$t't''\widetilde{\mathbf{m}}^T\boldsymbol{\tau}_{21} - t'u''\widetilde{\mathbf{m}}^T\boldsymbol{\tau}_{23} - v't''\widetilde{\mathbf{m}}^T\boldsymbol{\tau}_{31} + v'u''\widetilde{\mathbf{m}}^T\boldsymbol{\tau}_{33} = 0 \tag{4.15}$$

$$-t'v''\widetilde{\mathbf{m}}^T\boldsymbol{\tau}_{21} + t'u''\widetilde{\mathbf{m}}^T\boldsymbol{\tau}_{22} + v'v''\widetilde{\mathbf{m}}^T\boldsymbol{\tau}_{31} - v'u''\widetilde{\mathbf{m}}^T\boldsymbol{\tau}_{32} = 0 \tag{4.16}$$

$$t't''\widetilde{\mathbf{m}}^T\boldsymbol{\tau}_{12} - t'v''\widetilde{\mathbf{m}}^T\boldsymbol{\tau}_{13} - u't''\widetilde{\mathbf{m}}^T\boldsymbol{\tau}_{32} + u'v''\widetilde{\mathbf{m}}^T\boldsymbol{\tau}_{33} = 0 \tag{4.17}$$

$$-t't''\widetilde{\mathbf{m}}^T\boldsymbol{\tau}_{11} + t'u''\widetilde{\mathbf{m}}^T\boldsymbol{\tau}_{13} + u't''\widetilde{\mathbf{m}}^T\boldsymbol{\tau}_{31} - u'u''\widetilde{\mathbf{m}}^T\boldsymbol{\tau}_{33} = 0 \tag{4.18}$$

$$t'v''\widetilde{\mathbf{m}}^T\boldsymbol{\tau}_{11} - t'u''\widetilde{\mathbf{m}}^T\boldsymbol{\tau}_{12} - u'v''\widetilde{\mathbf{m}}^T\boldsymbol{\tau}_{31} + u'u''\widetilde{\mathbf{m}}^T\boldsymbol{\tau}_{32} = 0 \tag{4.19}$$

$$-v't''\widetilde{\mathbf{m}}^T\boldsymbol{\tau}_{12} + v'v''\widetilde{\mathbf{m}}^T\boldsymbol{\tau}_{13} + u't''\widetilde{\mathbf{m}}^T\boldsymbol{\tau}_{22} - u'v''\widetilde{\mathbf{m}}^T\boldsymbol{\tau}_{23} = 0 \tag{4.20}$$

$$v't''\widetilde{\mathbf{m}}^T\boldsymbol{\tau}_{11} - v'u''\widetilde{\mathbf{m}}^T\boldsymbol{\tau}_{13} - u't''\widetilde{\mathbf{m}}^T\boldsymbol{\tau}_{21} + u'u''\widetilde{\mathbf{m}}^T\boldsymbol{\tau}_{23} = 0 \tag{4.21}$$

$$-v'v''\widetilde{\mathbf{m}}^T\boldsymbol{\tau}_{11} + v'u''\widetilde{\mathbf{m}}^T\boldsymbol{\tau}_{12} + u'v''\widetilde{\mathbf{m}}^T\boldsymbol{\tau}_{21} - u'u''\widetilde{\mathbf{m}}^T\boldsymbol{\tau}_{22} = 0 \,. \tag{4.22}$$

Consider for example equations (4.14), (4.15), (4.17) and (4.18). It is easy to see that the terms corresponding to $\boldsymbol{\tau}_{22}$, $\boldsymbol{\tau}_{21}$, $\boldsymbol{\tau}_{12}$ and $\boldsymbol{\tau}_{11}$ appear only in one of the four equations, which implies that the four equations are linearly independent in terms of the 27 elements of $\mathbf{K}$, $\mathbf{L}$ and $\mathbf{M}$. The remaining five equations are linear combinations of these four equations:

$$\begin{aligned}
&(4.16) = -(u''(4.14) + v''(4.15))/t''\\
&(4.19) = -(u''(4.17) + v''(4.18))/t''\\
&(4.20) = -(u'(4.14) + v'(4.17))/t'\\
&(4.21) = -(u'(4.15) + v'(4.18))/t'\\
&(4.22) = (u'u''(4.14) + u'v''(4.15) + v'u''(4.17) + v'v''(4.18))/(t't'') \,,
\end{aligned}$$

where the numbers in parenthesis represent the equation numbers. The fact that there exist 4 independent equations between three views is noted by Shashua [139]. The 9 equations (4.13) are linear in each coordinate of the triplet of points $(\mathbf{m}, \mathbf{m}', \mathbf{m}'')$, and are called the *trilinearity relationships* by Shashua. Hartley [57] calls $\mathbf{K}$, $\mathbf{L}$, and $\mathbf{M}$ the *trifocal tensor*. There are 27 elements, but are defined up to a *common* scale factor.

It is interesting to note that we have four independent equations *only* in the 27 elements of matrices $\mathbf{K}$, $\mathbf{L}$, and $\mathbf{M}$. Algebraically, there are only *three* independent equations. Indeed, from the pinhole model, a point in each image provides two equations with one unknown scalar, thus a triplet of points provide in total six equations which contain three unknowns. Removing the three unknowns yields three algebraically independent equations.

Trifocal Constraints for Pinhole Cameras

We consider the case that the three views are taken by pinhole cameras. As described in the last chapter, we can only reconstruct the scene up to an arbitrary projective transformation. Without loss of generality, we can assume that the first camera projection matrix is of the form $\mathbf{P} = [\mathbf{I} \mid \mathbf{0}]$, where $\mathbf{I}$ is a 3×3 identity matrix and $\mathbf{0}$ is a zero vector. Then, from the definitions of $\mathbf{B}'$, $\mathbf{B}''$, $\mathbf{b}'$ and $\mathbf{b}''$ given before, we say that the second and third camera projection matrices are defined by $\mathbf{P}' = [\mathbf{B}' \mid \mathbf{b}']$ and $\mathbf{P}'' = [\mathbf{B}'' \mid \mathbf{b}'']$. Therefore, the trifocal constraint for pinhole cameras, if they are expressed in the preceding form, has exactly the same equation as that of (4.13).

Trifocal Constraints for Affine Cameras

If the three views are taken by affine cameras, then the trifocal tensors can be simplified, and the trilinear equations become linear.

Recall that for affine cameras, all of $\mathbf{P}$, $\mathbf{P}'$ and $\mathbf{P}''$ are in the form of

$$\begin{bmatrix} \mathbf{p}_1^T & \\ \mathbf{p}_2^T & \mathbf{p}_4 \\ \mathbf{0}_3^T & \end{bmatrix},$$

$\mathbf{P}^+$ in the form of

$$\begin{bmatrix} & \mathbf{Q} & \\ 0 & 0 & \frac{1}{P_{34}} \end{bmatrix},$$

where $\mathbf{Q}$ is a 3×3 matrix, and $\mathbf{p}^\perp$ in the form of $[* \;\; * \;\; * \;\; 0]^T$, where $*$ indicates values that are generally non-zeros. Now it is easy to see that $\mathbf{B}'$ and

$\mathbf{B}''$ are in the form of

$$\begin{bmatrix} * & * & * \\ * & * & * \\ 0 & 0 & * \end{bmatrix},$$

and $\mathbf{b}'$ and $\mathbf{b}''$ in the form of $[* \;\; * \;\; 0]^T$. That is, the third elements of $\mathbf{b}'$, $\mathbf{b}''$, $\mathbf{b}'_1$, $\mathbf{b}'_2$, $\mathbf{b}''_1$ and $\mathbf{b}''_2$ are all zero. From this, we can show that $\mathbf{K}$ and $\mathbf{L}$ are both in the form of

$$\begin{bmatrix} * & * & 0 \\ * & * & 0 \\ 0 & 0 & 0 \end{bmatrix}$$

and $\mathbf{M}$ is in the form of

$$\begin{bmatrix} * & * & * \\ * & * & * \\ * & * & 0 \end{bmatrix}.$$

Substituting these for equations (4.14), (4.15), (4.17) and (4.18), we have

$$\begin{aligned} -t't''\tilde{\mathbf{m}}^T\boldsymbol{\tau}_{22} + t'v''tM_{23} + v't''tM_{32} &= 0\,, & (4.23)\\ t't''\tilde{\mathbf{m}}^T\boldsymbol{\tau}_{21} - t'u''tM_{23} - v't''tM_{31} &= 0\,, & (4.24)\\ t't''\tilde{\mathbf{m}}^T\boldsymbol{\tau}_{12} - t'v''tM_{13} - u't''tM_{32} &= 0\,, & (4.25)\\ -t't''\tilde{\mathbf{m}}^T\boldsymbol{\tau}_{11} + t'u''tM_{13} + u't''tM_{31} &= 0\,. & (4.26) \end{aligned}$$

Setting $t = t' = t'' = 1$ in the above equations, they become

$$\begin{aligned} -uK_{22} - vL_{22} + v''M_{23} + v'M_{32} - M_{22} &= 0\,, & (4.27)\\ uK_{21} + vL_{21} - u''M_{23} - v'M_{31} + M_{21} &= 0\,, & (4.28)\\ uK_{12} + vL_{12} - v''M_{13} - u'M_{32} + M_{12} &= 0\,, & (4.29)\\ -uK_{11} - vL_{11} + u''M_{13} + u'M_{31} - M_{11} &= 0\,. & (4.30) \end{aligned}$$

It is easy to see that all the 4 equations are linear in the image coordinates. There are 16 unknowns in the 4 equations. Thus, given 4 triples of point matches, all the trifocal tensors can be recovered.

Actually, from these 4 equations, we can represent (u, v) by linear combinations of (u', v') and (u'', v''), if the trifocal tensors are given. This means that, if we are given two images and the trifocal tensor, then we can generate a *virtual image*. This can be applied to image synthesis or to 3D object recognition.

Actually, this can be understood as an extension of Ullman and Basri's linear combination theory from orthographic projection or weak perspective projection to affine cameras, a broader class of projection. For application of using linear combination of model views to 3D object recognition, see Chapter 8.

4.3.2 Trifocal Constraints for Line Correspondences

We now consider line correspondences between three views, denoted by $\{\mathbf{l}_i, \mathbf{l}'_i, \mathbf{l}''_i\}$ $(i = 1, \ldots, n)$. A 2D line $\mathbf{l}$ can be represented by a point $\mathbf{p}$ and the direction vector $\mathbf{u}$. Then, any point on line $\mathbf{l}$ can be expressed in parametric form as $\mathbf{m} = \mathbf{p} + \alpha\mathbf{u}$, where α is the parameter which controls the position of the represented point on the line. Another representation for a line is to use a 3D vector $\mathbf{n}$ such that $\mathbf{n}^T\widetilde{\mathbf{m}} = 0$ for any point $\mathbf{m}$ on the line. It is evident that vector $\mathbf{n}$ is only defined up to a scalar factor. In particular, we have $\mathbf{n} = \widetilde{\mathbf{p}} \times \breve{\mathbf{u}}$ where $\mathbf{p}$ is any point on the line and $\breve{\mathbf{u}} = [u_1, u_2, 0]^T$ if $\mathbf{u} = [u_1, u_2]^T$ [9].

For a given line correspondence $(\mathbf{l}, \mathbf{l}', \mathbf{l}'')$ (we omit here the subscript to simplify the notation), consider a point $\mathbf{p}$ on line $\mathbf{l}$. Its corresponding points on $\mathbf{l}'$ and $\mathbf{l}''$ should be $\mathbf{m}'$ and $\mathbf{m}''$, which are given by

$$\mathbf{m}' = \mathbf{p}' + \alpha'\mathbf{u}' \quad \text{and} \quad \mathbf{m}'' = \mathbf{p}'' + \alpha''\mathbf{u}''$$

for some undeterminate real numbers α' and α''. From (4.13), we have

$$[\widetilde{\mathbf{p}}' + \alpha'\breve{\mathbf{u}}']_\times [(\mathbf{K}, \mathbf{L}, \mathbf{M}) * \widetilde{\mathbf{p}}][\widetilde{\mathbf{p}}'' + \alpha''\breve{\mathbf{u}}'']_\times = \mathbf{O} .$$

Since α' and α'' are unknowns we eliminate them by pre- and post-multiplying the above equation by $\breve{\mathbf{u}}'^T$ and $\breve{\mathbf{u}}''$:

$$(\widetilde{\mathbf{p}}' \times \breve{\mathbf{u}}')^T [(\mathbf{K}, \mathbf{L}, \mathbf{M}) * \widetilde{\mathbf{p}}](\widetilde{\mathbf{p}}'' \times \breve{\mathbf{u}}'') = 0 ,$$

or

$$\mathbf{n}'^T [(\mathbf{K}, \mathbf{L}, \mathbf{M}) * \widetilde{\mathbf{p}}]\mathbf{n}'' \equiv \begin{bmatrix} \mathbf{n}'^T\mathbf{K}\mathbf{n}'' \\ \mathbf{n}'^T\mathbf{L}\mathbf{n}'' \\ \mathbf{n}'^T\mathbf{M}\mathbf{n}'' \end{bmatrix}^T \widetilde{\mathbf{p}} = 0 , \tag{4.31}$$

where $\mathbf{n}'$ and $\mathbf{n}''$ are the vectors representing lines $\mathbf{l}'$ and $\mathbf{l}''$. We have thus one scalar equation.

[9]Note the difference in notation between $\widetilde{}$ and $\breve{}$. The former is used for a point by adding 1 as the last element, while the latter is used for a direction vector by adding 0 as the last element.

However, equation (4.31) must be true for every point $\mathbf{m} = \mathbf{p} + \alpha\mathbf{u}$ on line $\mathbf{l}$. This implies

$$\begin{bmatrix} \mathbf{n}'^T\mathbf{K}\mathbf{n}'' \\ \mathbf{n}'^T\mathbf{L}\mathbf{n}'' \\ \mathbf{n}'^T\mathbf{M}\mathbf{n}'' \end{bmatrix}^T (\tilde{\mathbf{p}} + \alpha\tilde{\mathbf{u}}) = 0 \qquad \forall\alpha\,.$$

Therefore, the first vector is perpendicular to all vector $(\tilde{\mathbf{p}} + \alpha\tilde{\mathbf{u}})$, which is equivalent to say that it is parallel to vector $\mathbf{n} = \tilde{\mathbf{p}} \times \tilde{\mathbf{u}}$. Its cross product with $\mathbf{n}$ must be zero:

$$\begin{bmatrix} \mathbf{n}'^T\mathbf{K}\mathbf{n}'' \\ \mathbf{n}'^T\mathbf{L}\mathbf{n}'' \\ \mathbf{n}'^T\mathbf{M}\mathbf{n}'' \end{bmatrix} \times \mathbf{n} = \mathbf{0}\,. \tag{4.32}$$

This is a vector equation that is equivalent to three scalar ones, only two of which are independent.

4.3.3 Linear Estimation of K, L, and M Using Points and Lines

As described previously, each point match provides four scalar equations, and each line correspondence provides two scalar equations. Although they are nonlinear in terms of the camera parameters, they are linear in the 27 elements of the three 3×3 matrices $\mathbf{K}$, $\mathbf{L}$, and $\mathbf{M}$. We arrange the elements of these three matrices in a 27×1 vector $\mathbf{x}$ which is the column vector of the unknowns. For the ith point match, we can construct a 9×27 matrix of coefficients of the unknowns in (4.13), denoted by $\mathbf{N}_i$, and equation (4.13) becomes $\mathbf{N}_i\mathbf{x} = \mathbf{0}$. For the jth line correspondence, we can construct a 3×27 matrix of coefficients of the unknowns in (4.32), denoted by $\mathbf{Q}_j$, and equation (4.13) becomes $\mathbf{Q}_j\mathbf{x} = \mathbf{0}$. These two types of equations can be combined together to estimate the vector $\mathbf{x}$. Remember that $\mathbf{x}$ is defined up to a scale factor, and we can impose the constraint $\|\mathbf{x}\| = 1$. Let m and n be the number of point matches and that of line correspondences. If we have $4m+2n \geq 26$, we can estimate $\mathbf{x}$ by minimizing

$$\mathcal{F}(\mathbf{x}) = \sum_{i=1}^{m} w_{pi}\mathbf{x}^T\mathbf{N}_i^T\mathbf{N}_i\mathbf{x} + \sum_{j=1}^{n} w_{lj}\mathbf{x}^T\mathbf{Q}_j^T\mathbf{Q}_j\mathbf{x} \equiv \mathbf{x}^T\mathbf{N}\mathbf{x}$$

subject to $\|\mathbf{x}\| = 1$. Here $\mathbf{N} = \sum_i w_{pi}\mathbf{N}_i^T\mathbf{N}_i + \sum_j w_{lj}\mathbf{Q}_j^T\mathbf{Q}_j$ is a 27×27 symmetric matrix, and w_p and w_l are the weights for points and lines. The solution is readily obtained using, for example, Lagrange multipliers (see e.g. Sect. 3.1.2), and is the unit eigenvector of $\mathbf{N}$ associated to the smallest eigenvalue.

The minimum number of points and lines, based on $4m + 2n \geq 26$, required for a unique solution of **K**, **L**, and **M** is given in Table 4.1. We thus need at least 13 lines if there are no point matches, and at least 7 points if points are used alone.

Table 4.1 Minimum point and line matches required for solving trifocal tensors

m	0	1	2	3	4	5	6	7
n	13	11	9	7	5	3	1	0

Note that although this technique is simple, it is not the best in the presence of noise. Indeed, $\mathbf{x}$ is not any vector having unit norm. Its elements are those of the matrices **K**, **L**, and **M** which are all singular.

One open question is how to choose the appropriate weights w_p and w_l. This problem needs to be studied carefully before developing a real optimal solution. Part of the answer can be found in [146].

4.3.4 Determining Camera Projection Matrices

We discuss in this section the techniques for recovering the camera projection matrices from matches between three views.

Closed-Form Solution

We assume that matrices **K**, **L**, and **M** have been estimated by the technique described in the last section. The presentation is largely based on the work described in [145, 59].

Recovering $\mathbf{b}'$ and $\mathbf{b}''$. The matrices **K**, **L**, and **M** are all singular, and equations (4.9) to (4.11) can be rewritten as

$$\mathbf{K} = [\mathbf{b}' \ \mathbf{b}'_1] \begin{bmatrix} \mathbf{b}_1''^T \\ -\mathbf{b}''^T \end{bmatrix}$$

$$\mathbf{L} = [\mathbf{b}' \ \mathbf{b}'_2] \begin{bmatrix} \mathbf{b}_2''^T \\ -\mathbf{b}''^T \end{bmatrix}$$

$$\mathbf{M} = [\mathbf{b}' \ \mathbf{b}'_3] \begin{bmatrix} \mathbf{b}_3''^T \\ -\mathbf{b}''^T \end{bmatrix} .$$

Assume that there is only one zero singular value for each matrix **K**, **L**, and **M**. Through SVD, we obtain the left and right null-spaces of these matrices, which are denoted by $(\mathbf{u}_K, \mathbf{v}_K)$, $(\mathbf{u}_L, \mathbf{v}_L)$, and $(\mathbf{u}_M, \mathbf{v}_M)$, respectively. Following the definition of null-space, we have $\mathbf{K}\mathbf{v}_K = \mathbf{0}$, which implies that $\mathbf{v}_K$ is perpendicular to $\mathbf{b}_1''$ and $\mathbf{b}''$ (i.e. $\mathbf{b}_1''^T\mathbf{v}_K = 0$, $\mathbf{b}''^T\mathbf{v}_K = 0$). In the same way, $\mathbf{v}_L$ is perpendicular to $\mathbf{b}_2''$ and $\mathbf{b}''$, and $\mathbf{v}_M$ is perpendicular to $\mathbf{b}_3''$ and $\mathbf{b}''$. This leads to the following observation: $\mathbf{b}''$ must be the common perpendicular to $\mathbf{v}_K, \mathbf{v}_L$ and $\mathbf{v}_M$. Let

$$\mathbf{V} = [\mathbf{v}_K \quad \mathbf{v}_L \quad \mathbf{v}_M] .$$

In the noise-free case, we must have $\mathbf{V}^T\mathbf{b}'' = \mathbf{0}$. Therefore, we can estimate $\mathbf{b}''$ by minimizing $\|\mathbf{V}^T\mathbf{b}''\|^2$. Since $\mathbf{b}''$ is defined up to a scale factor, we can impose $\|\mathbf{b}''\| = 1$. And the solution for $\mathbf{b}''$ is then the unit eigenvector of $\mathbf{V}\mathbf{V}^T$ associated to the smallest eigenvalue. Similarly, $\mathbf{b}'$ can be estimated as the unit eigenvector of $\mathbf{U}\mathbf{U}^T$ associated to the smallest eigenvalue, where $\mathbf{U} = [\mathbf{u}_K, \mathbf{u}_L, \mathbf{u}_M]$ is composed of the left null-spaces of **K**, **L**, and **M**. The reader is referred to [145, appendix II] for the discussion when one of the three matrices has a second singular value that is zero.

Recovering the remaining entries. Fron (4.9) to (4.11), it is not difficult to see that if we replace $\mathbf{b}'_i$ by $\mathbf{b}'_i + \alpha_i\mathbf{b}'$ and $\mathbf{b}''_i$ by $\mathbf{b}''_i + \alpha_i\mathbf{b}''$ for the same α_i $(i = 1, 2, 3)$, then we get the same matrices **K**, **L**, and **M**. To remove this ambiguity, we can impose that $\mathbf{b}_i'^T\mathbf{b}' = 0$ for $i = 1, 2, 3$[10]. Multiplying $\mathbf{K}^T$ by $\mathbf{b}'$ from the right gives $\mathbf{b}_1''$. Similarly, we have $\mathbf{b}_2'' = \mathbf{L}^T\mathbf{b}'$ and $\mathbf{b}_3'' = \mathbf{M}^T\mathbf{b}'$. This means that

$$\mathbf{P}'' = [\mathbf{K}^T\mathbf{b}' \quad \mathbf{L}^T\mathbf{b}' \quad \mathbf{M}^T\mathbf{b}' \ \mathbf{b}'] .$$

Now substituting $\mathbf{b}_1'' = \mathbf{K}^T\mathbf{b}'$ into (4.9) gives $\mathbf{b}'_1\mathbf{b}''^T = (\mathbf{b}'\mathbf{b}'^T - \mathbf{I})\mathbf{K}$. Finally, multiplying it by $\mathbf{b}''$ yields $\mathbf{b}'_1 = (\mathbf{b}'\mathbf{b}'^T - \mathbf{I})\mathbf{K}\mathbf{b}''$. Similarly, we can compute

[10] We can impose this constraint on $\mathbf{b}_i''$. But it can be only imposed once, either on $\mathbf{b}'_i$ or $\mathbf{b}''_i$, because the coefficient α_i should be the same for $\mathbf{b}'_i$ and $\mathbf{b}''_i$.

$\mathbf{b}'_2$ and $\mathbf{b}'_3$. In summary, we have

$$\mathbf{P}' = [-\mathbf{QKb}'' \quad -\mathbf{QLb}'' \quad -\mathbf{QMb}''] ,$$

where $\mathbf{Q} = \mathbf{I} - \mathbf{b}'\mathbf{b}'^T$.

Nonlinear Optimization

Although the closed-form solution is simple and nice, it is not the best in the presence of noise. In Sect. 4.3.3, we estimate a 27-vector $\mathbf{x}$, but it is not just any vector. Its elements are those of three singular matrices. The closed-form solution is only possible by ignoring these constraints. Hartley [59] reported that the closed-form solution is very unstable in the presence of noise.

One way is to estimate directly the camera projection matrices $\mathbf{P}'$ and $\mathbf{P}''$ by minimizing (4.13) and (4.32). This is a nonlinear minimization and requires an initial guess which can be provided by the closed-form solution. As described previously, $\mathbf{P}'$ and $\mathbf{P}''$ can not be determined uniquely, so the minimization should be conducted subject to the following constraints: $\|\mathbf{b}'\| = 1$, $\|\mathbf{b}''\| = 1$, and $\mathbf{b}_i'^T\mathbf{b}' = 0$ for $i = 1, 2, 3$.

4.3.5 Image Transfer

Besides the importance of the trifocal constraints (4.13) and (4.32) in estimating the camera projection matrices and performing the 3D reconstruction, they are very useful in image synthesis. Consider (4.13). It is equivalent to

$$\tilde{\mathbf{m}}'' \cong [\tilde{\mathbf{m}}']_\times[(\mathbf{K}, \mathbf{L}, \mathbf{M}) * \tilde{\mathbf{m}}] .$$

Given one point in two images ($\mathbf{m}$ and $\mathbf{m}'$), the above equation allows us to determine the position of the corresponding point in a third *virtual* image by choosing an appropriate camera projection matrix $\mathbf{P}''$ (which determines matrices $\mathbf{K}$, $\mathbf{L}$, and $\mathbf{M}$). It is thus possible to generate a new image from two given images using the trilinearities. The reader can find such applications in [84].

Similarly, (4.32) can be rewritten as

$$\mathbf{n} \cong \begin{bmatrix} \mathbf{n}'^T\mathbf{K}\mathbf{n}'' \\ \mathbf{n}'^T\mathbf{L}\mathbf{n}'' \\ \mathbf{n}'^T\mathbf{M}\mathbf{n}'' \end{bmatrix} .$$

Knowing two images of a line, one can deduce the projection of this line in the third image. Thus, the trilinearities allow us to transfer lines, too.

The trifocal constraints play also a key role in *trinocular* stereo matching.

4.4 SUMMARY

Line segments are very important geometric primitives in Computer Vision. However, they are traditionally treated as straight lines, because their endpoints are not reliable due to a variety of sources. Nevertheless, the endpoints still contain rich information, which tell us the longitudinal position of the observed line segment along the infinite support line. We have shown in the first part of this chapter that two views are generally enough to recover the motion and structure of the scene. The only assumption we used is that two matched line segments contain the projection of a *common part* of the corresponding line segment in space (and we say that the two 2D line segments *overlap*). Indeed, this assumption is minimal, and is what we already use in matching line segments between different views. The analysis has been done for calibrated images, but it applies equally well to uncalibrated images. This is because the overlap assumption has been implemented through the use of the epipolar geometry, which is of course unknown. Both synthetic and real data have been used to test the proposed algorithm, and excellent results have been obtained with real data containing about one hundred line segments. The results are comparable with those obtained with stereo calibration.

The second part of this chapter deals with the geometric constraints between three views, which are called the trifocal constraints. The motivation is that three views are required to recover the motion and structure if we ignore the endpoints of line segments, that is, if we use infinite support lines. However, we have derived trifocal constraints for both points and lines, and have presented a unified algorithm for determining the epipolar geometry between three views from either points alone, or straight lines alone, or combination of points and lines. The trifocal constraints are not only important in motion and structure recovery, but also useful in image synthesis and stereo matching.

5

REDEFINING STEREO, MOTION AND OBJECT RECOGNITION VIA EPIPOLAR GEOMETRY

In this chapter, we review correspondence problems in stereo, motion, and object recognition from the epipolar geometry point of view, and the analysis shows that once the epipolar geometry is recovered, all can be redefined as a 1D correspondence search problem plus a segmentation problem which can be solved simultaneously.

5.1 CONVENTIONAL APPROACHES TO STEREO, MOTION AND OBJECT RECOGNITION

In this section, we review conventional approaches to these three old problems.

5.1.1 Stereo

Stereo is one of the earliest problems treated in computer vision [97, 50]. The epipolar constraint is well known from the beginning and was used to ease the difficulty of matching. Most algorithms assume that the epipolar lines are given *a priori*, and thus pose the stereo matching problem as a 1D search problem.

The classical technique for obtaining the epipolar geometry is to estimate the perspective projection matrix for each camera using an apparatus with known shape and size, and then compute the epipolar geometry from the projection matrices [40, 163, 160, 41]. This is a fastidious task. More often, two cameras are mechanically set to have parallel optical axes such that the epipolar lines are horizontal in both images. By employing this camera configuration, one can be relieved from thinking about the somewhat troublesome calibration problem, and can be concentrated on the handsome correspondence problem. For this, most stereo algorithms assume this imaging geometry [50] and most standard stereo image pairs in well-known and widely accessed databases are taken by carefully arranging the camera(s) this way.

Here we would like to point out that, first, human eyes do not work this way [42], rather, they do correspondence using *vergence*vergence, thus the angle between the two eyes is not zero and is constantly changing. Second, even if we try carefully to arrange the imaging geometry that way, there is still error, and usually the corresponding points are not strictly on the same horizontal lines.

Therefore, in the general case, calibration is necessary to recover the epipolar geometry accurately. As we have seen in the previous chapters, the epipolar geometry can be recovered from two uncalibrated images alone without use of any calibration apparatus, and once this is done, we say that the two images are *weakly calibrated* [126, 186]. Recently, an automatic and robust technique has been developed [188], and we will present it in Chap. 6.

Now let us have a look at the correspondence problem given the epipolar lines. The problem can be posed in several ways. One can find matches for edge points, or for outstanding feature points, or for edge lines and contours, or for every image pixel which is visible in both images.

The most troublesome problem here, which is actually everywhere in any correspondence problem, is matching ambiguity, i.e. for a given feature in one image, there are multiple candidates in the other which can be paired with this feature. Though in different forms, all algorithms use one common constraint, that is disparity *continuity* or *smoothness*, to choose one from among the multiple candidates [48, 118].

5.1.2 Motion

Motion is also one of the oldest problems in computer vision. Ullman's pioneering work on this problem has great influence on later research. [164]. The problem has been divided into two smaller ones. The first is *correspondence*, which decides how points "flow" between consecutive images in the image sequence. The second is called *structure-from-motion*, which determines the 3D structure and motion of the objects from matched points in the images. Most of the past research tries to solve one of the above two problems.

For the correspondence problem, the difficulty is considered to be the *aperture problem*aperture problem, that is, in motion, different from stereo, the displacement is not *a priori* considered constrained to be one-dimensional, rather, it is two-dimensional. The array of displacements is called *optical flow*optical flow. Thus, if we see things locally with a small aperture, then there are indefinite number of possible solutions. Solutions are uniquely determined only at those structures like corners, junctions, spots, etc., which are characterisitic enough to be uniquely localized and distinguished. Only if we see things with apertures large enough to include these structures, can we propagate their flows as constraints to determine flows for other points which otherwise would not have unique solutions.

This idea leads to a whole family of algorithms which impose a 2D smoothness constraint on the flow field [69, 65]. For points which happen to have unique solutions, their flows are so determined. For points which do not have unique structures, their flows are determined by minimizing a global sum of derivatives of the flow field.

This family of approaches have three common problems. Firstly, correspondence determined this way is not guaranteed to produce correct result. For example, in the case of rigid motion, optical flow obtained by imposing a 2D smoothness constraint does not satisfy the rigidity constraint in general. Secondly, the smoothness constraint is not valid along flow discontinuities. Unfortunately, these discontinuities are not known *a priori*. Thus, applying this constraint *blindly* not only gives wrong answers along these discontinuities, but these wrong answers can also propagate to neighboring areas, whose influence can reach very far from those discontinuities. Thirdly, minimizing a global functional of 2D flows is very computationally costly. Even the deepest descent method, which is the least costly algorithm available currently, still consumes a lot of computation power. Let alone those algorithms that seek global minimums.

5.1.3 Object Recognition

Object recognition is another hardest problem in vision. Actually it is called the final objective of vision. Marr claims, "Vision is the process of knowing what is where through seeing" [96]. "knowing what" is exactly the problem of object recognition.

Object recognition is actually not a single problem. It can be further divided into the problem of "recognizing a particular object" and the problem of "recognizing a generic object". The former has been intensively studied, and the later received attention only recently. In this book we only treat the former problem.

The approaches to object recognition have been deeply influenced by Marr's philosophy [96]. Marr claims that to recognize 3D objects, one must have enough 3D information about the objects to be recognized, thus only when the full 3D shape is recovered from images, can the process of recognition start. Under his influence, most early approaches to object recognition assume that 3D object models are given *a priori*, and the task is then to find a particular transformation that projects the model onto a particular part of the image, usually segmented from the background in advance. The problem with this approach is that 3D models are not easily available. It is not easy to determine shape of objects from input images through stereo, or motion, or other visual cues.

Recently, there are a number of attempts to avoid 3D object models, but instead to use 2D model views. Ullman proposes to use linear combination of matched model views as object models. [165]. Poggio proposes to use a number of matched views to train an approximation network whose internal representation functions as object model [115]. Xu proposes to use only one image as object model [177].

Anyway, if the problem of object recognition is posed as one of matching model views with the input image, which not only includes the target objects but also background, then correspondence between the model views, and between model view(s) and input view, becomes essential and necessary. Actually we believe that correspondence, localization and recognition are essentially the same process.

5.2 CORRESPONDENCE IN STEREO, MOTION AND OBJECT RECOGNITION AS 1D SEARCH

In this section, we redefine the three problems from the perspective of epipolar geometry. As will be seen when reading through this section, the correspondence problem in both motion and object recognition is changed to be a 1D search problem, similar to that in stereo, if the epipolar geometry underlying the images is recovered.

5.2.1 Stereo Matching

It is straightforward and conventional wisdom that the stereo correspondence problem is a 1D search problem, if the epipolar geometry is known *a priori*. Here it suffices to only stress that the recovery of epipolar geometry is necessary if the two cameras are not carefully displaced in a special manner. If the scene is stationary, there is no difference between two motion images and a pair of stereo images. The epipolar geometry underlying the stereo images is the same as that underlying the two motion images caused by a moving camera. Thus, the algorithm to recover epipolar geometry should also be the same.

5.2.2 Motion Correspondence and Segmentation

What differs between a general motion problem and a general stereo problem, is that in motion problem, we can have different motions simultaneously. The camera can move, the scene can move in a different way, and more importantly, there can be multiple objects moving independently in the scene, while in stereo, only the camera moves. Thus, between general motion images, there can be multiple epipolar geometries while in stereo there can only be one.

If we can somehow recover all the epipolar geometries underlying two motion images, then the search for correspondence is reduced from 2-dimensional to 1-dimensional. Thus, the aperture problem no longer exists, though ambiguity still exists, or multiple candidates still exist along the epipolar lines, as in stereo matching.

5.2.3 3D Object Recognition and Localization

The problem of using model view(s) for 3D object recognition can be further divided into two cases. One is of using only one model view, and the other is of using multiple views. Here we stress that in either case, the correspondence between the model and input views, and between model views is essential. Matching model views is nothing but matching two uncalibrated stereo images. It is also basically the same thing as matching the model view with the input view.

If the object in one model view is the same as the object in the input view, then they should satisfy the epipolar geometry, and the points in model view(s) can be matched against the points in the input view. If they are not the same, then we cannot find effective epipolar geometry between the images and cannot find smooth correspondences between the images.

The problem is then divided into one of finding possible epipolar geometry between a model view and the input view, and one of finding legitimate correspondence of image points based on the recovered epipolar geometry between the model view and the input view. If it is successful, then the two images are very likely to represent the same object; otherwise, they are not.

If the input view also includes other objects and background, finding the corresponding points to the model is also localizing the object from other objects and background.

The part of recovering epipolar geometry between a model view and an input view is essentially the same as that in uncalibrated stereo. The difference is that usually there are other objects and background in the input view, which, as described later, brings difficulty to epipolar equation recovery.

For how to recover the epipolar geometries in stereo, motion and object recognition, read the individual chapters that follow this one.

5.3 DISPARITY AND SPATIAL DISPARITY SPACE

Assuming that the epipolar geometry is recovered, then finding correspondence between motion images and between a model view and an input view is similar to that in stereo matching. Here we can also define disparity as the displacement of a pair of corresponding points along the epipolar lines in the two images.

5.3.1 Disparity under Full Perspective Projection in the Parallel Camera Case

In the parallel camera case, as shown in Fig.5.1, disparity is defined as the difference between the horizontal coordinates of the two corresponding points, as they have the same vertical coordinates. Let the two points be (u, v) and (u', v'). Disparity d is defined as

$$d = u - u', \tag{5.1}$$

And we know

$$v = v'. \tag{5.2}$$

The epipoles are both at infinity. Disparity is inversely proportional to the depth Z, the distance from the object to the base line linking the two optical centers:

$$d = u - u' = \frac{fb}{Z}, \tag{5.3}$$

where f is the focal length, and b is the length of the base line. It is clear that disparity approaches zero when depth approaches infinity.

This camera model is widely used by most stereo algorithms, since it is the simplest geometry and the relation between image displacement and depth is straightforward. As will be seen in the next subsection, the relation is much more complex for the general case under the full perspective projection.

5.3.2 Disparity under Full Perspective Projection in the General Case

In the general case, different from the parallel camera case, the corresponding points are not defined in a common reference frame such that the disparity can be measured directly. The idea here is to *rectify* the two images such that they look like being taken by two parallel cameras, the case which we discussed in the last subsection.

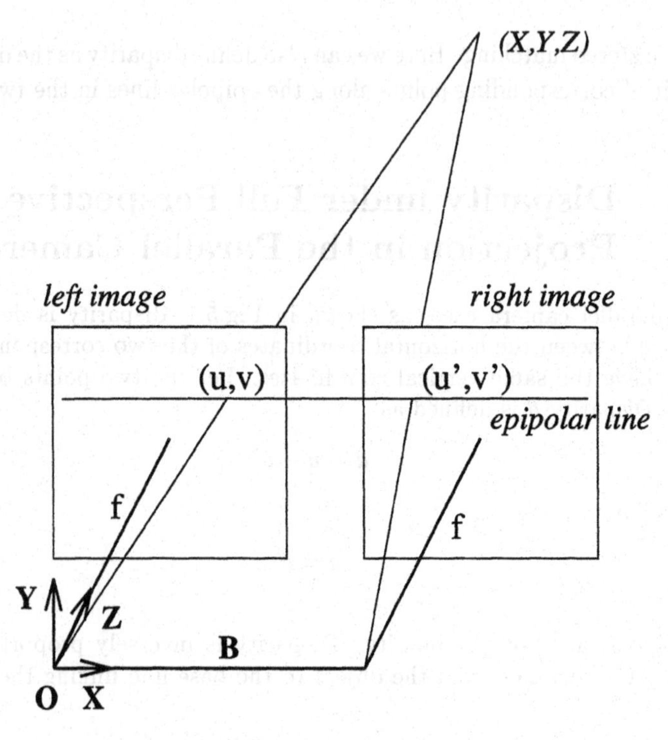

Figure 5.1 The stereo geometry for parallel cameras

Recall that given fundamental matrix $\mathbf{F}$, we can compute the epipoles $\widetilde{\mathbf{e}}$ for the first image and $\widetilde{\mathbf{e}}'$ for the second image, respectively, up to a scale factor, from

$$\mathbf{F}\widetilde{\mathbf{e}}' = 0\,, \tag{5.4}$$

$$\mathbf{F}^T\widetilde{\mathbf{e}} = 0\,. \tag{5.5}$$

See (2.66) and (2.68) in Chapter 2 for details.

Given a point $\widetilde{\mathbf{m}}'$ in the second image, the corresponding point $\widetilde{\mathbf{m}}$ in the first image can be expressed as

$$s\widetilde{\mathbf{m}} = s'\mathbf{H}\widetilde{\mathbf{m}}' + \widetilde{\mathbf{e}}\,. \tag{5.6}$$

where $\mathbf{H}$ is a 3×3 matrix. This equation is equivalent to (2.88).

From (2.89), we know

$$\mathbf{F} = [\widetilde{\mathbf{e}}]_\times \mathbf{H} \, . \tag{5.7}$$

All the three matrices are 3×3. But since $[\widetilde{\mathbf{e}}]_\times$ is not invertible, $\mathbf{H}$ cannot be uniquely determined from $\mathbf{F}$. The ambiguity in $\mathbf{H}$ will be discussed later.

Now let us choose $\hat{\mathbf{H}}$ such that

$$\hat{\mathbf{H}}\widetilde{\mathbf{e}} = \kappa[1,0,0]^T \, , \tag{5.8}$$

where κ is any nonzero scalar. It is evident that $\hat{\mathbf{H}}$ cannot be uniquely determined from this equation only. Multiplying $\hat{\mathbf{H}}$ to the two sides of (5.6) yields

$$s\hat{\mathbf{H}}\widetilde{\mathbf{m}} = s'\hat{\mathbf{H}}\mathbf{H}\widetilde{\mathbf{m}}' + \kappa[1,0,0]^T \, . \tag{5.9}$$

Let us define

$$\hat{\mathbf{m}} = [\hat{u}, \hat{v}, 1]^T \, , \tag{5.10}$$

$$\hat{\mathbf{m}}' = [\hat{u}', \hat{v}', 1]^T \, , \tag{5.11}$$

$$\hat{s}\hat{\mathbf{m}} = s\hat{\mathbf{H}}\widetilde{\mathbf{m}} \, , \tag{5.12}$$

$$\hat{s}'\hat{\mathbf{m}}' = s'\hat{\mathbf{H}}'\widetilde{\mathbf{m}}' = s'\hat{\mathbf{H}}\mathbf{H}\widetilde{\mathbf{m}}' \, . \tag{5.13}$$

We call the new images defined by Eqs. (5.10 -5.13) *rectified images*, and call this whole process *rectification* [41, 127].

Now (5.9) can be rewritten as

$$\hat{s}\hat{\mathbf{m}} = \hat{s}'\hat{\mathbf{m}}' + \kappa[1,0,0]^T \, . \tag{5.14}$$

Eq.(5.14) shows that the corresponding points $\hat{m}$ in the first rectified image and $\hat{m}'$ in the second rectified image have the same vertical coordinates, that is,

$$\hat{v} = \hat{v}' \, . \tag{5.15}$$

This means that the two rectified images look like a pair of images taken by a pair of parallel cameras. Disparity can be defined as the difference in the horizontal coordinates:

$$d = \hat{u} - \hat{u}' \, . \tag{5.16}$$

Geometrically, rectification is a process of projecting the two images onto another pair of retinal planes which are parallel to each other and to the line connecting the two optical centers. There are, however, an infinite number of such planes. This is the ambiguity in determining the two matrices $\hat{\mathbf{H}}$ and $\mathbf{H}$. This ambiguity can be used to optimize some physically meanful quantities. One such example is to minimize distortion between the original images and the rectified images [127].

Once $\hat{\mathbf{H}}$ and $\mathbf{H}$ are determined, given a point in the first image and a disparity d, we can compute a point in the second image. In general, both $\hat{\mathbf{H}}^{-1}$ and $(\hat{\mathbf{H}}')^{-1}$ exist. We first determine $\hat{\mathbf{m}}$ by

$$\hat{s}\hat{\mathbf{m}} = s\hat{\mathbf{H}}\widetilde{\mathbf{m}} ,$$

then obtain $\hat{\mathbf{m}}'$ as

$$\hat{\mathbf{m}}' = \hat{\mathbf{m}} - [d, 0, 0]^T ,$$

and finally determine the point in the second image as

$$s'\widetilde{\mathbf{m}}' = \hat{s}'(\hat{\mathbf{H}}')^{-1}\hat{\mathbf{m}}' .$$

5.3.3 Disparity under Weak Perspective and Paraperspective Projections

It is easier to define disparity for images taken under the weak perspective projection and paraperspective projection models. Given the coefficients of the epipolar equation, we can define the following transformation

$$\begin{bmatrix} \bar{u} \\ \bar{v} \end{bmatrix} = \begin{bmatrix} \cos\alpha & \sin\alpha \\ -\sin\alpha & \cos\alpha \end{bmatrix} \begin{bmatrix} u \\ v \end{bmatrix} \tag{5.17}$$

and

$$\begin{bmatrix} \bar{u}' \\ \bar{v}' \end{bmatrix} = \rho \begin{bmatrix} \cos\gamma & \sin\gamma \\ -\sin\gamma & \cos\gamma \end{bmatrix} \begin{bmatrix} u' \\ v' \end{bmatrix} + \begin{bmatrix} 0 \\ -\lambda \end{bmatrix} , \tag{5.18}$$

such that every pair of corresponding points in the two new images have the same vertical coordinates.

$$\bar{v} = \bar{v}' \tag{5.19}$$

This situation is the same as the standard stereo images in the parallel camera case. It is thus easy to define the disparity. It is simply the difference between the horizontal coordinates between the corresponding points in the two new images,

$$d = \bar{u} - \bar{u}' \tag{5.20}$$

For a point (u, v), given disparity d, its correspoding point in the other image (u', v') can be directly computed as

$$\begin{bmatrix} u' \\ v' \end{bmatrix} = \frac{1}{\rho} \begin{bmatrix} \cos\gamma & -\sin\gamma \\ \sin\gamma & \cos\gamma \end{bmatrix} \begin{bmatrix} \cos\alpha & \sin\alpha \\ -\sin\alpha & \cos\alpha \end{bmatrix} \begin{bmatrix} u \\ v \end{bmatrix} + \frac{1}{\rho} \begin{bmatrix} \cos\gamma & -\sin\gamma \\ \sin\gamma & \cos\gamma \end{bmatrix} \begin{bmatrix} -d \\ \lambda \end{bmatrix} \tag{5.21}$$

For the other direction, given a point (u', v') and d, the corresponding point (u, v) is computed by the following equation,

$$\begin{bmatrix} u \\ v \end{bmatrix} = \rho \begin{bmatrix} \cos\alpha & -\sin\alpha \\ \sin\alpha & \cos\alpha \end{bmatrix} \begin{bmatrix} \cos\gamma & \sin\gamma \\ -\sin\gamma & \cos\gamma \end{bmatrix} \begin{bmatrix} u' \\ v' \end{bmatrix} + \begin{bmatrix} \cos\alpha & -\sin\alpha \\ \sin\alpha & \cos\alpha \end{bmatrix} \begin{bmatrix} d \\ -\lambda \end{bmatrix} \tag{5.22}$$

Note that d is a function of the unknown rotation angle β and unknown depth Z which can only be determined up to a scale factor. Without loss of generality, here assume that it has the same scale as the images. From

$$\bar{u} = \bar{u}' \cos\beta + Z \sin\beta\ , \tag{5.23}$$

we have

$$d = \bar{u}'(\cos\beta - 1) + Z \sin\beta\ . \tag{5.24}$$

Knowing d cannot uniquely determine depth Z and rotation angle β, thus cannot uniquely determine structure and motion. One more view with different β is required.

5.3.4 Spatial Disparity Space and Smoothness

Looking for correspondence for each point is the same as determining disparity for each point. This problem can be intuitively represented as a search in a 3D space with the image as the first and second dimensions and disparity as the third dimension. This space (called *Spatial Disparity Space*)spatial disparity space is originally proposed by Yang it et al. for stereo matching [178].

Figure 5.2 gives an illustration of the space. Let us explain what SDS represents and how the constraints can be rephrased in terms of SDS.

A point $(u - v - d)$ means a pair of correspondence between (u, v) in the first and (u', v') in the second image which has the disparity of d. The uniqueness constraint simply means that in SDS, each column (u, v) can have only one active point, that, one disparity. If each column has only one active point, then they form a surface. The contrinuity constraint can be understood as a requirement that the neighboring columns should have continuous disparities. If there is a jump in disparity, it means a discontinuity.

In stereo, to find correspondence for each point in the left image is the same as finding a surface in SDS with its $u - v$ coordinates being identical with the left image. Usually we impose the smoothness constraint on matching, which is the same as requiring the surface to be as smooth as possible (there are different mathematical representations for quantitizing smoothness [151].)

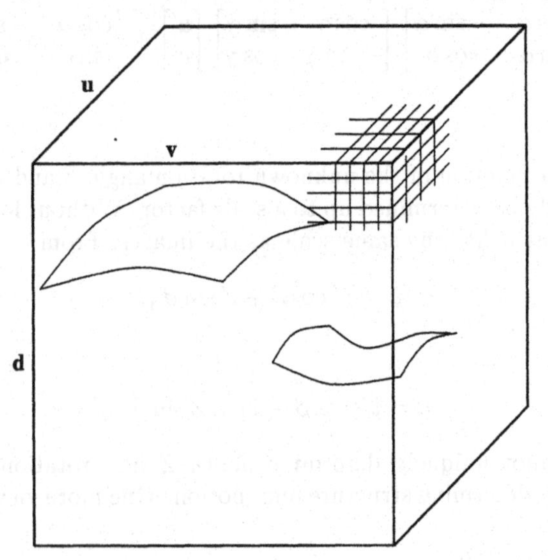

Figure 5.2 Spatial Disparity Space for stereo matching

In motion, when there are different rigid motions, as long as their epipolar equations have been recovered, we can add as many blocks of SDS as motions, so that each block represents one motion. The uniqueness constraint applies again. That is, each column can have only one active node along it, no matter how many blocks there are, and no matter in which block the active node resides.

We call this multi-block Spatial Disparity Space *Extended Spatial Disparity Space*, or ESDS.

Figure 5.3 gives an illustration of ESDS. Note that there are inactive insulator layers between every two neighboring blocks so that local operations do not cover two different blocks. Also note that for 2D affine motions, each block has only one layer, which does not represent a range of possible disparities but only represents whether that point belongs to a particular 2D affine motion.

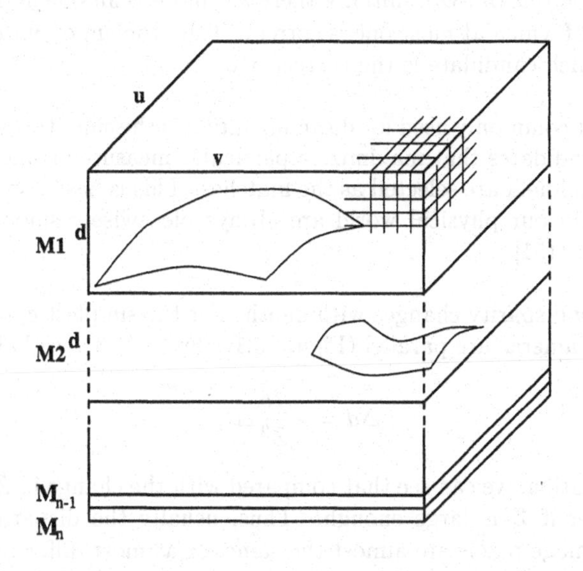

Figure 5.3 The Extended Spatial Disparity Space for multiple motions

In object recognition, once possible epipolar equation is recovered, the SDS can be established in the same manner as in stereo. Each node represents a match between a point in the model image and a point in the input image.

If the $u - v$ coordinates of the ESDS is identical with the model image, then the task is then to find which node is the true match. In this case, there is only one block.

If the $u - v$ coordinates of the ESDS is identical the input image, the ESDS can represent the problem of finding identities for all the parts of the input image in the model database. Each part of the input image is matched against a particular model under a particular epipolar equation. The final result is a set of models that match individual parts of the input image. In this case the problem is similar to that in multiple motions

5.3.5 Correspondence as Search for Surfaces and Contours in SDS

As always is the problem in matching, there are multiple candidates for correspondence. In terms of SDS, initially there are more than one active nodes for each column, of which at most one is correct. The problem of matching is then to find out which candidate is the correct one.

The constraint commonly used for disambiguation is the smoothness constraint, that is, the candidates that maximize a particular measure of smoothness over the whole visual field are selected as the matches. This is based on observations that surfaces in our physical world are always piecewisely smooth except at discontinuites. [151]

Let us see how disparity changes with depth. For the simplest case, we assume that the two cameras are parallel (Figure 5.1.) From (5.3), we have

$$\Delta d = -\frac{fb}{Z^2}\Delta Z \tag{5.25}$$

From this equation, we can see that compared with the change in Z, the change in d is smaller if Z is large enough. Thus, usually the disparity values for neighboring image pixels are almost the same, or at most differ by 1 pixel.

We can have similar analysis for the weak perspective projection model, though the analysis for the full perspective projection in the general case is more difficult.

Strictly speaking, the "smoothest" surfaces are not always guaranteed to be the true ones. But stochastically, they are the most likely ones to be true.

There are many ways to define and measure smoothness. In SDS, we can define smoothness for surfaces in SDS, and in some cases contours if only edges are used for matching.

In the former case, the first-order differentials $d_u^2 + d_v^2$ are extensively used [69]. Some others argue that second-order differentials are also necessary. [151]

For contours, both the first-order differential d_s^2 and the second-order differential d_{ss}^2 are widely used [65, 77, 176]. But as the order of points on the edges is not necessarily clear, d_s^2 and $d_{ss}^2$are not always available. In this case, one way is to define smoothness stochastically. We will use a stochastic measure of smoothness in later chapters for matching edge points.

5.4 SUMMARY

In this chapter, we have reexamined the problems of correspondence and segmentation in stereo, motion and object recognition from the epipolar geometry point of view. We have shown that once the epipolar geometry is recovered, all can be redefined as a 1D correspondence search problem plus a segmentation problem which can be solved simultaneously. This analysis provides a unified framework to do correspondence and segmentation in all multiple view problems.

We have also shown how to define disparity for both full perspective projections and affine cameras. We have further proposed to represent all possible correspondences along the epipolar lines in a three-dimensional representation called *Extended Spatial Disparity Space*, with one of the images as the first and second dimensions and disparity as the third dimension. Determining unique correspondences can be posed as finding the smoothest surfaces or smoothest contours in this space.

6

IMAGE MATCHING AND UNCALIBRATED STEREO

Matching different images of a single scene remains one of the bottlenecks in computer vision. A large amount of work has been carried out during the last 15 years, but the results are, however, not satisfactory. The only geometric constraint we know between two images of a single scene is the *epipolar constraint*. However, when the motion between the two images is unknown, the epipolar geometry is also unknown. The methods reported in the literature all exploit some heuristics in one form or another, for example, intensity similarity, which are not applicable to most cases. The difficulty is partly bypassed by taking long sequences of images over short time interval [27, 183]. Indeed, as the time interval is small and object velocity is constrained by physical laws, the interframe displacements of objects are bounded, i.e., the correspondence of a token at the subsequent instant must be in the neighborhood of the first. However, in many cases, such as a pair of uncalibrated stereo images, this technique does not apply. Developing a robust image matching technique is thus very important.

Over the years numerous algorithms for image matching have been proposed. They can roughly be classified into two categories:

1. **Template matching.** In this category, the algorithms attempt to correlate the grey levels of image patches in the views being considered, assuming that they present some similarity [9, 47, 52, 20, 43]. The underlying assumption appears to be a valid one for relatively textured areas and for image pairs with small difference; however it may be wrong at occlusion boundaries and within featureless regions.
2. **Feature matching.** In this category, the algorithms first extract salient primitives from the images, such as edge segments or contours, and match

them in two or more views. An image can then be described by a graph with primitives defining the nodes and geometric relations defining the links. The registration of two maps becomes the mapping of the two graphs: *subgraph isomorphism.* Common techniques are tree searching, relaxation, maximal clique detection, etc. Some heuristics such as assuming affine transformation between the two images are usually introduced to reduce the complexity. These methods are fast because only a small subset of the image pixels are used, but may fail if the chosen primitives cannot be reliably detected in the images. The following list of references is by no means exhaustive: [164, 136, 10, 19, 122, 94, 68]

The approaches we have proposed in [188, 175] aims at exploiting the only geometric constraint, i.e., the epipolar constraint, to establish robustly correspondences between two perspective or affine images of a single scene. However, in order to reduce the complexity of the algorithm, we still exploit heuristic techniques to find an initial set of matches. We first extract high curvature points and then match them using a classical correlation technique followed by a new fuzzy relaxation procedure or clustering technique. More precisely, our algorithm consists of three steps:

- Establish initial correspondences using some classical techniques,
- Estimate robustly the epipolar geometry, and
- Establish correspondences using estimated epipolar geometry as in classical stereo matching.

The basic idea is first to estimate robustly the epipolar geometry, and then reduce the general 2D image matching problem to 1D stereo matching. The robust technique for estimating the epipolar geometry is described in Sect. 3.1.6 and Sect. 3.3.

In the following, Sect. 6.1 describes detection of feature points like corners, and establishing initial correspondences between two sets of feature points. Sect. 6.2 describes using relaxation to reduce ambiguity and using robust estimation of epipolar geometry under perspective projection to determine unique correspondences between two sets of feature points. Sect. 6.3 describes how to apply clustering technique to the initial matches for affine images. It produces unique matches and robustly estimates affine epipolar geometry. Sect. 6.4 describes two simple techniques to match corners and edges using recovered epipolar geometry. The following two sections give one example for matching two perspective images, and one example for matching two affine images.

6.1 FINDING MATCH CANDIDATES BY CORRELATION

Before recovering the epipolar geometry, we must establish a few point matches between images. To this end, a corner detector is first applied to each image to extract high curvature points. A classical correlation technique is then used to establish match candidates between the two images. This technique is used for both perspective images and affine images.

6.1.1 Extracting Points of Interest

First, feature points corresponding to high curvature points are extracted from each image. A great deal of effort has been spent by the computer vision community on this problem, and several approaches have been reported in the literature in the last few years. They can be broadly divided into two groups. The first group consists in first extracting edges as a chain code, and then searching for points having maxima curvature [26, 4, 101] or performing a polygonal approximation on the chains and then searching for the line segment intersections [67]. The second group works directly on a grey-level image. The large number of techniques that have been proposed within this group are generally based on the measurement of the gradients and of the curvatures of the surface (see [29] for a review).

In our application, we use the corner detector [56], which is a slightly modified version of the Plessey corner detector [55, 110]. It is based on the following operator:

$$R(x,y) = \det[\hat{C}] - k\,\mathrm{trace}^2[\hat{C}]\,, \tag{6.1}$$

where $\hat{C}$ is the following matrix:

$$\hat{C} = \begin{bmatrix} \hat{I}_x^2 & \widehat{I_x I_y} \\ \widehat{I_x I_y} & \hat{I}_y^2 \end{bmatrix}\,, \tag{6.2}$$

where $\hat{I}$ denotes the smoothing operation on the grey level image $I(x,y)$. I_x and I_y indicate the x and y directional derivatives respectively.

We use a value of k equal to 0.04 to provide discrimination against high contrast pixel step edges. After that, the operator output is thresholded for the corner detection. It should be pointed out that this method allows us to recover a corner position up to pixel precision. In order to recover the corner position up to sub-pixel position, one uses the model based approach we have already developed and presented in [25], where corners are extracted directly from the image by searching the parameters of the parametric model that best approximates the observed grey level image intensities around the corner position detected. One can notice that such an approach is well adapted for scenes containing polyhedral objects, but not for most outdoor scenes.

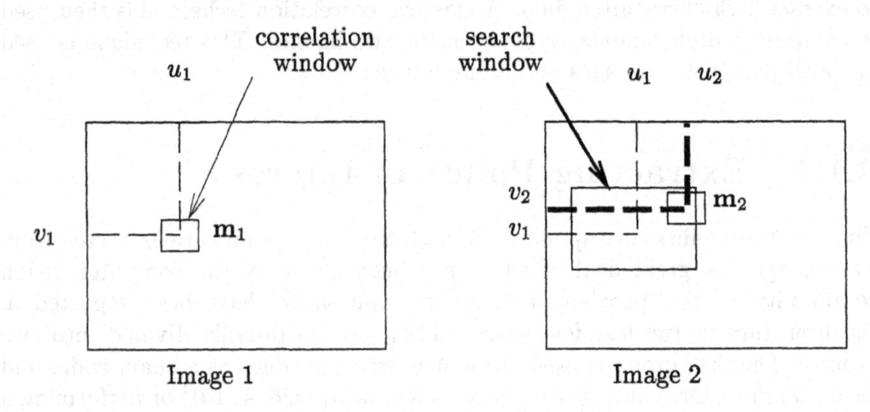

Figure 6.1 Correlation

6.1.2 Matching Through Correlation

Since the epipolar geometry is not yet known, the search for a correspondence should be performed theoretically in the whole image. Given a high curvature point $\mathbf{m}_1$ in image 1, we use a correlation window of size $(2n+1) \times (2m+1)$ centered at this point. We then select a rectangular search area of size $(2d_u + 1) \times (2d_v + 1)$ around this point in the second image, and perform a correlation operation on a given window between point $\mathbf{m}_1$ in the first image and all high curvature points $\mathbf{m}_2$ lying within the search area in the second image. The search window reflects some *a priori* knowledge about the disparities between the matched points. This is equivalent to reducing the search area for a corresponding point from the whole image to a given window. If no such knowledge is available, then we have to search the whole image. The correlation score is defined as

$$Score(\mathbf{m}_1, \mathbf{m}_2) = \frac{\sum_{i=-n}^{n} \sum_{j=-m}^{m} [I_1(u_1+i, v_1+j) - \overline{I_1(u_1, v_1)}] \times [I_2(u_2+i, v_2+j) - \overline{I_2(u_2, v_2)}]}{(2n+1)(2m+1)\sqrt{\sigma^2(I_1) \times \sigma^2(I_2)}}, \quad (6.3)$$

where $\overline{I_k(u,v)} = \sum_{i=-n}^{n} \sum_{j=-m}^{m} I_k(u+i, v+j)/[(2n+1)(2m+1)]$ is the average at point (u, v) of I_k $(k = 1, 2)$, and $\sigma(I_k)$ is the standard deviation of the image I_k in the neighborhood $(2n+1) \times (2m+1)$ of (u, v), which is given by:

$$\sigma(I_k) = \sqrt{\frac{\sum_{i=-n}^{n} \sum_{j=-m}^{m} I_k^2(u,v)}{(2n+1)(2m+1)} - \overline{I_k(u,v)}^2} \, .$$

The score ranges from -1, for two correlation windows which are not similar at all, to 1, for two correlation windows which are identical.

A constraint on the correlation score is then applied in order to select the most consistent matches: For a given couple of points to be considered as a match candidate, the correlation score must be higher than a given threshold. If the above constraint is fulfilled, we say that the pair of points considered is self-consistent and forms a *match candidate*. For each point in the first image, we thus have a set of match candidates from the second image (the set is possibly nil); and in the same time we have also a set of match candidates from the first image for each point in the second image.

In our implementation, $n = m = 7$ for the correlation window, and a threshold of 0.8 on the correlation score is used.

6.1.3 Rotating Correlation Windows

The above correlation score is defined without considering the possible large rotation of the local image patterns. In uncalibrated stereo, the images may have been taken by cameras in different poses, thus large image torsions are possible.

To deal with this problem, we have to allow the image windows to rotate. It is sufficient to rotate image windows in only one of images. Since the two images play a symmetric role, rotating either of them is the same. Let us rotate windows in image 1. Also, if we allow large image torsions, then the search window has to be enlarged to include the whole image.

There is no definite answer to the question on resolution of rotation angles. Empirical data show that 16 angles are usually sufficient. Each angle represents 22.5 degrees. Let us call them $\alpha_k, k = 1, ..., 16$. Now for each feature point in image 1, we can prepare 16 windows.

As we allow the images to rotate, it is possible that one point on the left side of the image 1 matches one point on the right side of the other image. Thus, basically we have to examine every pairing of all feature points in one image with all feature points in the other image. If there are M feature points in image 1 and N feature points in image 2, then we have to compute the matching scores for $16MN$ times.

Equation (6.3) can now be modified to be

$$Score(\mathbf{m}_{1i}, \mathbf{m}_{2j}, \alpha_k) = \tag{6.4}$$

$$\frac{\sum_{x=-n}^{n} \sum_{y=-m}^{m} I_1'(u_{1i} + x, v_{1i} + y) \times I_2'(u_{2j} + \Delta u(\theta_k, x, y), v_{2j} + \Delta v(\theta_k, x, y))}{(2n+1)(2m+1)\sqrt{\sigma^2(I_1) \times \sigma^2(I_2)}},$$

$$\text{for } i = 1, ..., M; j = 1, ..., N; k = 1, ..., 16;$$

where

$$\begin{bmatrix} \Delta u(\theta_k, x, y) \\ \Delta v(\theta_k, x, y) \end{bmatrix} = \begin{bmatrix} \cos\theta_k & \sin\theta_k \\ -\sin\theta_k & \cos\theta_k \end{bmatrix} \begin{bmatrix} x \\ y \end{bmatrix};$$

$$I_1'(u_{1i} + x, v_{1i} + y) =$$
$$I_1(u_{1i} + x, v_{1i} + y) - \frac{\sum_{x=-n}^{n} \sum_{y=-m}^{m} I_1(u_{1i} + x, v_{1i} + y)}{(2n+1)(2m+1)};$$

$$I_2'(u_{2j} + \Delta u(\theta_k, x, y), v_{2j} + \Delta v(\theta_k, x, y)) =$$
$$I_2(u_{2j} + \Delta u(\theta_k, x, y), v_{2j} + \Delta v(\theta_k, x, y))$$
$$- \frac{\sum_{x=-n}^{n} \sum_{y=-m}^{m} I_2(u_{2j} + \Delta u(\theta_k, x, y), v_{2j} + \Delta v(\theta_k, x, y))}{(2n+1)(2m+1)}.$$

Since here we assume that the scene is stationary, it is not possible to have feature points being matched with largely different rotation angles. This is especially true for the affine projections, as the image torsions α and γ are everywhere identical in the images.

This property implies that if the image is rotated for the correct angle, then matching scores should be high for all the corresponding feature points with that rotation angle. Under the weak or para- perspective projections, this angle corresponds to θ. If we rotate image 2 by θ, then a simple correlation defined in (6.4) will give high matching scores for all corresponding feature points with that rotation angle θ. This suggests a histogram algorithm for evaluation of the rotation angle. As in the last subsection, if the score is higher than a predetermined threshold, that pair of feature points is considered a *match candidate*, with rotation angle θ_k. We can define an indicator $S(\mathbf{m}_{1i}, \mathbf{m}_{2j}, \theta_k)$ for the pair of points $\mathbf{m}_{1i}, \mathbf{m}_{2j}$ with angle θ_k, which is 1 if the score is higher than the threshold, and 0 otherwise. Now we can further define a counter for θ_k,

$$C(\theta_k) = \sum_{i=1,M} \sum_{j=1,N} S(\mathbf{m}_{1i}, \mathbf{m}_{2j}, \theta_k); \; k = 1, ...16.$$

Once a single outstanding peak is found, only the match candidates with this particular rotation angle are used for later processes. Usually, ambiguity remains to be cleared.

6.2 UNIQUE CORRESPONDENCE BY RELAXATION AND ROBUST ESTIMATION OF EPIPOLAR GEOMETRY FOR PERSPECTIVE IMAGES

Using the correlation technique described above, a point in the first image may be paired to several points in the second image (which we call *match candidates*), and vice versa. Several techniques exist for resolving the matching ambiguities. The technique we use falls into the class of techniques known as *relaxation techniques*. The idea is to allow the match candidates to reorganize themselves by propagating some constraints, such as continuity and uniqueness, through the neighborhood.

6.2.1 Measure of the Support for a Match Candidate

Consider a match candidate $(\mathbf{m}_{1i}, \mathbf{m}_{2j})$ where $\mathbf{m}_{1i}$ is a point in the first image and $\mathbf{m}_{2j}$ is a point in the second image. Let $\mathcal{N}(\mathbf{m}_{1i})$ and $\mathcal{N}(\mathbf{m}_{2j})$ be, respectively, the neighbors of $\mathbf{m}_{1i}$ and $\mathbf{m}_{2j}$ within a disc of radius R. If $(\mathbf{m}_{1i}, \mathbf{m}_{2j})$ is a good match, we will expect to see many matches $(\mathbf{n}_{1k}, \mathbf{n}_{2l})$, where $\mathbf{n}_{1k} \in \mathcal{N}(\mathbf{m}_{1i})$ and $\mathbf{n}_{2l} \in \mathcal{N}(\mathbf{m}_{2j})$, such that the position of $\mathbf{n}_{1k}$ relative to $\mathbf{m}_{1i}$ is similar to that of $\mathbf{n}_{2l}$ relative to $\mathbf{m}_{2j}$. On the other hand, if $(\mathbf{m}_{1i}, \mathbf{m}_{2j})$ is a bad match, we will expect to see only few matches, or even not any at all, in their neighborhood.

More formally, we define a measure of support for a match, which we call the *strength of match* (SM for abbreviation), as

$$S_M(\mathbf{m}_{1i}, \mathbf{m}_{2j}) = c_{ij} \sum_{\mathbf{n}_{1k} \in \mathcal{N}(\mathbf{m}_{1i})} \left[\max_{\mathbf{n}_{2l} \in \mathcal{N}(\mathbf{m}_{2j})} \frac{c_{kl}\, \delta(\mathbf{m}_{1i}, \mathbf{m}_{2j};\ \mathbf{n}_{1k}, \mathbf{n}_{2l})}{1 + dist(\mathbf{m}_{1i}, \mathbf{m}_{2j};\ \mathbf{n}_{1k}, \mathbf{n}_{2l})} \right],$$

where c_{ij} and c_{kl} are the goodness of the match candidates $(\mathbf{m}_{1i}, \mathbf{m}_{2j})$ and $(\mathbf{n}_{1k}, \mathbf{n}_{2l})$, which can be the correlation scores given in the last section, $dist(\mathbf{m}_{1i}, \mathbf{m}_{2j};\ \mathbf{n}_{1k}, \mathbf{n}_{2l})$ is the average distance of the two pairings, i.e.,

$$dist(\mathbf{m}_{1i}, \mathbf{m}_{2j};\ \mathbf{n}_{1k}, \mathbf{n}_{2l}) = [d(\mathbf{m}_{1i}, \mathbf{n}_{1k}) + d(\mathbf{m}_{2j}, \mathbf{n}_{2l})]/2$$

with $d(\mathbf{m}, \mathbf{n}) = \|\mathbf{m} - \mathbf{n}\|$, the Euclidean distance between $\mathbf{m}$ and $\mathbf{n}$, and

$$\delta(\mathbf{m}_{1i}, \mathbf{m}_{2j};\ \mathbf{n}_{1k}, \mathbf{n}_{2l}) = \begin{cases} e^{-r/\varepsilon_r} & \text{if } (\mathbf{n}_{1k}, \mathbf{n}_{2l}) \text{ is a match candidate and } r < \varepsilon_r \\ 0 & \text{otherwise} \end{cases}$$

where r is the relative distance difference given by

$$r = \frac{|d(\mathbf{m}_{1i}, \mathbf{n}_{1k}) - d(\mathbf{m}_{2j}, \mathbf{n}_{2l})|}{dist(\mathbf{m}_{1i}, \mathbf{m}_{2j};\ \mathbf{n}_{1k}, \mathbf{n}_{2l})}$$

and ε_r is a threshold on the relative distance difference. The above definition of the strength of a match is similar in the form to that used in the PMF stereo algorithm [118].

Several remarks can be made regarding our measure of matching support.

- Firstly, the strength of a match actually counts the number of match candidates found in the neighborhoods, but only those whose positions relative to the considered match are similar are counted.

- Secondly, the test of similarity in relative positions is based on the relative distance (the value of r). Indeed, the similarity in relative positions is justified by the hypothesis that an affine transformation can approximate the change between the neighborhoods of the match candidate being considered. This assumption is reasonable only for a small neighborhood. Thus we should allow larger tolerance in distance differences for distant points, and this is exactly what our criterion does.

- Thirdly, the contribution of a match candidate $(\mathbf{n}_{1k}, \mathbf{n}_{2l})$ to the strength of the match $(\mathbf{m}_{1i}, \mathbf{m}_{2j})$ is the exponential of the negative relative error r, which is strictly monotonically decreasing function of r. When r is very big, then $\exp(-r/\varepsilon_r) \rightarrow 0$, and the match candidate can be ignored. When $r \rightarrow 0$, i.e., the difference is very small, then $\exp(-r/\varepsilon_r) \rightarrow 1$, and the candidate will largely contribute to the match $(\mathbf{m}_{1i}, \mathbf{m}_{2j})$.

- Fourthly, if a point in the left image has several match candidates in the right image, only the one which has smallest distance difference is accounted for, which is done by the "max" operator.

- Lastly, the contribution of each match candidate in the neighborhood is weighted by its distance to the match. The addition of '1' is only to prevent the over weight for very close points. In other words, a close match candidate gives more support to the match being considered than a distant one. This is also connected to the fact that an affine approximation is only reasonable for a small neighborhood.

The measure of matching support defined above, however, is not symmetric. That is, the strength of a match is possibly not the same if we reverse the role of the two images, i.e., possibly we have $S_M(\mathbf{m}_{1i}, \mathbf{m}_{2j}) \neq S_M(\mathbf{m}_{2j}, \mathbf{m}_{1i})$. This occurs when several points $\mathbf{n}_{1k} \in \mathcal{N}(\mathbf{m}_{1i})$ are candidate matches of a single point $\mathbf{n}_{2l} \in \mathcal{N}(\mathbf{m}_{2j})$, as illustrated in Fig. 6.2 where $\mathbf{n}_{11}$ and $\mathbf{n}_{12}$ share the same point $\mathbf{n}_{21}$ as their match candidate. In our implementation, we have made the following modification in order to achieve the symmetry. Before computing the summation, if several points $\mathbf{n}_{1k} \in \mathcal{N}(\mathbf{m}_{1i})$ score the maximal value with the same point $\mathbf{n}_{2l} \in \mathcal{N}(\mathbf{m}_{2j})$, then only the point which gives the largest value is counted. This assures that the same pairing will be counted if we reverse the role of the two images.

Other heuristics can be integrated into the computation of the strength of a match. For example, if the angle of the rotation in the image plane is assumed to be less than Θ, then we can impose the following constraint: the angle between $\overrightarrow{\mathbf{m}_{1i}\mathbf{n}_{1k}}$ and $\overrightarrow{\mathbf{m}_{2j}\mathbf{n}_{2l}}$ must be less than Θ. In other words, for a match candidate

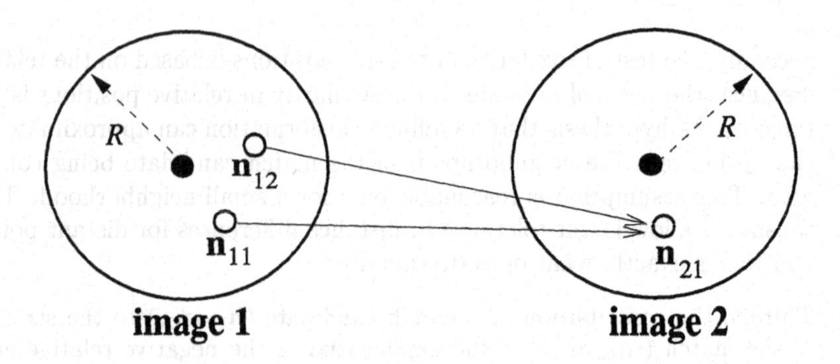

Figure 6.2 Illustration of the non-symmetric problem of the matching support measure

$(\mathbf{n}_{1k}, \mathbf{n}_{2l})$ which does not satisfy the above constraint, its $\delta(\mathbf{m}_{1i}, \mathbf{m}_{2j};\ \mathbf{n}_{1k}, \mathbf{n}_{2l})$ takes the value of zero.

In our implementation, R = one eighth of the image width, $c_{ij} = 1$, $\varepsilon_r = 0.3$ and $\Theta = 90°$.

6.2.2 Relaxation Process

If we define the energy function as the sum of the strengths of all match candidates, i.e.,

$$\mathcal{J} = \sum_{(\mathbf{m}_{1i}, \mathbf{m}_{2j})} S_M(\mathbf{m}_{1i}, \mathbf{m}_{2j})$$

then the problem of disambiguating matches is equivalent to minimizing the energy function $\mathcal{J}$. The relaxation scheme is one approach to it. It is an iterative procedure, and can be formulated as follows:

iterate {

- compute the matching strength for each match candidate
- update the matches by minimizing the total energy

} **until the convergence of the energy**

After the correlation procedure, for each point in the first image, we have a set of match candidates from the second image (the set is possibly nil); and in the same time we have also a set of match candidates from the first image for each point in the second image. The last subsection has already explained how to compute the SM for each match candidate. As the definition of SM is now symmetric, we only need to compute SMs for the list of match candidates in the left image and assign the values to the match candidates in the right image, thus saving half of the computation.

There are several strategies for updating the matching in order to minimize the total energy. The first is the "*winner-take-all*", as exploited by Rosenfeld et al. [128], Zucker et al. [194], and Pollard et al. [118]. The method works as follows. At each iteration, any matches which have the highest matching strengths for both of the two image points that formed them are immediately chosen as "correct". That is, a match $(\mathbf{m}_{1i}, \mathbf{m}_{2j})$ is selected if its points (either $\mathbf{m}_{1i}$ or $\mathbf{m}_{2j}$) have no higher matching-strength scores with any other matches they can form. Then, because of the uniqueness constraint, all other matches associated with the two points in each chosen match are eliminated from further consideration. This allows further matches, that were not previously either selected or eliminated, to be selected as correct provided they now have the highest matching strengths for both constituent points. This method proceeds as a steepest-descent approach, and is thus fast. However, it may get stuck easily at a bad local minimum.

The second is the "*looser-take-nothing*" [88]. The method works as follows. For each point in the first image, the candidate which has gained the weakest matching strength is eliminated. The process suppresses at most one candidate at each iteration until one and only one candidate is left for each point, finally achieving an unambiguous set of matches. Since the suppressed matches have gained the weakest support, they are very possibly not among the correct matches. This method thus proceeds as a slowest-descent approach, and is not efficient if a point has many match candidates. Furthermore, this method is not symmetric for the two images: reversing the role of the two images may give different result.

We have developed a new update strategy, which we would like to call "*some-winners-take-all*". It differs from "winner-take-all", which is in fact *all-winners-take-all*, and works as follows. As with "winner-take-all", we consider all matches which have the highest matching strength for both of the two image points that formed them. We shall call such matches the *potential matches*, and denote them by $\{\mathcal{P}_i\}$. For $\{\mathcal{P}_i\}$, we construct two tables. The first, denoted by $\mathcal{T}_{SM}$, saves the matching strength of each $\mathcal{P}_i$, and is then sorted in decreasing order. The second, denoted by $\mathcal{T}_{UA}$, saves a value which indicates how unambiguous each $\mathcal{P}_i$ is. This is defined as

$$U_A = 1 - S_M^{(2)} / S_M^{(1)} ,$$

where $S_M^{(1)}$ is the SM of $\mathcal{P}_i$, and $S_M^{(2)}$ is the SM of the second best match candidate. Thus U_A is ranging from 1 (unambiguous) to 0 (ambiguous). The table $\mathcal{T}_{UA}$ is also sorted in decreasing order. Finally, those potential matches $\mathcal{P}_i$ which are among both the first q percent of matches in $\mathcal{T}_{SM}$ and the first q percent of matches in $\mathcal{T}_{UA}$ are selected as correct matches. Thus, ambiguous potential matches will not be selected even they have high SM, and those having weak SM will not selected even they are unambiguous. We have therefore prevented the problem of evolve-too-soon-ness with "winner-take-all" while maintaining computational efficiency. If a match candidate does not receive any support ($S_M = 0$), it will be eliminated from further consideration. If $q = 100$, i.e., one hundred percent selection case, our method becomes "winner-take-all". Not that q must be larger than 50 in order to assure that at least one potential match will be selected at each iteration if there exist several potential matches. If $q < 50$, a premature stop may occur. In our implementation, q is set to 60.

Our algorithm necessarily converges, because if during one iteration there is no match selected, then the total energy will remain the same at the next iteration. The number of selected matches is evidently limited because the number of match candidates is limited.

6.2.3 Detection of False Matches

The set of matches established by the correlation technique, or improved by the relaxation technique, usually contains false matches, because only heuristics are used. The only geometric constraint between two images is the epipolar constraint. If two points are correctly matched, they must satisfy this constraint, which is however unknown in the general case. The LMedS technique described in Sect. 3.1.6 allows us to detect the false matches and estimate in the same time the epipolar geometry robustly.

6.3 UNIQUE CORRESPONDENCE BY ROBUST ESTIMATION OF EPIPOLAR GEOMETRY FOR AFFINE IMAGES

Using the correlation technique described in Sect. 6.1, a point in the first image is usually paired to more than one points in the second image (which we call *match candidates*), and vice versa. For images that can be modelled as weak perspective or paraperspective images, there are only 4 degrees of freedom in the epipolar geometry. We can apply the clustering technique described in Sect. 3.3 to robustly estimate the underlying epipolar geometry. The clustering technique is also good at rejecting outliers. To produce a hypothesis of epipolar geometry, we need 4 pairs of point matches, which project onto the parameter space as a pair of points. To limit the number of points away from the clusters, we need to further reduce the ambiguity in the initial matches. The essence of the techniques used here is the same as described in the last section, that is, relative distances among neighboring points do not change drastically between images.

6.3.1 Discarding Unlikely Match Candidates

Suppose that we are given a list of match candidates $\mathbf{u}_k = [u_k, v_k, u'_k, v'_k]^T, k = 1, ..., n$ between point $[u_k, v_k]^T$ in the first image and $[u'_k, v'_k]^T$ in the second image. We now define the relative distance difference between $\mathbf{u}_k$ and $\mathbf{u}_l$ as

$$r_{kl} = \frac{\sqrt{(u_k - u_l)^2 + (v_k - v_l)^2} - \sqrt{(u'_k - u'_l)^2 + (v'_k - v'_l)^2}}{\sqrt{(u_k - u_l)^2 + (v_k - v_l)^2} + \sqrt{(u'_k - u'_l)^2 + (v'_k - v'_l)^2}} . \tag{6.5}$$

To measure consistency of a particular match candidate $\mathbf{u}_k$ with other match candidates, we compute

$$R_k = \sum_{l=1, l \neq k}^{n} r_{kl} . \tag{6.6}$$

As illustrated in Fig.6.3, if a particular match candidate is inconsistent with other match candidates, R tends to be very large. If it is larger than a threshold, then it is discarded from the list without further consideration. By doing so, the number of match candidates can be further reduced. In general, however, matches are still not unique.

In our current implementation, the threshold is the average of R for all match candidates.

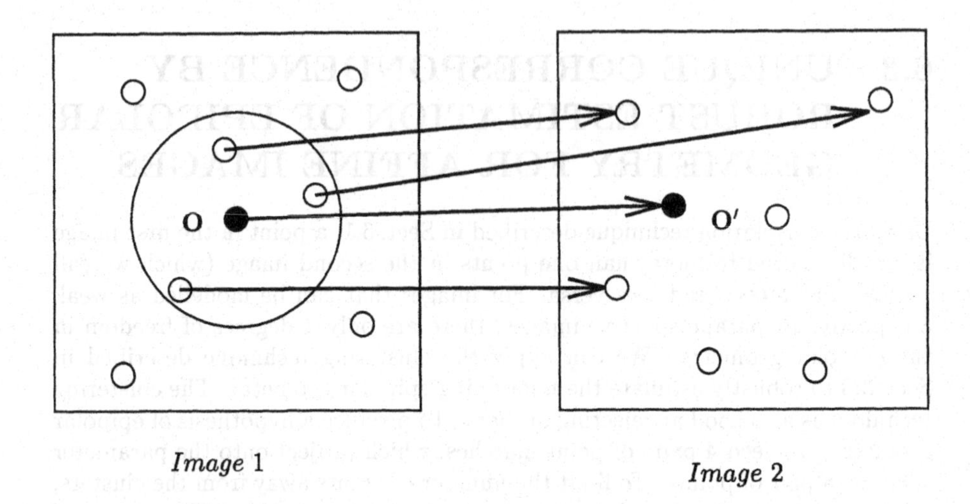

Figure 6.3 If a particular match candidate is inconsistent with other match candidates, it can be removed by computing the sum of relative distance differences with respect to all other match candidates.

6.3.2 Generating Local Groups of Point Matches for Clustering

To generate one hypothesis of epipolar geometry, we need 4 point matches. If we generate groups of 4 point matches *randomly*, the number of generated hypotheses can become very large, if the number of feature points is over 10 or 15. The strategy we use to limit the number of hypotheses is to only generate local groups of point matches. For each pair of match candidates, we only generate groups by finding the 3 closest neighbors.

As illustrated in Fig.6.4, for a pair of points O and O′, we find the 3 closest neighbors for O. As the neighboring points do not necessarily have unique matches, there are usually more than one groups of 4 point matches. For each group, we compute the relative distance difference r using (6.5). If one of the match candidates causes inconsistency, then this group is discarded. If no group remains for the current 3 neighbors, then the trouble-making neighbor is removed, and the next closest neighbor is chosen to form new groups, till at least one group passes the consistency check.

For each group that passes the consistency check, we compute one epipolar geometry, and project two points corresponding to the same epipolar geometry

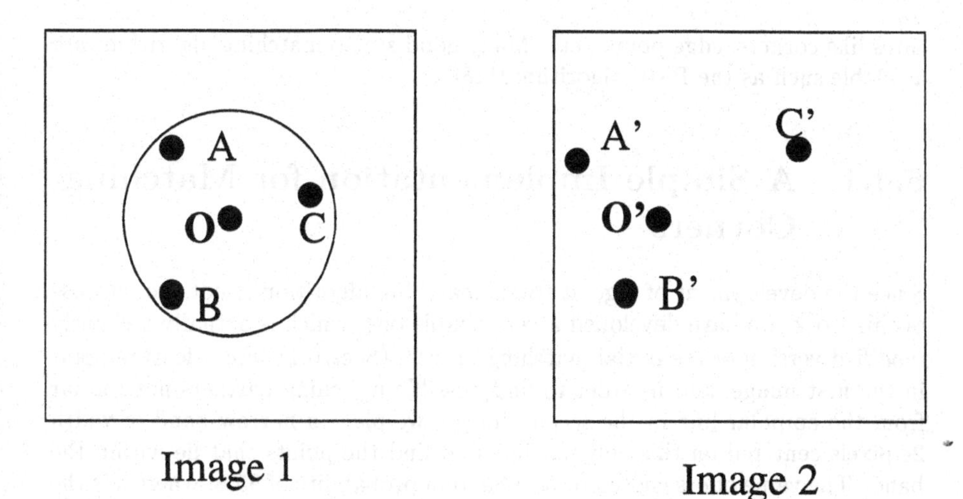

Figure 6.4 If a particular match candidate in the neighborhood is inconsistent with other match candidates, it is removed and the next closest neighbor is chosen.

onto the parameter space. And after this, the clustering algorithm described in Sect. 3.3 is applied to find clusters in the parameter space. The corresponding epipolar equations are determined by substituting the motion parameters of the cluster centers for (2.113).

Once the epipolar equations are determined, we can check which initial match candidates satisfy the epipolar equation while keeping consistent with other matches. By doing this, the individual match that did not have chance to form correct groups can also be found.

Note that we do not use the global relaxation technique because the above technique using local information can be applied to finding multiple clusters for multiple motions.

6.4 IMAGE MATCHING WITH THE RECOVERED EPIPOLAR GEOMETRY

Once the fundamental matrix has been determined robustly, we can use the recovered epipolar constraint to determine correspondence for other image features like corners, edge points, etc. Many good stereo matching algorithms are available such as the PMF algorithm [118].

6.4.1 A Simple Implementation for Matching Corners

Since the development of a good stereo matching algorithm is not the purpose of this work, we have developed a very simple one, which is actually a slightly modified version of the initial matching process (Sect. 6.1). For a feature point in the first image, and in order to find possible matching partners not too far from the epipolar line in the second image, we place a narrow band of width 2ϵ pixels centered on this epipolar line and find the points that lie within the band. The value of ϵ is chosen to be $3.8\bar{d}$ for a probability of 95%, where $\bar{d}$ is the root of mean squares of distances between the points and their epipolar lines defined by the recovered fundamental matrix, i.e., $\bar{d} = \sqrt{\sum_i w_i r_i^2 / \sum_i w_i}$. The same constraints as in Sect. 6.1 are then applied to select the most consistent matches, except that the constraint on the disparity (defined by the search window) is replaced by the epipolar constraint just described.

If needed, we can refine the estimation of the fundamental matrix using all correspondences established at this point. The number of correspondences found in this step is usually larger.

6.4.2 A Simple Implementation for Matching Edges

Under the weak or para- perspective projections, once the epipolar equations are recovered, disparity can be defined according to (5.20). Then the matching of edge images can be greatly simplified by making use of this constraint. As described in Chap. 5, we can construct the Spatial Disparity Space, in which all possible matches are represented as an active node. We need not search over all the image, once the feature points like corners have been matched. The disparities for the matched feature points are computed and used as a rough estimate of the disparity range we need to search over.

Next, the task is to choose at most one match for each edge point and delete all others by maximizing the smoothness of the surfaces or contours in the SDS.

A stochastic measure of smoothness can be defined as

$$E(u,v,d) = \sum_u \sum_v \sum_{i=-\Delta u}^{\Delta u} \sum_{j=-\Delta v}^{\Delta v} \frac{(\Delta d_{min}(u+j, v+j))^2}{i^2+j^2} , \tag{6.7}$$

where $\Delta d_{min}(u+j, v+j)$ is the possible minimal difference between $d(u,v)$ and other active nodes for $(u+j, v+j)$.

There are different algorithms to minimize this energy. We found in our examples that minimizing the energy independently for each edge point gives sufficiently good results, because the search range is quite limited due to knowledge of feature point matching. This algorithm is the simplest among all possible minimization algorithms. It is also used in our approaches to the motion correspondence and segmentation problem and to the object recognition and localization problem, as described in the following chapters.

6.5 AN EXAMPLE OF MATCHING UNCALIBRATED PERSPECTIVE IMAGES

The proposed algorithm has been implemented in C programming language with X/Motif interface. It can be used either in interactive mode or in batch mode. It is made available on the internet since early 1994, and is used by a number of laboratories all over the world. It has been tested on different types of images, such as indoor, rocks, road, and textured dummy scenes. Good results have been obtained. Here, we just provide one example. The reader is referred to [188] for more examples. The interested reader is encouraged to try his/her own images. As of this writing, the program is available through the WWW page:

`http://www.inria.fr/robotvis/personnel/zzhang/zzhang-eng.html`

The image pair is the same as that used in Sect. 3.1.8, and was taken by a stereo system in a room containing a stack of rocks. The stereo calibration parameters are not used in the experiment reported here; that is, the two images are treated as being uncalibrated. The Harris corner detector is applied to each image, and we obtain 566 points for the left image and 465 points for the right. They are indicated by a cross in Fig. 6.5.

Figure 6.5 Image pair of a rock scene overlayed with points detected by Harris corner detector.

In order to show the performance of the relaxation procedure, we also provide the matching results given by the correlation technique. The correlation results are obtained as follows. We perform the correlation twice by reversing the roles of the two images (i.e., from left to right, and then from right to left) and consider as valid only those matches for which the reverse correlation has fallen on the initial point in the first image. More precisely, for a given point $\mathbf{m}_1$ in the left image, let the match candidate with the highest correlation score be $\mathbf{m}_2$ through a left-to-right correlation. Before validating the match, we perform a right-to-left correlation. If the match candidate with the highest correlation score for $\mathbf{m}_2$ is again $\mathbf{m}_1$, then this match will be validated; otherwise, it will be rejected. The two images thus play a symmetric role. This validity test allows us to reduce greatly the probability of error.

Figure 6.6 shows the matching result with the correlation technique. Matched points in the two images are given the same number. Because the rocks have a nice texture, the correlation works quite well. Among 242 matches found, only 52 are false. After carrying out the relaxation procedure, the result improves (see Fig. 6.7). Several false matches have been corrected, and more matches have been found. The total number of matches is 301, and there are 46 false matches. In this example, the number of ambiguous matches due to correlation is not very high. If the number of ambiguous matches is high (such as in a scene containing repetitive patterns, see [188] for examples), the improvement in the matching result due to relaxation is impressive thanks to the use of contextual (neighboring) information.

Figure 6.6 Matching result with the correlation technique.

Figure 6.7 Matching result with the relaxation technique. The estimated epipolar geometry without false match detection is also shown.

If we now estimate the epipolar geometry with the nonlinear technique using all matched points by relaxation, the result is not correct, as shown by the epipolar lines in Fig. 6.7, which should be compared with the "ideal" epipolar geometry shown in Fig. 3.5. The average distance between a point and its

Figure 6.8 Outliers detected by the LMedS method. All 46 false matches, plus 11 not very well localized points

epipolar line is 15.1 pixels, which is very large. If we use the measure introduced in Sect. 3.1.8 on page 116, the difference between this estimate and the "ideal" epipolar geometry is as high as 109.2 pixels. Numerically, the two fundamental matrices are given by

$$\mathbf{F}_{\text{relax}} = \begin{bmatrix} 1.0698 \times 10^{-6} & -1.5101 \times 10^{-5} & 0.00573 \\ 1.7159 \times 10^{-5} & 1.8818 \times 10^{-6} & -0.00542 \\ -0.00814 & 0.00301 & 0.99993 \end{bmatrix},$$

$$\text{and} \quad \mathbf{F}_{\text{calib}} = \begin{bmatrix} -5.8629 \times 10^{-6} & -8.8801 \times 10^{-6} & -0.03424 \\ 1.1702 \times 10^{-5} & 3.4835 \times 10^{-6} & -0.02307 \\ 0.03392 & 0.02067 & 0.99836 \end{bmatrix}.$$

They have been, respectively, normalized by their Frobenius norm.

The LMedS technique has detected all 46 false matches. 11 good, but not well localized, matches were also identified by LMedS as outliers. All these are shown in Fig. 6.8. The epipolar geometry estimated by LMedS was already shown in Fig. 3.8. The average distance between points and epipolar lines is now 0.87 pixels. One can observe the significant change in the epipolar geometry. Indeed, the estimation of the epipolar geometry is a procedure very sensitive to false matches. We have noticed that even with one or two false matches the epipolar geometry estimated without outlier rejection may be completely different from the true one. Since M-estimators depend tightly on the initial estimate, use of robust techniques with high breakpoint like LMedS becomes indispensable in practice.

6.6 AN EXAMPLE OF MATCHING UNCALIBRATED AFFINE IMAGES

Using the techniques described in Sect. 3.2, Sect. 3.3 and Sect. 6.3, we have successfully found the epipolar equation of the following pair of uncalibrated stereo images and matched the two images using the recovered epipolar equation.

Figure 6.9 Two uncalibrated views of a Mac computer

Figure 6.9 shows two uncalibrated images of a Mac computer. Assuming the two images are taken by a weak perspective or paraperspective camera, we first extract feature points from the images, find possible matches by rotated template matching, and then use the inconsistency check to exclude inconsistent matches. From each pair of match, the closest 3 neighbors in the first image are located, which form groups of 4 pairs of point matches. For each group, if the 4 pairs of matches are consistent with each other, an epipolar geometry is determined. If no group remains after the consistency check, then the trouble-making neighbor is found and removed, and the next closest neighbor is included. This process repeats until we can determine epipolar geometry. The computed epipolar geometries are then projected onto the motion parameter space, and clustering technique is used to find clusters in it. The result is shown in Fig.6.10. The concentration of the clusters for the true epipolar geometry is several times larger than other "clusters". The feature points that satisfy the recovered epipolar equation are numbered and shown in Fig.6.11. The epipolar equation is then recomputed using these matched points as

$$0.003u - 0.591v - 0.056u' + 0.576v' + 10.0 = 0\ .$$

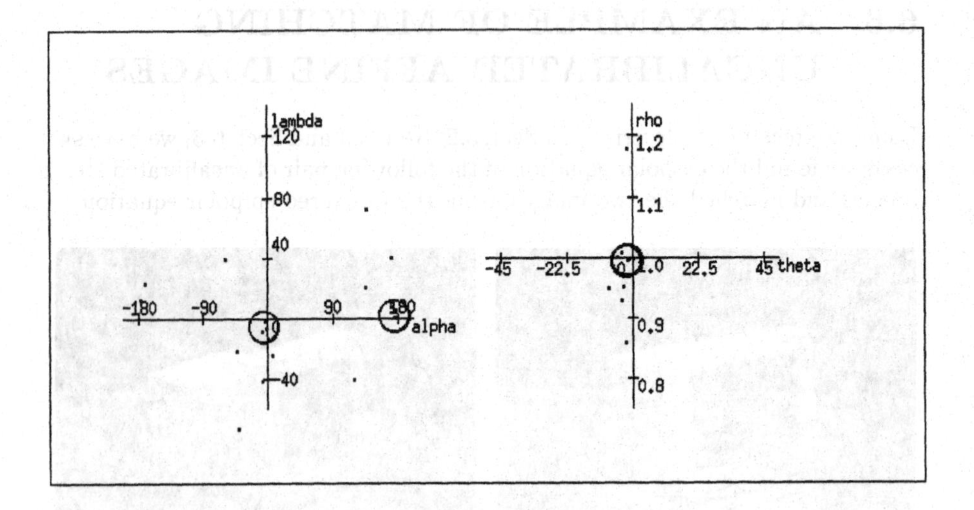

Figure 6.10 Clusters found for the two views of Mac Computer: $\alpha_1 = -6.553879$, $\alpha_2 = 175.980881$, $\theta_1 = -2.845327$, $\theta_2 = -2.244064$, $\rho_1 = 0.996819$, $\rho_2 = 0.999406$, $\lambda_1 = -5.645007$, and $\lambda_2 = 2.335877$

The edge images (Fig.6.12) are then matched using the recovered epipolar equation. The matched edge points in image 2 with respect to image 1 are shown in Fig.6.13. It can be seen that most edge points are correctly matched. The existence of unmatched points near the upper left and right bottom corners is due to the mispositioning of epipolar lines for a few pixels along the perpendicular direction.

The matched edge images can be used as model views for recognizing and localizing the Mac computer in other images. For how to do this, see Chap. 8.

Figure 6.11 Matched feature points are marked by the same numbers.

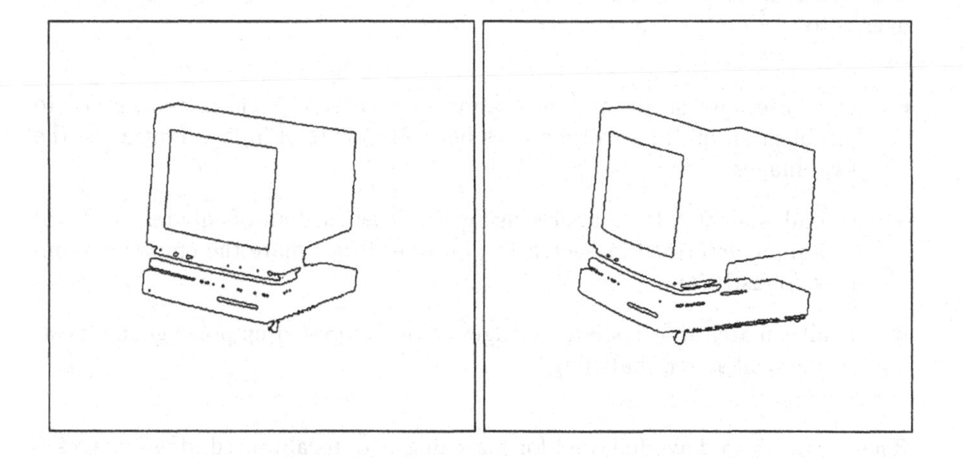

Figure 6.12 Edges in the two images

6.7 SUMMARY

In this chapter, we have described two complete systems for matching two uncalibrated perspective images and matching two uncalibrated affine images by recovering their epipolar geometry.

Figure 6.13 Edge points matched in image 1 with respect to image 2

The approach we have followed for matching two uncalibrated perspective images is to

- first, use correlation and relaxation or consistency check techniques to establish an initial set of matches between points of high curvature in the two images,
- second, detect false matches using the least-median-of-squares (LMedS) technique described in Sect. 3.1.6, in order to estimate the epipolar geometry robustly, and
- finally, match more points or edges using estimated epipolar geometry as in classical stereo matching.

The approach we have followed for matching two uncalibrated affine images is to

- first, use correlation and consistency check techniques to establish an initial set of matches between points of high curvature in the two images,
- second, detect false matches using the clustering technique described in Sect. 3.3, in order to estimate the epipolar geometry robustly, and

- finally, match more points or edges using estimated epipolar geometry as in classical stereo matching.

The difference between the two approaches is only in the second step.

The first step is not crucial, and other techniques can be used. It is important mainly from a pratical point of view. The objective is to reduce the complexity by restricting the parameter space using some valuable heuristics. The second step is necessary because the set of matches established by heuristic techniques usually contain false matches, which, sometimes even one, may perturb the estimation of epipolar geometry completely if least-squares techniques are used.

7

MULTIPLE RIGID MOTIONS: CORRESPONDENCE AND SEGMENTATION

Motion has been one of the main research topics in computer vision [96]. Traditionally, people have divided the motion problem into correspondence and structure-from-motion [164]. Actually, however, there is another problem, that is, segmentation of motion images into different rigid objects. Segmentation is very important itself, because in many real world tasks like target following segmentation is a precondition, perhaps more frequently required than accurate shape recovery. Also, from a computational point of view, it has to be solved in the case of multiple motions for both correct correspondence and structure-from-motion computation.

7.1 PROBLEMS OF MULTIPLE RIGID MOTIONS

In matching two motion images, the *aperture problem* has to be solved [179, 63]. The conventional wisdom is to impose the smoothness constraint on the 2D flow field so that we can obtain a solution for every point in the image [69, 65]. One problem with this is that smoothing the whole motion field uniformly does not give correct answers in many cases. As has been seen before, corresponding points can only lie on unknown 1D epipolar lines. 2D smoothing, however, does not in any way guarantee this property. Another problem with this approach is that imposing smoothness along motion boundaries gives wrong answers to those boundary points and their neighbors. Here segmentation and motion boundary detection is vital to correct solutions.

In structure-from-motion algorithms, since the fundamental assumption is *rigidity* of objects, segmenting images into independent rigid motions is a precondition for applying those algorithms [164, 90, 161, 174, 154, 190, 8]. To avoid the segmentation problem, many of the papers assume that there is only one moving object against the static background or that the camera is moving so that the environment can be regarded as a single rigid body.

Compared with the correspondence problem and structure-from-motion problem, segmentation has received less attention. Direct segmentation has been difficult because image flow arises from both object motion and camera motion. The work reported so far can be classified into two categories. The first category relies on continuity and discontinuity in the 2D flow field [1, 106, 32, 21]. The second category relies on examining if flows share or violate the epipolar and rigidity constraints [108, 142, 109, 168, 15].

We take an approach which explicitly uses the epipolar constraint for segmentation, that is, whether epipolar equations are shared or not is explicitly examined. Since we are not given segmented images, we have to take a generate-and-test approach to segmentation. Hypotheses of epipolar equations are generated and if there are clusters of the hypotheses, they are regarded as independent rigid motions.

We start from matching feature points like corners and junctions detected by a corner detector and matching them by template correlation. They are then grouped by the epipolar constraint. The obtained epipolar equations are then used for correspondence, as the epipolar constraint reduces the search space from 2 dimensions to 1 dimension, thus eliminating the "aperture" problem. It looks like a problem of stereo matching, but it differs from that in the sense that there are more than one set of epipolar lines in the images and the boundary is not given.

It is somewhat surprising that this approach has not been tried so far. The apparent reasons seem to be that the epipolar geometry between arbitrary object and view motions has not been well understood and that the recovery of multiple epipolar geometries has been a hard problem.

We present an algorithm for segmentation and correspondence between edge images with known epipolar constraint. We also present an algorithm to detect transparent rigid motions.

7.2 DETERMINING EPIPOLAR EQUATIONS FOR MULTIPLE RIGID MOTIONS

The recovery of epipolar geometry for multiple motions through matching feature points is a vital step [109, 144, 168].

7.2.1 The Algorithm

Now given two images, we first match feature points by template correlation techniques. Secondly, to find the epipolar equations underlying the two images, we take the generate-and-test strategy, that is, we generate hypotheses of groups of matched points, compute epipolar equations for them, and then see which are shared by many points. The clustering algorithm is discussed in detail in Sect. 3.3.

The procedure is:

1. find corners, junctions, and other feature points in two images;
2. establish correspondence between the two sets of feature points by computing correlation between local image patterns around feature points;
3. for each matched point in image 1, find the $k (k \geq 3)$ (in the current implementation, k is 6) closest neighbors, and form a group for each combination of 3 neighbors;
4. for each group, compute the epipolar equation and the corresponding motion parameters; if the points undergo a 2D affine motion (if the third eigenvalue of $\mathbf{W}$ is smaller than 1.0 in this implementation), then determine the 2D affine motion equations and motion parameters; if the computed motion parameters are too large, discard the group;
5. find clusters in the motion parameter space;
6. for each cluster, compute the epipolar equation;
7. merge individual matches that satisfy the epipolar motion equations.

In the current implementation, k is 6. Due to the limited number of feature points in each image, the computation is not that terrible as one might imagine, even though we have to determine the equations for each combination of local groups. Moreover, much of the computation can be done in parallel. The 2D affine motions, if any, can be found in the same way.

7.2.2 Experimental Results

We have implemented the algorithm to recover the multiple epipolar equations from feature matching in motion images. Two image sequences are used. One is taken by a moving camera of a scene including a static background and a moving soccer ball. Since the camera is moving, every point in the image is apparently moving. The second image sequence is taken by a static camera of a scene in which two soccer balls move independently in front of a static background. Thus there are three motions, the two balls undergoing rigid motions represented by epipolar equations and the background represented by a 2D affine motions.

First, feature points are detected by a modified feature detection operator proposed by [28, 78]. They are matched through the image sequence (as long as possible). Since the motion between successive images may not be sufficient for clusters to emerge in the motion parameter space, the decision is prolonged till the motion flows are long enough to determine reliable epipolar equations.

Figure 7.1 Motion image 1

Figure 7.2 Motion image 2

Figs. Fig. 7.1,Fig. 7.2 and Fig. 7.3 show two original motion images, the detected feature points and their flows. The clusters for the ball are found in the motion parameter space (Fig. 7.4). Their concentration measures are 3 times larger than other "clusters". The next figures in Figs. Fig. 7.5 and Fig. 7.6 show the segmentation results, one for points on the ball and the other for points on the background. Nearly all points are correctly classified.

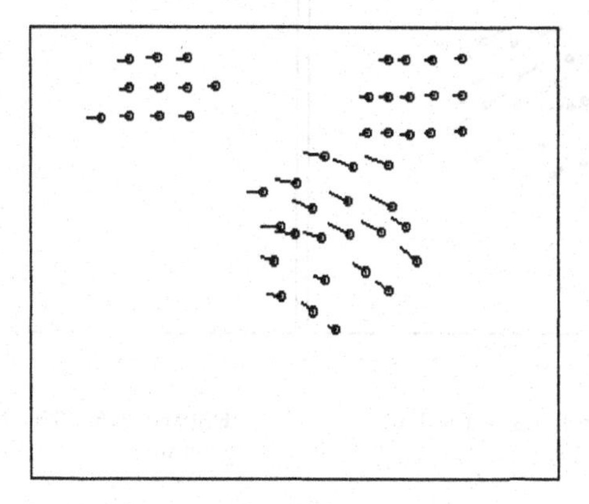

Figure 7.3 The optical flow of feature points

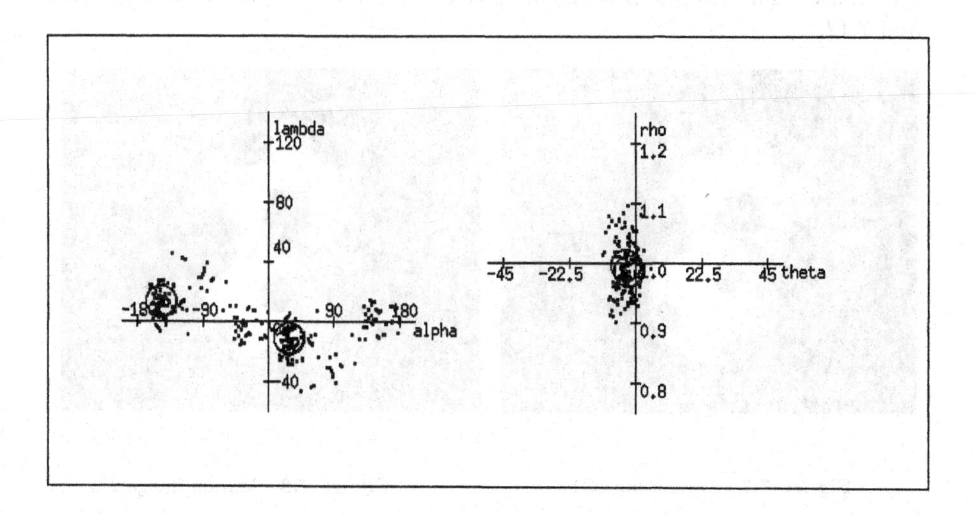

Figure 7.4 The clusters found for the ball in motion parameter space. The background points are excluded due to rank deficiency. $\alpha_1 = 28.282593$, $\alpha_2 = -149.29328$, $\theta_1 = -2.999845$, $\theta_2 = -3.839111$, $\rho_1 = 0.988856$, $\rho_2 = 0.998571$, $\lambda_1 = -11.611038$, and $\lambda_2 = 14.337216$

Figs. 7.7, 7.8 and 7.9 show another example, where there are two balls moving independently. The clusters for the two balls are located in the motion param-

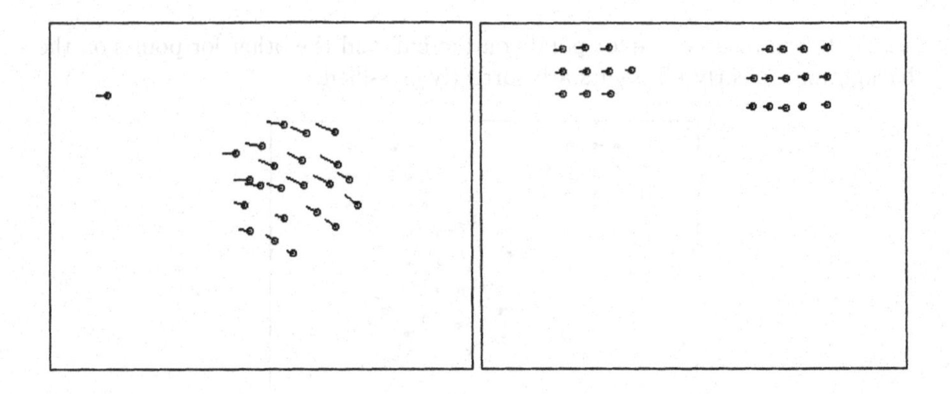

Figure 7.5 The flows of the ball

Figure 7.6 The flows of the background

eter space. The results of segmentation of the flows are shown in Figs. 7.11 and 7.12.

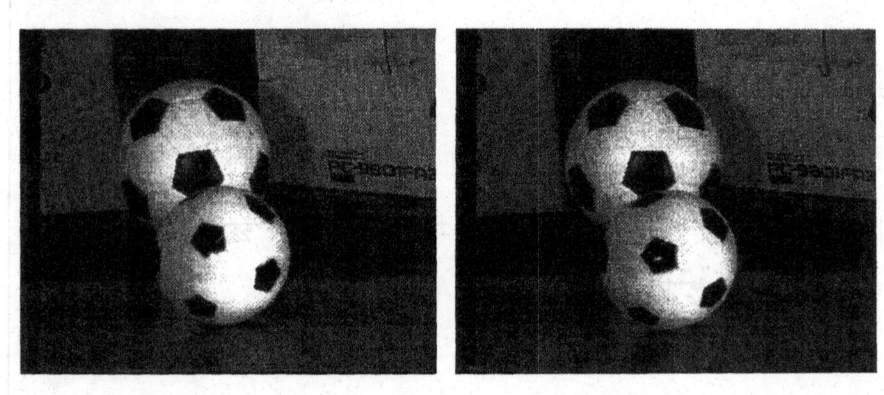

Figure 7.7 Motion image 1

Figure 7.8 Motion image 2

Note that the clustering is not necessary for every pair of consecutive images. Actually, once the feature points are classified into different motions and objects, the groupings can be kept through the image sequence as long as they are visible.

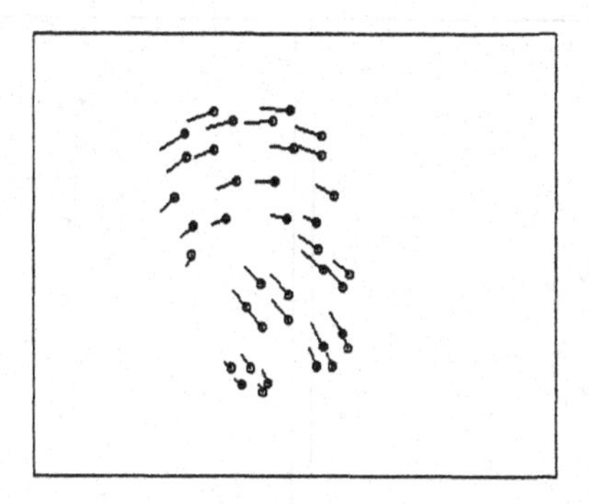

Figure 7.9 The optical flow of feature points

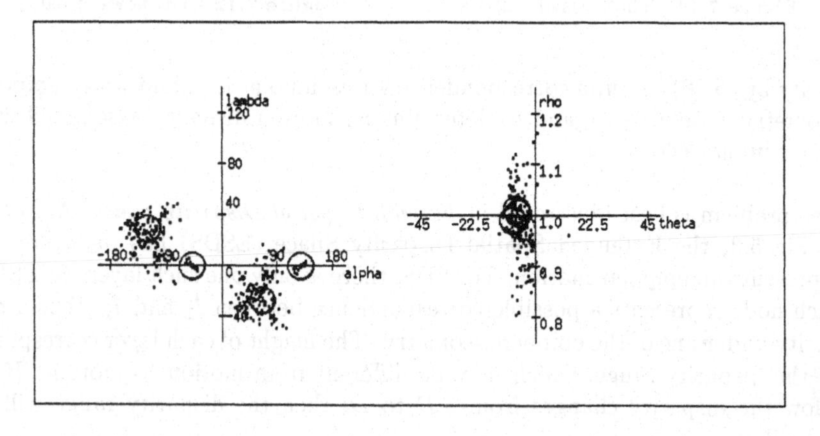

Figure 7.10 The clusters for the 2 balls in motion parameter space. There are too few points from the background to form a cluster. For ball 1, $\alpha_1 = 66.773567$, $\alpha_2 = -122.562492$, $\theta_1 = -7.010643$, $\theta_2 = -7.681648$, $\rho_1 = 1.007331$, $\rho_2 = 1.013413$, $\lambda_1 = -27.145382$, and $\lambda_2 = 29.840086$. And for ball 2, $\alpha_1 = 129.210052$, $\alpha_2 = -50.059856$, $\theta_1 = -9.476761$, $\theta_2 = -9.445133$, $\rho_1 = 0.992618$, $\rho_2 = 0.990300$, $\lambda_1 = -0.544330$, and $\lambda_2 = -0.357600$.

7.3 MATCHING AND SEGMENTING EDGE IMAGES WITH KNOWN EPIPOLAR EQUATIONS

7.3.1 Representing the Problem in ESDS

Once the epipolar equation is known, by rotating the two images so that the corresponding points always lie on the same horizontal lines, we can define dis-

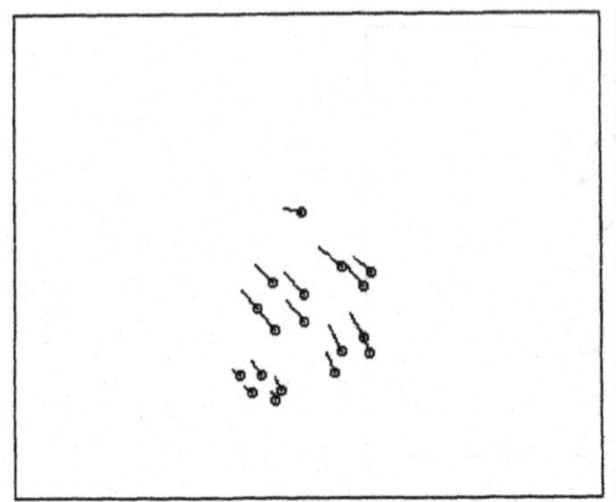

Figure 7.11 The flows of ball 1

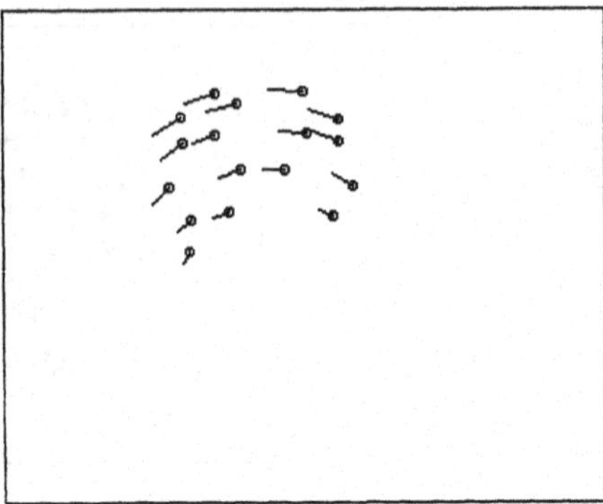

Figure 7.12 The flows of ball 2

parity by (5.20). To find correspondence for each point is to find which epipolar geometry it belongs to, and to determine its motion disparity associated with that epipolar geometry.

The problem is best illustrated in *Extended Spatial Disparity Space.* As shown in Fig. 5.3, the Extended Spatial Disparity Space (ESDS) has m layers. F_i represents an epipolar motion. For SDS, there is only one such layer. In ESDS, each node represents a possible correspondence between I_1 and I_2, thus a disparity under one of the epipolar geometry. The height of each layer corresponds to the disparity range, which may be different from motion to motion. If we allow the disparity changes from $-D$ to D, then the disparity range will be $2D + 1$.

For the 2D affine motions, the thickness of each layer is only 1 pixel, because the motion is completely determined.

For each edge point in the first frame, we find out the possible corresponding edge points in the second frame, for which the corresponding disparity, or corresponding nodes in the ESDS, are marked *active.* The operation is illustrated in Fig. 7.13.

For each column, we usually have multiple active nodes (in Fig. 7.13 there are two epipolar lines and 4 active nodes), thus making it necessary to use the smoothness constraint to choose only one from among the 4 candidates. It can be easily seen from Fig. 7.13 that correct matching does have smoother disparity changes.

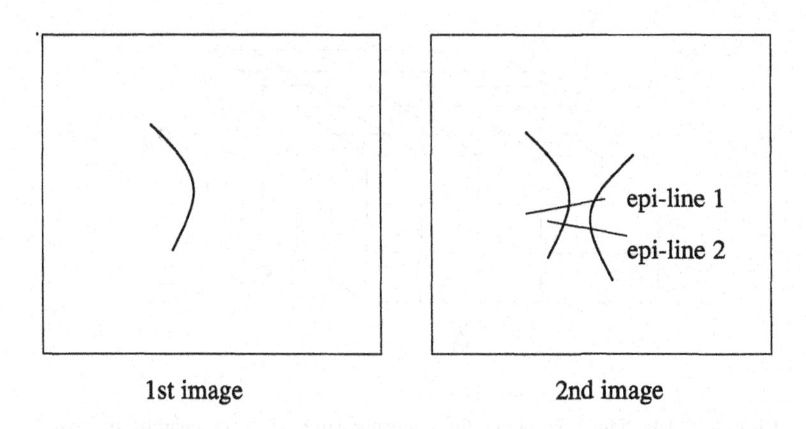

Figure 7.13 Disparities defined along multiple epipolar lines

7.3.2 A Support Measure for Selection from Multiple Candidates

The criterion for selection is the smoothness constraint. That is, we select the disparity that is least different from the disparity values of its neighbors. For each disparity, we define a box centered at that node, with the size to be S_x, S_y and S_d (Fig. 7.14). Each active node, corresponding to a possible disparity, within the box, is counted, according to a point system, which gives 2 points if there is no disparity difference, 1 point if the disparity difference is equal to 1 pixel, and 0 point otherwise. This is a kind of support for the disparity from neighbors. The higher the points are, the more neighboring points have similar disparities. This, as a smoothness measure, is different from the derivatives defined for continuous field. Here, for discrete edge points, we can only define a kind of stochastic measure for smoothness like this one.

The total score for a node $C(u, v, d)$ is defined as

$$C(u, v, d) = \sum_{y \in S_y} \sum_{x \in S_x} \max_{d \in S_d} P(x, y, d) \ , \tag{7.1}$$

where $P(x, y, d)$ is the score that the node (x, y, d') contributes to $C(u, v, d)$:

$$P(x, y, d) = \begin{cases} 2 & d - d' = 0 \ , \\ 1 & d - d' = \pm 1 \ , \\ 0 & otherwise \ . \end{cases}$$

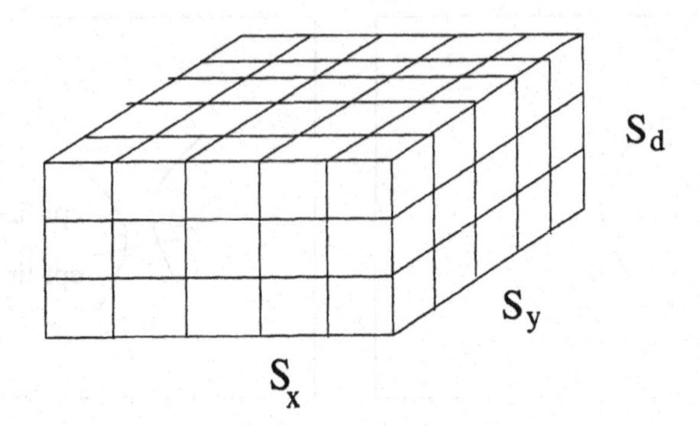

Figure 7.14 The box range for counting support from neighboring edge points

Along each column in the ESDS, or for each image point (u, v), the disparity node which has the largest count is selected as the true match, and the motion that this disparity represents is the motion that this edge point belongs to.

As the disparity value cannot change very much between successive images, this simple measure gives good estimate of the disparity variation. Experimental results show astonishingly good correspondences and correct segmentation between edge images.

7.3.3 Experimental Results

We have tested the above algorithm to a number of motion images.

Figs. Fig. 7.15 and Fig. 7.16 are two edge images taken out from the first image sequence described in Sect. 7.2.2. The epipolar equation for the moving ball and the motion equations for the background are computed and used for matching the two edge images. The results are shown in Fig. 7.17 for the ball and in Fig. 7.18 for the background.

Figure 7.19 and Fig. 7.20 are another pair of edge images taken out from the second image sequence described in Sect. 7.2.2. The epipolar equations for the two moving balls and the motion equations for the background are computed and used for matching the two edge images. The results are shown in Fig. 7.21 for ball 1, Fig. 7.22 for ball 2 and Fig. 7.23 for the background.

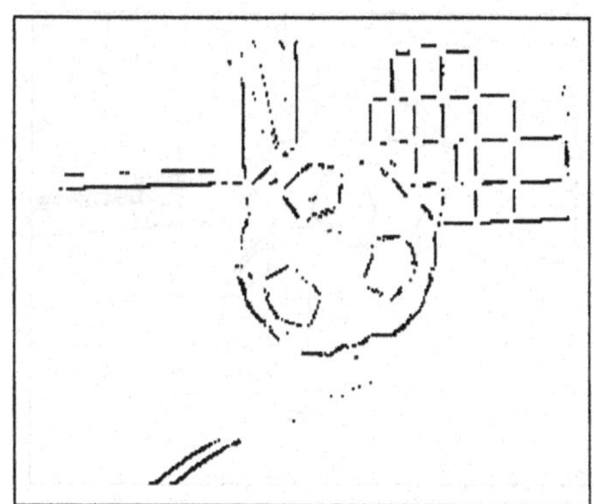

Figure 7.15 Edge image 1

Figure 7.16 Edge image 2

Figure 7.17 Segmented ball

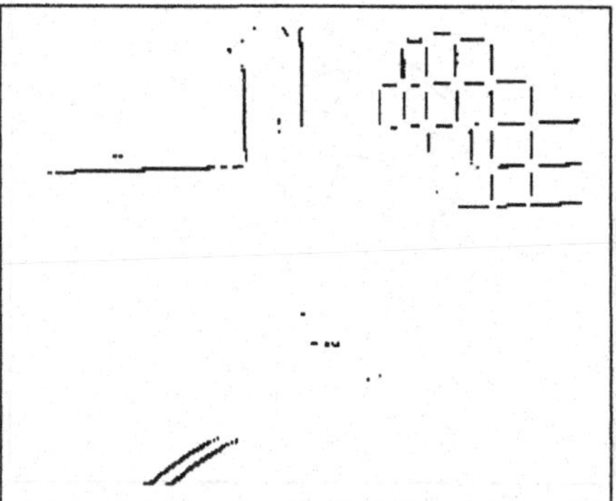

Figure 7.18 Segmented background

It can seen from these figures that most of edge points are correctly segmented. This is encouraging because we use only very simple local operations. Some of the misclassified edge points are due to poor performance of the edge detectors which fail to detect some edge points at important locations.

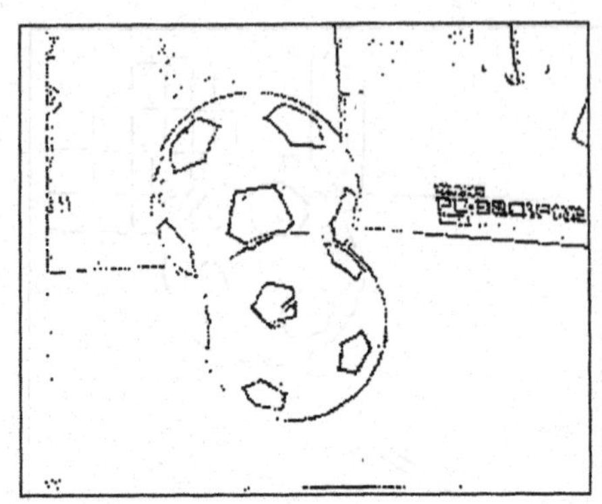

Figure 7.19 Motion image 1

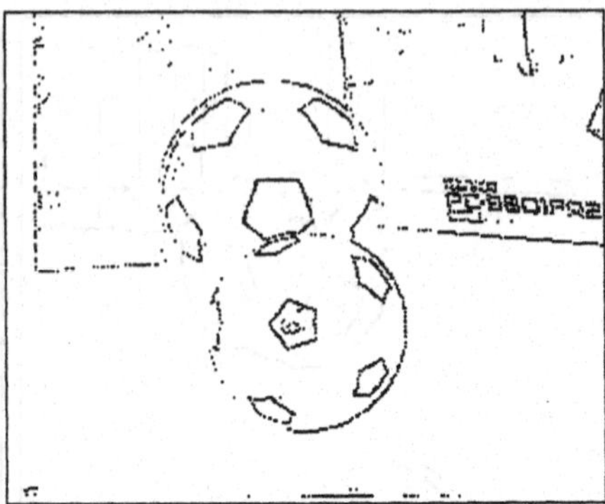

Figure 7.20 Motion image 2

Figure 7.21 Segmented ball 1

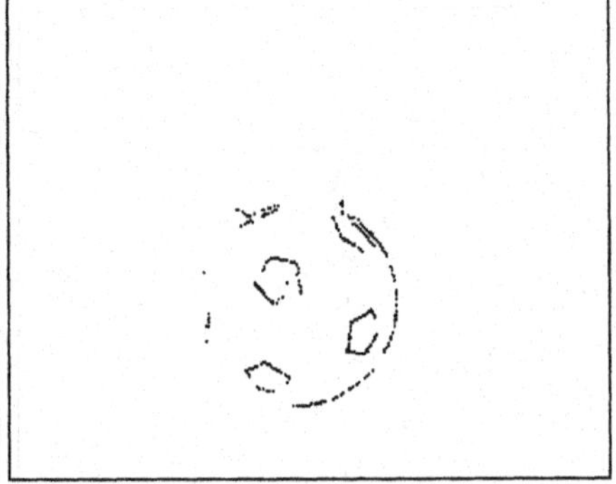

Figure 7.22 Segmented ball 2

7.4 TRANSPARENT MULTIPLE RIGID MOTIONS

The above algorithm can be modified to deal with transparent motions [141]. Transparent motions can be understood as superimposition of object motions behind moving transparent object like glass, transparencies, or clouds. The objective here is to segment the feature points into groups that each represents a separate motion.

As we assume that the motions are rigid, they must also satisfy the epipolar constraint. What differs between transparent motions and non-transparent

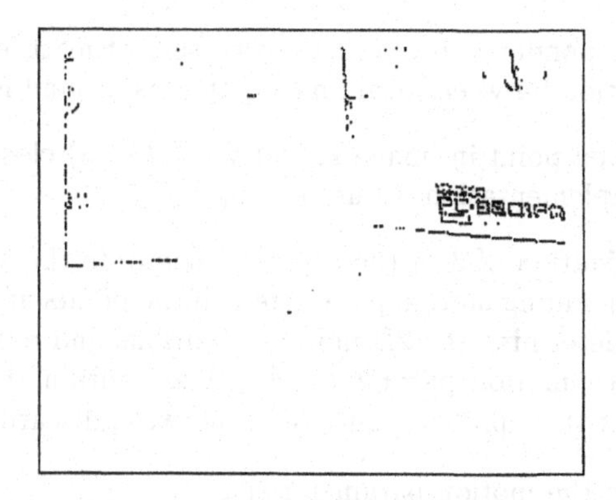

Figure 7.23 The matched edge points for the background

motions is that the flows of different motions in the former case are spatially mixed, while in the latter case the flows of different motions are separated by objects' boundaries. This difference brings difficulty in finding clusters in the motion parameter space, because within a local neighborhood, it is likely to have feature points from other objects. We can enlarge the neighborhood to include more points, thus having a larger probability of including 3 other points from the same object. This of course leads to more combinations of 4 point groups, for each of which an epipolar equation is computed.

Computation cost is one minor problem; the major concern is that more random points in the motion parameter space might blur the clusters.

Fortunately, if the motions are different, then a combination of flows from different motions would lead to an epipolar equation corresponding to strangely large image torsions, or strangely large scale changes, strangely large translations, which are not likely thus not allowed in successive motion images. By this constraint on temporal continuity, we can reject combinations of flows which produce such motion parameters, so that the number of epipolar equations registered in the motion parameter space for clustering is greatly reduced and limited.

The procedure for transparent motions is given below, which is modified from that in Sect. 7.2.1:

1. find corners, junctions, and other feature points in two images;

2. establish correspondence between the two sets of feature points by computing correlation between local image patterns around feature points;
3. for each matched point in image 1, find the $k (k \geq 3)$ closest neighbors (in the current implementation we use $k = 6$;
4. for each combination of 3 of the k points, compute the epipolar equation and the corresponding motion parameters; if the points undergo a 2D affine motion, then determine the 2D motion equations and motion parameters; if the computed motion parameters θ, ρ, λ are within the predetermined ranges, register this epipolar equation; otherwise, discard this group;
5. find clusters in the motion parameter space;
6. for each cluster, compute the epipolar equation;
7. merge individual matches that satisfy the epipolar motion equations.

We have used the above procedure for a set of synthesized data which include two rigid motions, each of which has 30 matched points. Each point is added an isotropic Gaussian noise with deviation of 1 pixel. The motion parameters used for synthesis are: for motion 1, $\alpha = 10.0$, $\theta = 8.0$, $\rho = 1.1$, and for motion 2, $\alpha = -45.0$, $\theta = -9.0$ and $\rho = 0.9$. The mixed flows are shown in Fig. 7.24, and the clusters are shown in Fig. 7.25 and the segmented flows are shown separately in Fig. 7.27 and Fig. 7.29. If one compares the results with the given two motions (Fig. 7.26 and Fig. 7.28), it is found that one flow failed to be correctly segmented in motion 1 and two flows failed in motion 2. In general, the performance is influenced by the difference between the motions. The larger the difference, the easier the segmentation.

7.5 SUMMARY AND DISCUSSIONS

We have in this chapter presented a new approach to motion correspondence and segmentation via epipolar geometry. Firstly, feature points like corners are detected and matched between successive motion images. Then local groups of 4 flows are generated and for each group an epipolar geometry is determined. The epipolar geometry hypotheses are projected onto the motion parameter space for clustering. Once clusters are found, they are used to classify the individual flows for feature points.

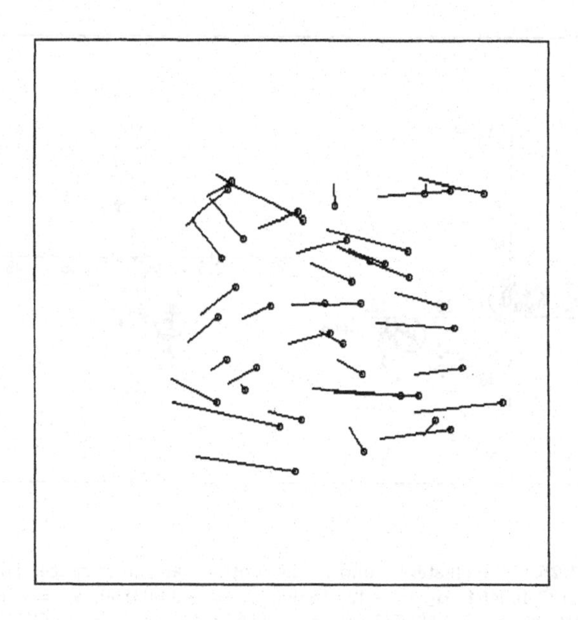

Figure 7.24 Flows generated for two different rigid motions are mixed.

The recovered epipolar equations can then be used to match edge images. Now we search for correspondences along multiple epipolar lines. The Spatial Disparity Space for stereo matching is extended to allow more than one epipolar geometries. Once the Extended Spatial Disparity Space is constructed, the search for unique correspondence from among multiple candidates can be done in a similar way as in stereo matching, using a measure of local smoothness. The final result determines not only the disparity but also which epipolar line each match satisfies.

We have also discussed the case of transparent motions. Basically the same algorithm can be applied to transparent motions. Good results are obtained.

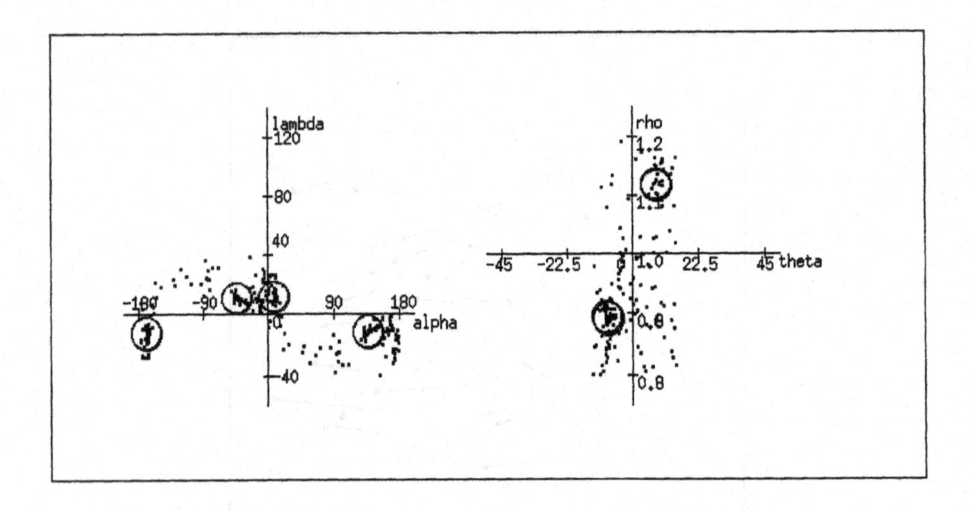

Figure 7.25 The clusters found in the motion parameter space. For motion 1, $\alpha_1 = 137.460464$, $\alpha_2 = -42.655895$, $\theta_1 = -9.003800$, $\theta_2 = -8.993267$, $\rho_1 = 0.894926$, $\rho_2 = 0.895087$, $\lambda_1 = -11.565264$, and $\lambda_2 = 11.624498$. And for motion 2, $\alpha_1 = 9.602901$, $\alpha_2 = -170.052139$, $\theta_1 = 8.066992$, $\theta_2 = 8.072153$, $\rho_1 = 1.115633$, $\rho_2 = 1.115886$, $\lambda_1 = 11.836720$, and $\lambda_2 = -11.770472$.

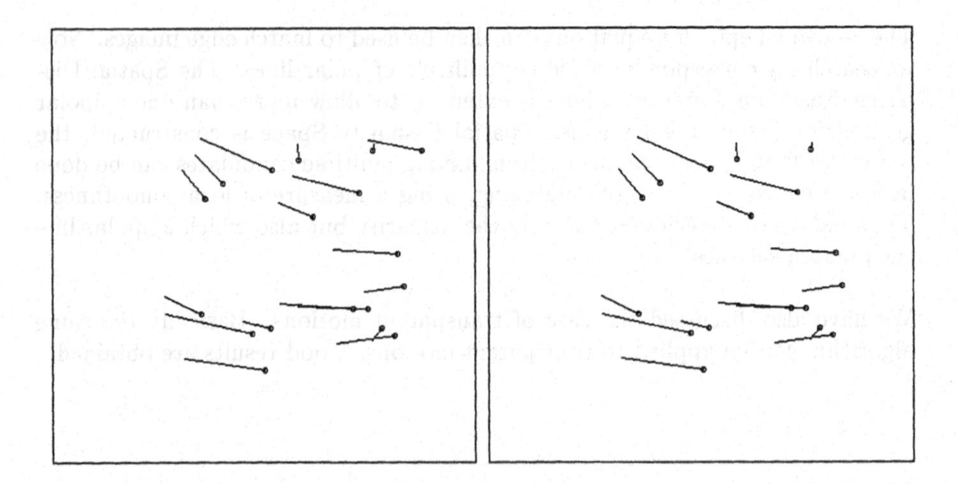

Figure 7.26 Flows belonging to cluster 1

Figure 7.27 Flows found belonging to cluster 1

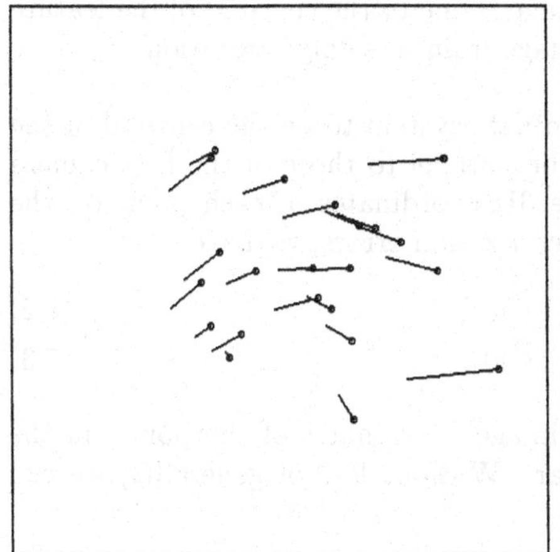

Figure 7.28 Flows belonging to cluster 2

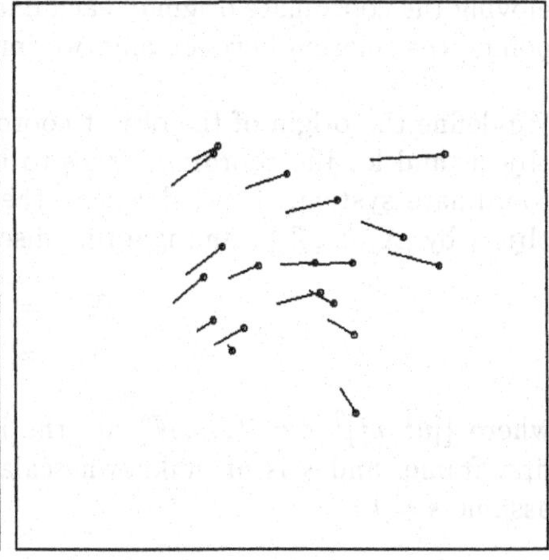

Figure 7.29 Flows found belonging to cluster 2

7.6 APPENDIX: SVD ALGORITHM FOR STRUCTURE AND MOTION RECOVERY WITH KNOWN EPIPOLAR EQUATIONS

Tomasi and Kanade proposed an effective and elegant algorithm using singular value decomposition to determine motion and object structure simultaneously from image points [155]. With known epipolar equations the algorithm can be modified and simplified.

As the image change is most significantly evidenced along the epipolar direction, we only look at the coordinates of points along the epipolar lines.

Suppose that we are given N matched points in F frames of image, and we have recovered the epipolar equation between the first frame and each of the other frames. Using (5.18), we can compute the new coordinate along the epipolar direction for each point in each image, whose size is normalized in accordance with the first image.

Let the new coordinate along the epipolar lines be $\bar{u}_i^j$, where the subscript i indicates i^{th} point, and the superscript j indicates the jth image frame. By moving the coordinate origin of each image frame to the centroid of the feature points, the relation between any two image frames is only a rotation.

We define the origin of the object coordinate system to be the centroid of the object, and let the coordinate axes to be parallel to those of the first camera coordinate system. Now, denoting the 3D coordinates of each point on the object by $[X_i, Y_i, Z_i]^T$ and ignoring discretization errors, we have

$$X_i = s\, u_i^1 , \tag{7.2}$$

$$Y_i = s\, v_i^1 , \tag{7.3}$$

where $\{[u_i^1, v_i^1]^T, i = 1, ..., N\}$ are the image coordinates of the points in the first frame, and s is an unknown scalar. Without loss of generality, we can assume $s = 1$.

The horizontal coordinates computed for each frame except for the first frame using (5.18) are

$$\bar{u}_i^j = \mathbf{m}^{jT} \begin{bmatrix} X_i \\ Y_i \\ Z_i \end{bmatrix} , \quad j \neq 1 , \tag{7.4}$$

where $\mathbf{m}^j$ is the first row vector of the jth rotation matrix which determines the rotation between the first and j^{th} image frames, satisfying $\mathbf{m}^{jT}\mathbf{m}^j = 1$. For the first frame, we just use the horizontal coordinates. Thus, $\mathbf{m}^{1T} = (1, 0, 0)^T$.

If we put all the points and the vectors representing motion into matrix form as

$$\mathbf{H} = \begin{bmatrix} \bar{u}_1^1 & \cdots & \bar{u}_N^1 \\ \vdots & \vdots & \vdots \\ \bar{u}_1^F & \cdots & \bar{u}_N^F \end{bmatrix} ,$$

$$\mathbf{S} = \begin{bmatrix} X_1 & \cdots & X_N \\ Y_1 & \cdots & Y_N \\ Z_1 & \cdots & Z_N \end{bmatrix} ,$$

and

$$\mathbf{M} = \begin{bmatrix} \mathbf{m}^{1T} \\ \vdots \\ \mathbf{m}^{FT} \end{bmatrix} ,$$

then the equation is

$$\mathbf{H} = \mathbf{MS} \ . \tag{7.5}$$

It can be proven in a similar way that the matrix $\mathbf{H}$ is of rank 3. Now the *singular value decomposition* of $\mathbf{H}$ yields

$$\mathbf{H} = \mathbf{U}\Sigma\mathbf{V}^T \ ,$$

where, $\mathbf{U}$ is a $F \times N$ matrix composed of orthonormal unit vectors, $\mathbf{V}$ is a $N \times N$ matrix composed of orthonormal unit vectors, and Σ is a $N \times N$ diagonal matrix composed of singular values sorted in a non-increasing manner ($\sigma_1 \geq \cdots \geq \sigma_N$). $\mathbf{U}$ and $\mathbf{V}$ satisfy $\mathbf{U}^T\mathbf{U} = \mathbf{V}^T\mathbf{V} = \mathbf{V}\mathbf{V}^T = \mathbf{I}$.

Since the rank of $\mathbf{H}$ is only 3, we can retain the largest 3 singular values and those left and right singular vectors associated with them, and treat other singular values as noise. Note that it can be proven [46] that this is the best possible rank-3 approximation.

Let the upper left 3×3 part of Σ be Σ', the first three columns of $\mathbf{U}$ be $\mathbf{U}'$, and the first three columns of $\mathbf{V}$ be $\mathbf{V}'$. Then we have

$$\mathbf{H}' = \mathbf{U}'\Sigma'\mathbf{V}'^T \ , \tag{7.6}$$

where $\mathbf{H}'$ is the optimal estimate of the ideal noise-free measurement matrix.

If we define

$$\mathbf{M}' = \mathbf{U}'[\Sigma']^{\frac{1}{2}} \ ,$$

and

$$\mathbf{S}' = [\Sigma']^{\frac{1}{2}}\mathbf{V}'^T \ ,$$

where

$$[\Sigma']^{\frac{1}{2}} = \begin{bmatrix} \sigma_1^{\frac{1}{2}} & 0 & 0 \\ 0 & \sigma_2^{\frac{1}{2}} & 0 \\ 0 & 0 & \sigma_3^{\frac{1}{2}} \end{bmatrix} ,$$

then (7.6) can be rewritten as

$$\mathbf{H}' = \mathbf{M}'\mathbf{S}' \ . \tag{7.7}$$

The two matrices $\mathbf{M}'$ and $\mathbf{S}'$ are of the same size as the desired rotation and shape matrices $\mathbf{M}$ and $\mathbf{S}$, but generally they are not the same, as the decomposition is not unique. Given any invertible 3×3 matrix $\mathbf{C}$, $\mathbf{M}'\mathbf{C}$ and $\mathbf{C}^{-1}\mathbf{S}'$ are also a valid decomposition. Actually, we can find the desired rotation and shape matrices by properly choosing $\mathbf{C}$.

As the row vectors of $\mathbf{M}$ are unit vectors, we can derive metric constraints for $\mathbf{C}$. Let

$$\mathbf{M} = \mathbf{M}'\mathbf{C}\,, \tag{7.8}$$

and

$$\mathbf{S} = \mathbf{C}^{-1}\mathbf{S}'\,. \tag{7.9}$$

Since

$$\mathbf{m}^{jT}\mathbf{m}^{j} = 1\,, \qquad j = 1, ..., F$$

we can define $\mathbf{Q} = \mathbf{C}\mathbf{C}^T$, and obtain F linear equations for the 6 components of the symmetric $\mathbf{Q}$:

$$\mathbf{m}'^{jT}\mathbf{Q}\mathbf{m}'^{j} = 1\,, \qquad j = 1, ..., F. \tag{7.10}$$

If $F \geq 6$, then $\mathbf{Q}$ can be determined uniquely in the least-square sense.

$\mathbf{Q}$ has only 6 unknowns, but $\mathbf{C}$ has 9 unknowns. Thus we cannot uniquely determine $\mathbf{C}$ from $\mathbf{Q}$. But fortunately, the remaining ambiguity is only a rotation of the whole reference system which has 3 degrees of freedom. We will first determine $\mathbf{C}'$ using eigenvalue decomposition, and then determine the rotation $\mathbf{R}_0$. $\mathbf{C}$ is determined as

$$\mathbf{C} = \mathbf{C}'\mathbf{R}_0\,. \tag{7.11}$$

Since $\mathbf{Q}$ is symmetric and positive-definite, it is straightforward to divide it into $\mathbf{C}'\mathbf{C}'^T$ by eigenvalue decomposition. Let the eigenvalues of $\mathbf{Q}$ be $\lambda_1, \lambda_2, \lambda_3$, and the respective associated eigenvectors be $\mathbf{e}_1, \mathbf{e}_2, \mathbf{e}_3$. Then we have

$$\mathbf{Q} = \begin{bmatrix}\mathbf{e}_1 & \mathbf{e}_2 & \mathbf{e}_3\end{bmatrix}\begin{bmatrix}\lambda_1 & 0 & 0\\ 0 & \lambda_2 & 0\\ 0 & 0 & \lambda_3\end{bmatrix}\begin{bmatrix}\mathbf{e}_1 & \mathbf{e}_2 & \mathbf{e}_3\end{bmatrix}^T.$$

It is straightforward to see that

$$\mathbf{C}' = \begin{bmatrix}\mathbf{e}_1 & \mathbf{e}_2 & \mathbf{e}_3\end{bmatrix}\begin{bmatrix}\lambda_1^{\frac{1}{2}} & 0 & 0\\ 0 & \lambda_2^{\frac{1}{2}} & 0\\ 0 & 0 & \lambda_3^{\frac{1}{2}}\end{bmatrix}.$$

Once $\mathbf{C}'$ is determined, we can rewrite (7.8) and (7.9) as

$$\mathbf{M} = \mathbf{M}'\mathbf{C}'\mathbf{R}_0, \tag{7.12}$$

and

$$\mathbf{S} = \mathbf{R}_0^T\mathbf{C}'^{-1}\mathbf{S}'. \tag{7.13}$$

Note that $\mathbf{R}_0^T = \mathbf{R}_0^{-1}$.

$\mathbf{R}_0$ can be determined by rotating $\mathbf{C}'^{-1}\mathbf{S}'$ such that (7.2) and (7.3) are satisfied. Let $\mathbf{R} = [\mathbf{r}_0^1 \quad \mathbf{r}_0^2 \quad \mathbf{r}_0^3]$. $\mathbf{r}_0^1$ is determined as

$$\mathbf{r}_0^1 = \mathbf{C}'^T\mathbf{S}'^+\mathbf{u}^1 \ , \tag{7.14}$$

$\mathbf{r}_0^2$ as

$$\mathbf{r}_0^2 = \mathbf{C}'^T\mathbf{S}'^+\mathbf{v}^1 \ , \tag{7.15}$$

where $\mathbf{u}^1 = [u_1^1, ..., u_N^1]^T$ and $\mathbf{v}^1 = [v_1^1, ..., v_N^1]^T$ are the horizontal and vertical coordinates of the points in the first image, respectively, and $\mathbf{r}_0^3$ as

$$\mathbf{r}_0^3 = \mathbf{r}_0^1 \times \mathbf{r}_0^2 \ .$$

Now both $\mathbf{M}$ and $\mathbf{S}$ can be computed from (7.12) and (7.13).

Further, once $\mathbf{m}^j$'s are determined, the rotation angles β^j's can also be computed. Together with the rotation angles α^j's and γ^j's determined from the epipolar equations, motion is thus completely determined in terms of Euler angles.

8

3D OBJECT RECOGNITION AND LOCALIZATION WITH MODEL VIEWS

8.1 INTRODUCTION: 3D MODEL VS 2D MODEL, AND SINGLE MODEL VIEW VS. MULTIPLE MODEL VIEWS

The conventional approaches to 3D object recognition are mostly based on 3D object models. They include model-based feature grouping [91], model-based geometrical reasoning [16], constrained search [49], model fitting guided by local feature [18], feature based geometric hashing [83], automatic generation of search trees [75]. All these approaches rely on explicit 3D data as object model in this or that way. Unfortunately, however, 3D data are not always available for every object. In the case of manufactured objects, the data used in designing may be available. If one has to obtain the data by vision, then the difficulty is that there is still no algorithm available that works in every kind of environment.

Therefore, to avoid 3D model is a natural choice. The problem with using 2D model views, that is, images, is that the same 3D object can give rise to an infinite number of different appearances, with varying distances, varying view points and varying viewing poses. It would not be feasible to prepare a database of infinite number of images for each object.

To tackle this problem, Ullman and Basri proposed using linear combination of a small number (2 or 3 for an aspect) of matched 2D views as object model [165], and Poggio and Edelman proposed a network that learns and recognizes 3D objects using a finite number of matched 2D views [115], thus removing the burden of building explicit 3D models. One thing that is not clear with these approaches, however, is how to efficiently match the (uncalibrated) model views and how to efficiently match the (uncalibrated) model and input views. Recently, the invariance school proposes to look for and compute projective invariants from 2D views of 3D objects and use them for object recognition and indexing [130]. It has the same advantage with the above approaches that 2D information can be directly used without explicitly recovering 3D shape, but has the same problem of feature correspondence in order to compute the invariants from the same sets of features in different views.

One other issue that has not been fully studied in the past research is "segmentation for recognition". Many approaches assume that the image part to be recognized is or can be segmented first by other techniques. However, it is actually very difficult. And we believe that recognition, and segmentation or localization, cannot be "segmented", and should be treated together as the same problem. If the correspondences between the model and the input image can be established, then object localization and recognition are achieved at the same time.

In this chapter, we consider the problem of using model views for recognition and localization of 3D objects. Considering it as a problem of matching between the model view(s) and input view, a number of different situations should be examined. First, is it possible to use only one model view for identification? If possible, how? If not possible, how many model views are necessary? If multiple model views are needed, how are they matched with each other? And how are they matched and localized against the input view with also background and other objects?

The basic idea here is that no matter what the situation is, there is always the problem of matching between two images. If the two images include the same object, the epipolar geometry is always a valid constraint for use. We show that if the two views are projections of the same object and the only object (without other objects) from different view points, the problem is identical with that of matching uncalibrated stereo images, as described in Chapter 6. In that case, the epipolar constraint is sufficient to bring them into correspondence. This solves the problem of correspondence between model views.

If there is only one model view, and the input view has the same object but also other objects, then the correspondence between them may not be unique, depending on the ratio of the size of the target object to the size of other objects. If that ratio is close to 1, that is, almost the whole image is the target object, then the situation becomes close to matching two uncalibrated stereo images.

As that ratio decreases, there may exist more than one epipolar equations and more than one sets of correspondences associated with the epipolar equations. However, if the epipolar equations are wrong, then correspondence between edge images using the wrong epipolar equations is not smooth. This provides an indicator of whether or not the recovered epipolar equations are correct.

The other way to resolve the ambiguity is to use two or more model views. As discussed in Chapter 4, with two model views, by recovering the trifocal tensors, the image can be synthesized. By superimposing this synthesized image with the input view, we can judge whether or not the object is identical.

8.2 RECOGNITION AND LOCALIZATION WITH SINGLE MODEL VIEW

8.2.1 Is a Single View Sufficient for 3D Object Recognition?

Using only one model view for 3D object identification is essentially an under-constrained problem in two senses [177]. First, there is no guarantee to find a set of unique correspondences. Second, even if we can find a unique solution, there is no guarantee that the two views are of the same object. One extreme example is that some objects may appear completely the same in image but differ in the z-direction.

On the other hand, we believe that many things can be done with just one model view. Though the case of same-apprearance-different-shape is possible in theory, this is a rare case in practice. It is excluded from consideration in vision from the standpoint of general viewpoint. Our hope is that our approach can at least reduce the possible matches in a large object database to a manageable number, sometimes to only one. In the case of multiple solutions, we can then use a second model view to dissolve the ambiguity. This is described in next section.

8.2.2 Matching Model View with Input View as Uncalibrated Stereo Images

The problem of matching model view with input view is similar to that of matching uncalibrated stereo images. However, there are 2 major differences between them. First, since in gerenal there are also other objects than the target object, there is inherently more ambiguity in the data. Secondly, while in object recognition, the lighting conditions may change significantly, in stereo, we can assume that the two images are taken at the same time, thus the lighting conditions do not change very drastically between them. Therefore, the problem of matching model and input views is generally more difficult.

Basically we use the same procedure as that used for matching uncalibrated stereo images.

- Find corners, junctions and other high curvature points as feature points;
- Find match candidates between the two sets of feature points by correlation technique while allowing identical image torsions;
- Form groups of neighboring 4 pairs of matches whose spatial relations are preserved;
- Estimate epipolar equation for each such group, and find clusters in the motion parameter space;
- Match edge points using estimated epipolar equation.

For details of each step, see Chapter 6.

Since there are feature points from other objects in the input image, there is a higher possibility that the true match is not found or not included in match candidates. And this possibility increases as the number of feature points from other objects increases. Since we use the local image pattern correlation to find match candidates, this possibility, of course, also depends on how the the image patterns of target object and other objects resemble. If they look different, then the possibility does not increase. However, if they look similar, then the possibility increases.

8.2.3 An Example

The above algorithm was tested with the example of an office scene. The Mac computer was placed in an office, and an input image (Fig. 8.1) was taken with the background included. The two images used in Chapter 6 are used here as model views to locate the Mac computer in the input image. The feature points in all these images are first extracted. Using the procedure described above, the feature points in Mac image 1 and in the input image are matched. The clusters found are shown in Fig. 8.2, and the matched points are numbered and shown in Fig. 8.3 and Fig. 8.4.

Figure 8.1 Edge in Input view with background

The edges in the input image are shown in Fig. 8.5. Using the recovered epipolar geometry, the edge points corresponding to those in the model image are found in the input image and shown in Fig. 8.6.

The same things can be done between the second Mac image and the input image. The clusters found are shown in Fig. 8.7, and the matched points are numbered and shown in Fig. 8.8 and Fig. 8.8.

Using the recovered epipolar equation, edge images can also be matched. The edge points corresponding to those in the second model image are found in the input image and shown in Fig. 8.11, together with the original input edge image Fig. 8.10.

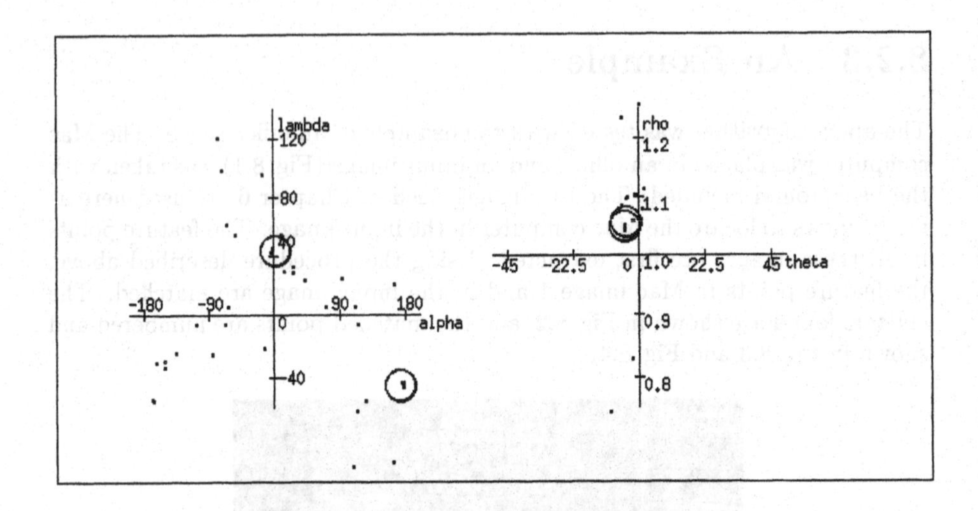

Figure 8.2 Cluster found between the first model view and the input view of the Mac computer: $\alpha_1 = 3.400936$, $\alpha_2 = 173.576248$, $\theta_1 = -4.769201$, $\theta_2 = -5.203446$, $\rho_1 = 1.059555$, $\rho_2 = 1.051757$, $\lambda_1 = 44.560696$, and $\lambda_2 = -45.441090$

Figure 8.3 The matched points are marked by numbers in Model Image 1.

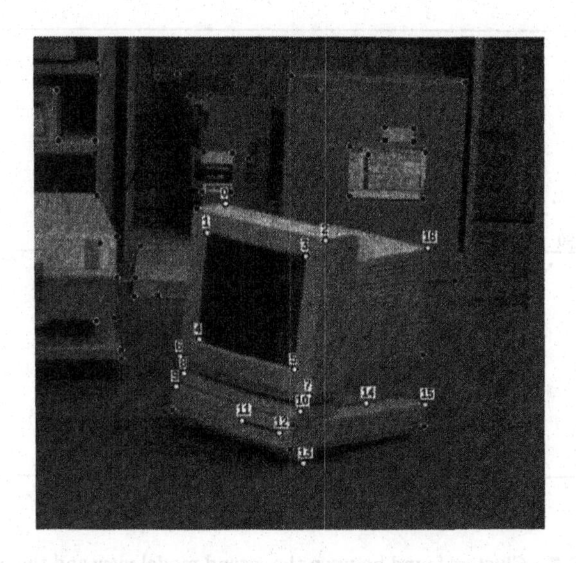

Figure 8.4 The matched points are marked by numbers in the Input image.

Figure 8.5 Edges in the input image

Figure 8.6 Edges in the input image matched with respect to Model image 1

To see how the epipolar equation affects the matching of edge images, we matched the model edge image and the input edge image by the same matching algorithm using a wrong epipolar equation. The matched edge points in the

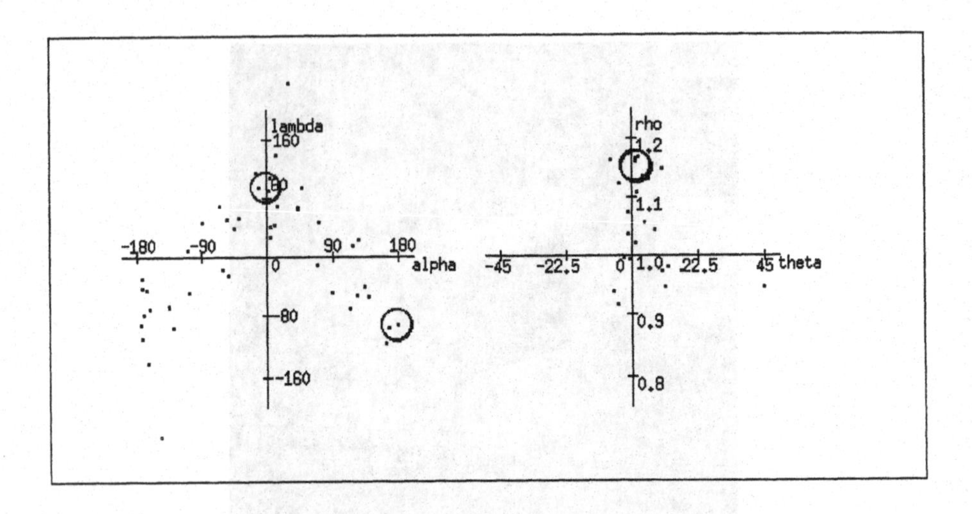

Figure 8.7 Clusters found between the second model view and the input view of the Mac computer $\alpha 1 = -2.713166$, $\alpha 2 = 177.041748$, $\theta 1 = 1.943668$, $\theta 2 = 1.194336$, $\rho 11.154302 =$, $\rho 2 = 1.156942$, $\lambda 1 = 96.984818$, and $\lambda 2 = -93.706825$

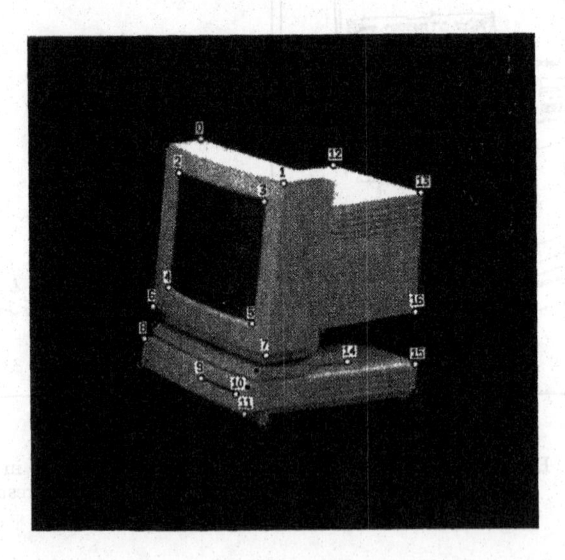

Figure 8.8 The matched points are marked by numbers in Model Image 2

input edge image are shown in Fig.8.13 together with the input edge image. As expected, the edge points are not smooth.

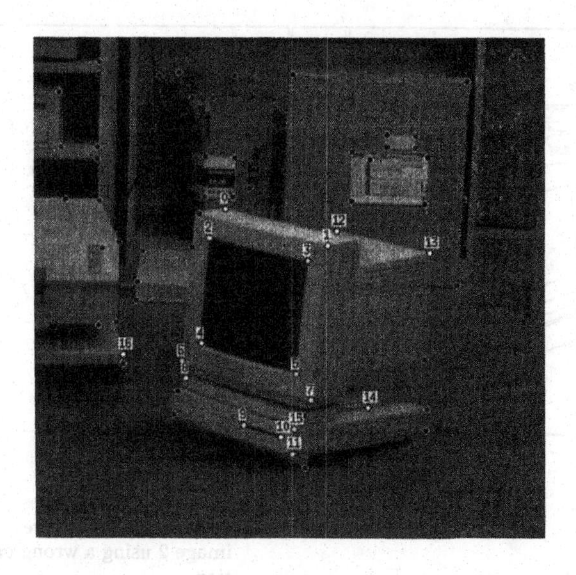

Figure 8.9 The matched points are marked by numbers in the input image.

Figure 8.10 Edges in the input image

Figure 8.11 Edges in the input image matched with respect to Model Image 2

To compare with the result using the correct epipolar equation, we compute the energy defined in (6.7) for the matching result using correct epipolar equation and that using wrong epipolar equation.

Figure 8.12 Edges in the input image

Figure 8.13 Edges in the input view matched with respect to Model image 2 using a wrong epipolar equation

Table 8.1 Comparison between matching using correct and wrong epipolar equations

	Average Smoothness Energy
Model 1 vs input, correct epi. eq.	0.348
Model 2 vs input, correct epi. eq.	0.259
Model 1 vs input, wrong epi. eq.	1.05

8.3 RECOGNITION AND LOCALIZATION WITH MULTIPLE MODEL VIEWS

While recognizing 3D object with one model view is inherently underconstraining, using two model views (which must be different) is sufficient. This conclusion can be obtained from the linear combination theorem by Ullman and Basri [165], and from trifocal tensor theory (see Chapter 4) [140]. In the following we first show that the intersection of epipolar lines is not an appropriate representation, and derive the linear combination expression from representing the image coordinates as basis vectors, from which we can easily determine how to choose basis vectors.

8.3.1 Intersection of Epipolar Lines

Intuitively, with two model views, we can draw two epipolar lines in the input image, and the corresponding point should coincide with their intersection.

Suppose we have two model views. We denote points in the first and second model views by (u', v') and (u'', v''), respectively, and a point in the input view as (u, v). If the corresponding points in the three views have been identified, then we have three epipolar equations

$$P_1 u + Q_1 v + S_1 u' + T_1 v' + C_1 = 0, \quad (8.1)$$
$$P_2 u + Q_2 v + S_2 u'' + T_2 v'' + C_2 = 0, \quad (8.2)$$
$$P_3 u' + Q_3 v' + S_3 u'' + T_3 v'' + C_3 = 0, \quad (8.3)$$

of which, only two are independent. To solve for the coefficients, we need at least 4 triples of points. Once the coefficients are determined, we can express a point in the input view by its corresponding points in the two model views,

$$\begin{bmatrix} u \\ v \end{bmatrix} = - \begin{bmatrix} P_1 & Q_1 \\ P_2 & Q_2 \end{bmatrix}^{-1} \begin{bmatrix} S_1 u' + T_1 v' + C_1 \\ S_2 u'' + T_2 v'' + C_2 \end{bmatrix} . \quad (8.4)$$

Visually, point (u, v) is the intersection of the two epipolar lines. It is thus natural, as can also be seen from the equations, if the two lines are parallel, or mathematically if the two equations (8.1) and (8.2) are not linear-independent, then increasing the number of model views does not solve the problem. Visually, this means that if the epipolar lines overlap each other, then there is an infinite number of solutions for (u, v). This corresponds to the situation that the viewing direction of the input view is coplanar with the viewing directions of the two model views (Figure 8.14). Unfortunately, for any two model views, such viewpoints always exist.

But if we take another approach to the derivation of the linear combination equations, then this problem does not occur.

8.3.2 Basis Vectors

Under the orthographic projection, the horizontal and vertical coordinates can be modelled as the inner products of two unit vectors with the 3D coordinates,

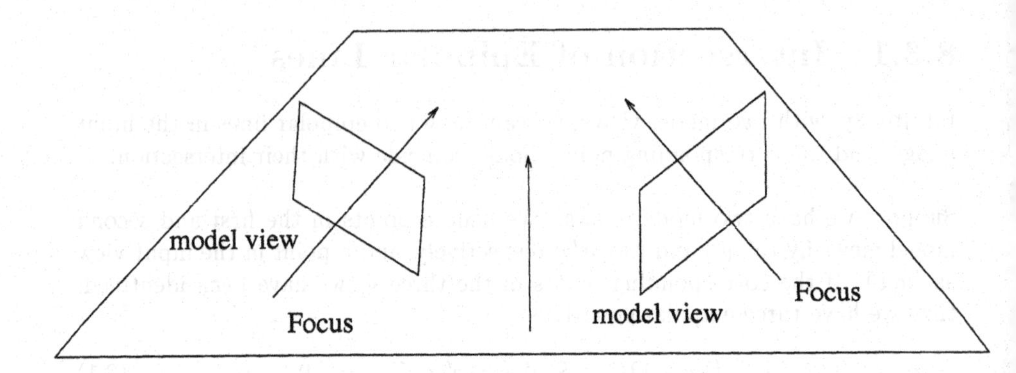

Figure 8.14 The epipolar line intersection approach collapses when the three viewing directions of two model views and the input view are coplanar.

respectively,

$$\begin{bmatrix} u \\ v \end{bmatrix} = \begin{bmatrix} \mathbf{r}_1^T \\ \mathbf{r}_2^T \end{bmatrix} \begin{bmatrix} X \\ Y \\ Z \end{bmatrix} + \begin{bmatrix} u_0 \\ v_0 \end{bmatrix} , \tag{8.5}$$

where $\mathbf{r}_1$ and $\mathbf{r}_2$ are the first two row vectors of the rotation matrix, which are unit vectors orthogonal to each other and from which the third row vector can be uniquely determined.

Suppose we have two model views. Then we have two rotation matrices whose mutual relations are fixed no matter what coordinate system we choose for the object. Let the first two row vectors of the two rotation matrices be $\mathbf{r}'_1$, $\mathbf{r}'_2$, and $\mathbf{r}''_1$, $\mathbf{r}''_2$, respectively. Then the problem of whether or not we can find a linear combination of the two model views for any image of the same object becomes the problem of whether or not we can model the first two row vectors of the unknown rotation matrix as a linear combination of $\mathbf{r}'_1$, $\mathbf{r}'_2$, $\mathbf{r}''_1$ and $\mathbf{r}''_2$, which we call *basis vectors*basis vectors.

It is then clear that such a linear combination does exist as far as the 4 basis vectors of the model views are not coplanar simultaneously. If the 4 basis vectors are coplannar, that is,

$$\mathbf{r}'_1 \times \mathbf{r}'_2 = \pm \mathbf{r}''_1 \times \mathbf{r}''_2 ,$$

then it is impossible to find a linear combination of them for an arbitrary unit vector. This corresponds to the situation where the optical axes of the two cameras are parallel, or the rotation between the two cameras is only within the image plane, or $\beta = 0$ in the Euler angle form.

If three of them are coplanar, but the 4th is not, then we can still find at least 1 group of 3 non-coplanar basis vectors.

Actually, even if the vectors are not strictly coplanar, the linear combination is actually difficult if they are close to a common plane. This closeness can be expressed as

$$C = |(\mathbf{r}_a \times \mathbf{r}_b) \cdot \mathbf{r}_c| \,. \tag{8.6}$$

This value will be exactly zero, if the three basis vectors are exactly coplanar, and it will grow up to one if the three vectors become orthogonal to each other. Note that two of them are already orthogonal.

It is intuitive that if the three basis vectors are orthogonal to each other, then an arbitrary unit vector can be modelled as the linear combination of them with all three coefficients being within the range from -1 to 1. But if C is small, which means that the three basis vectors are close to a common plane, then one has to have very large coefficients in order to model vectors of some orientations.

C depends on β in the Euler-angle form. Without loss of generality, let $\mathbf{r}_a$ and $\mathbf{r}_b$ be the basis vectors corresponding to the horizontal and vertical coordinates of one of the images. Then $\mathbf{r}_a \times \mathbf{r}_b$ corresponds to the optical axis of that image. Now $\mathbf{r}_c$ must come from the other image, corresponding to either the horizontal coordinates or the vertical coordinates.

Now let us consider when the inner product of vector $\mathbf{r}_a \times \mathbf{r}_b$ and vector $\mathbf{r}_c$ becomes maximal. That is when the angle between the two vectors becomes minimal. Since the rotation out of image plane is fixed (β), the problem is the rotation axis. This angle cannot be smaller than β. It is intuitive that when the rotation is within the plane determined by $\mathbf{r}_a \times \mathbf{r}_b$ and $\mathbf{r}_c$, the angle becomes minimal, that is, C becomes maximal. When the axis of the rotation out of image plane is $\mathbf{r}_c$, then no matter how large β is, the two vectors $\mathbf{r}_a \times \mathbf{r}_b$ and $\mathbf{r}_c$ are perpendicular to each other, that is, C is zero.

Let the angle between the image axis corresponding to $\mathbf{r}_c$ and projection of the rotation axis onto the image be α, then C can be determined as

$$C = \cos\alpha \sin\beta \,. \tag{8.7}$$

If $\mathbf{r}_a$ and $\mathbf{r}_b$ are from the other image, then C is

$$C = \cos\gamma \sin\beta \ . \tag{8.8}$$

Qualitatively speaking, the larger β is, the larger C is; and if β is zero, then C is zero, too.

If one has to choose 3 basis vectors from the 4, the C can be used as a criterion. There are 4 combinations. We can compute C for each of the combinations. We should choose the combination which give the largest C.

No matter how you choose the 3 basis vectors, two of them are already orthogonal. The remaining one must be one of vectors corresponding to the horizontal and vertical coordinates in the other image. Since β is fixed, we can only minimize α in (8.7). That is, of the horizontal and vertical axes, the one which has smaller angle to the epipolar direction should be chosen.

Since the discussion is symmetric with respect to the two images, the above discussion applies the other way. Therefore, what we can do is to find the coordinate axis which is closest to the epipolar directions in the two images, and use the coordinates corresponding to this axis and the two coordinates in the other image as the three basis vectors.

β cannot be determined given two images. However, it is qualitatively true that, the larger β is, the larger the disparities along the epipolar lines are. If β is zero, then disparity of every point is zero, too.

In the above discussion, we have assumed orthographic projection. However, even if we assume weak perspective projection, all the conclusions are valid except for the need to allow the basis vectors to have non-unit lengths.

8.3.3 Determining Coefficients

In this subsection we show how to determine the coefficients in the linear combination of basis vectors. There are two approaches. One is to determine them from the point matches, and the other is from recovered epipolar equations.

Given Point Matches

Suppose that we are given two model views, and we have chosen the basis vectors that maximize criterion C as in last subection. Without loss of generality, assume that they are $\mathbf{u}'$, $\mathbf{v}'$ and $\mathbf{u}''$, the horizontal and vertical coordinates of the first model image and the horizontal coordinates of the second model image. Note that choosing other combinations does not affect the following analysis. The general form of linear combination is

$$\mathbf{u} = a_1\mathbf{u}' + b_1\mathbf{v}' + c_1\mathbf{u}'' + e_1 , \tag{8.9}$$
$$\mathbf{v} = a_2\mathbf{u}' + b_2\mathbf{v}' + c_2\mathbf{u}'' + e_2 . \tag{8.10}$$

It is easy to see that with at least 4 triples of points the coefficients can all be determined as there are only 8 unknowns and 4 triples of points yield 8 linear equations.

Now suppose we are given $n (n \geq 4)$ triples of points. Let us use u_i for the horizontal coordinate of the i-th point, and similar notations for other coordinates. Then we can minimize

$$E = \sum_{i=1}^{n} \{(-u_i + a_1 u_i' + b_1 v_i' + c_1 u_i'' + e_1)^2 + (-v_i + a_2 u_i' + b_2 v_i' + c_2 u_i'' + e_2)^2\} \tag{8.11}$$

over the coefficients.

The solution can be obtained by doing the same analysis as for determining the epipolar equation given point matches. Let us define $u_0 = \frac{1}{n}\sum_{i=1}^{n} u_i$, $v_0 = \frac{1}{n}\sum_{i=1}^{n} v_i$, $u_0' = \frac{1}{n}\sum_{i=1}^{n} u_i'$, $v_0' = \frac{1}{n}\sum_{i=1}^{n} v_i'$, $u_0'' = \frac{1}{n}\sum_{i=1}^{n} u_i''$, $v_0'' = \frac{1}{n}\sum_{i=1}^{n} v_i''$. And further define, $\bar{u}_i = u_i - u_0$, $\bar{v}_i = v_i - v_0$, $\bar{u}_i' = u_i' - u_0'$, $\bar{v}_i' = v_i' - v_0'$, $\bar{u}_i'' = u_i'' - u_0''$, $\bar{v}_i'' = v_i'' - v_0''$. Then minimizing (8.11) is equivalent to solving

$$\begin{bmatrix} \bar{u}_1' & \bar{v}_1' & \bar{u}_1'' & -\bar{u}_1 \\ \vdots & \vdots & \vdots & \vdots \\ \bar{u}_n' & \bar{v}_n' & \bar{u}_n'' & -\bar{u}_n \end{bmatrix} \begin{bmatrix} a_1 \\ b_1 \\ c_1 \\ 1 \end{bmatrix} = \begin{bmatrix} 0 \\ \vdots \\ 0 \end{bmatrix} , \tag{8.12}$$

$$\begin{bmatrix} \bar{u}_1' & \bar{v}_1' & \bar{u}_1'' & -\bar{v}_1 \\ \vdots & \vdots & \vdots & \vdots \\ \bar{u}_n' & \bar{v}_n' & \bar{u}_n'' & -\bar{v}_n \end{bmatrix} \begin{bmatrix} a_2 \\ b_2 \\ c_2 \\ 1 \end{bmatrix} = \begin{bmatrix} 0 \\ \vdots \\ 0 \end{bmatrix} . \tag{8.13}$$

e_1 and e_2 can be obtained by

$$e_1 = a_1 u_0' + b_1 v_0' + c_1 u_0'' - u_0 , \tag{8.14}$$
$$e_2 = a_2 u_0' + b_2 v_0' + c_2 v_0'' - v_0 . \tag{8.15}$$

Let us call

$$\mathbf{A} = \begin{bmatrix} \bar{u}'_1 & \bar{v}'_1 & \bar{u}''_1 & -\bar{u}_1 \\ \vdots & \vdots & \vdots & \vdots \\ \bar{u}'_n & \bar{v}'_n & \bar{u}''_n & -\bar{u}_n \end{bmatrix},$$

and

$$\mathbf{B} = \begin{bmatrix} \bar{u}'_1 & \bar{v}'_1 & \bar{u}''_1 & -\bar{v}_1 \\ \vdots & \vdots & \vdots & \vdots \\ \bar{u}'_n & \bar{v}'_n & \bar{u}''_n & -\bar{v}_n \end{bmatrix}.$$

Then the solution of $[a_1 \quad b_1 \quad c_1 \quad 1]^T$ is the eigenvector associated with the smallest eigenvalue of $\mathbf{A}^T\mathbf{A}$, and scaled so that the last component is equal to 1. And the solution of $[a_2 \quad b_2 \quad c_2 \quad 1]^T$ is the eigenvector associated with the smallest eigenvalue of $\mathbf{B}^T\mathbf{B}$, and scaled so that the last component is equal to 1.

Actually, we need not to use only 3 basis vectors. Using all 4 coordinates as basis vectors is possible and sometimes even more advantageous. The algorithm described above can be extended straightforwardly for it.

Using 4 basis vectors, we have

$$\mathbf{u} = a_1\mathbf{u}' + b_1\mathbf{v}' + c_1\mathbf{u}'' + d_1\mathbf{v}'' + e_1\,, \tag{8.16}$$

$$\mathbf{v} = a_2\mathbf{u}' + b_2\mathbf{v}' + c_2\mathbf{u}'' + d_2\mathbf{v}'' + e_2\,, \tag{8.17}$$

Accordingly, define

$$\mathbf{A}' = \begin{bmatrix} \bar{u}'_1 & \bar{v}'_1 & \bar{u}''_1 & \bar{v}''_1 & -\bar{u}_1 \\ \vdots & \vdots & \vdots & \vdots & \vdots \\ \bar{u}'_n & \bar{v}'_n & \bar{u}''_n & \bar{v}''_n & -\bar{u}_n \end{bmatrix},$$

and

$$\mathbf{B}' = \begin{bmatrix} \bar{u}'_1 & \bar{v}'_1 & \bar{u}''_1 & \bar{v}'_1 & -\bar{v}_1 \\ \vdots & \vdots & \vdots & \vdots & \vdots \\ \bar{u}'_n & \bar{v}'_n & \bar{u}''_n & \bar{v}'_n & -\bar{v}_n \end{bmatrix}.$$

Then the solution of $[a_1 \quad b_1 \quad c_1 \quad d_1 \quad 1]^T$ is the eigenvector associated with the smallest eigenvalue of $\mathbf{A}'^T\mathbf{A}'$, and scaled so that the last component is equal to 1. And the solution of $[a_2 \quad b_2 \quad c_2 \quad d_2 \quad 1]^T$ is the eigenvector associated with the smallest eigenvalue of $\mathbf{B'}^T\mathbf{B'}$, and scaled so that the last component is equal to 1.

It is noted that using 4 basis vectors, the representation is not unique in theory. Now both $\mathbf{A'}^T\mathbf{A'}$ and $\mathbf{B'}^T\mathbf{B'}$ are of 5×5 size, but the rank is only 3. This means that the fourth and fifth smallest eigen values of the $\mathbf{A'}^T\mathbf{A'}$ and $\mathbf{B'}^T\mathbf{B'}$ do not differ very much. However, choosing the smallest eigen values and their associated eigen vectors does guarantee the best solution in terms of minimizing the following energy

$$\begin{aligned} E &= \textstyle\sum_{i=1}^{n}\{(-u_i + a_1 u_i' + b_1 v_i' + c_1 u_i'' + d_1 v_i'' + e_1)^2 \\ &\quad +(-v_i + a_2 u_i' + b_2 v_i' + c_2 u_i'' + d_2 v_i'' e_2)^2\} \,. \end{aligned} \tag{8.18}$$

Given Epipolar Equations

As described above, if the epipolar equations (8.1), (8.2) and (8.3) are not linear-independent, then the coefficients of the linear combination cannot be uniquely determined from the epipolar equations. However, if they are indeed linear-independent, then the coefficients of the linear combination can be uniquely determined from them.

It is worth mentioning that given the same basis vectors, the representation is unique. Then the coefficients determined from the point matches must be the same as those determined from the given epipolar equations.

For the equations (8.1 - 8.3) to be the same as (8.9-8.10), the following equations must be satisfied:

$$\begin{aligned} P_1 a_1 + Q_1 a_2 &= -S_1 \,, \\ P_2 a_1 + Q_2 a_2 &= \frac{T_2 P_3}{T_3} \,, \\ P_1 b_1 + Q_1 b_2 &= -T_1 \,, \\ P_2 b_1 + Q_2 b_2 &= \frac{T_2 Q_3}{T_3} \,, \\ P_1 c_1 + Q_1 c_2 &= 0 \,, \\ P_2 c_1 + Q_2 c_2 &= \frac{T_2 S_3}{T_3} - S_2 \,, \\ P_1 e_1 + Q_1 e_2 &= -C_1 \,, \\ P_2 e_1 + Q_2 e_2 &= \frac{T_2 C_3}{T_3} - C_2 \,. \end{aligned}$$

It is easy to find

$$
\begin{aligned}
\begin{bmatrix} a_1 \\ a_2 \end{bmatrix} &= \begin{bmatrix} P_1 & Q_1 \\ P_2 & Q_2 \end{bmatrix}^{-1} \begin{bmatrix} -S_1 \\ \frac{T_2 P_3}{T_3} \end{bmatrix}, \\
\begin{bmatrix} b_1 \\ b_2 \end{bmatrix} &= \begin{bmatrix} P_1 & Q_1 \\ P_2 & Q_2 \end{bmatrix}^{-1} \begin{bmatrix} -T_1 \\ \frac{T_2 Q_3}{T_3} \end{bmatrix}, \\
\begin{bmatrix} c_1 \\ c_2 \end{bmatrix} &= \begin{bmatrix} P_1 & Q_1 \\ P_2 & Q_2 \end{bmatrix}^{-1} \begin{bmatrix} 0 \\ \frac{T_2 S_3}{T_3} - S_2 \end{bmatrix}, \\
\begin{bmatrix} e_1 \\ e_2 \end{bmatrix} &= \begin{bmatrix} P_1 & Q_1 \\ P_2 & Q_2 \end{bmatrix}^{-1} \begin{bmatrix} -C_1 \\ \frac{T_2 C_3}{T_3} - C_2 \end{bmatrix},
\end{aligned}
\tag{8.19}
$$

provided

$$
\begin{vmatrix} P_1 & Q_1 \\ P_2 & Q_2 \end{vmatrix} \neq 0.
$$

8.3.4 An Example

Since we have two model views for the Mac computer, we have tried to use the matched edge points in the model views to synthesize the Mac image so that it superimposes with that part in the input view.

Using the techniques described in the above section, we can first match the feature points in the two model views and input view, and then match the edge points using the recovered epipolar equations.

Let us denote a point in the input view by (u, v), a point in model view 1 by (u', v'), and a point in model view 2 by (u'', v''). Then the recovered two epipolar equations for $(u, v) - (u', v')$, for $(u, v) - (u'', v'')$ and for $(u', v') - (u'', v'')$ are respectively

$$
\begin{aligned}
0.059751u' - 0.677430v' - 0.107834u + 0.725183v - 38.169512 &= 0, \\
0.038198u'' + 0.673137v'' - 0.064981u - 0.735666v + 56.323223 &= 0, \\
0.012576u' + 0.712034v' + 0.054180u'' - 0.699938v'' - 11.399557 &= 0.
\end{aligned}
$$

For the two model views, α and γ are $-1.011837°$ and $3.414448°$, respectively. If we choose 3 basis vectors, then $\mathbf{u}'$,$\mathbf{u}''$ and $\mathbf{v}''$ should be chosen.

Using (8.19), we can determine the equations for linear combination as

$$
\begin{aligned}
u &= 0.417222u' + 0.518943u'' - 0.013828v'' + 37.847492\,, \\
v &= -0.036853u' + 0.006085u'' + 0.916225v'' + 73.217812\,.
\end{aligned} \tag{8.20}
$$

Using eigen analysis to determine the coefficients from matched feature points, the equations for linear combination are

$$
\begin{aligned}
u &= 0.410758u' + 0.525500u'' - 0.012969v'' + 37.673837\,, \\
v &= -0.035980u' + 0.005223u'' + 0.916110v'' + 73.235615\,.
\end{aligned} \tag{8.21}
$$

Comparing (8.20) and (8.21), the equations are very close to each other.

Instead of using 3 basis vectors, we can also use 4 basis vectors. Using eigen analysis, the coefficients are determined from the eigen vectors associated with the 5th smallest eigen values. the equations are determined as

$$
u = 0.402295u' - 0.483757v' + 0.488607u'' + 0.462512v'' + 45.433265\,, \tag{8.22}
$$

$$
v = -0.019036u' + 0.954682v' + 0.077785u'' - 0.022360v'' + 57.95204 \tag{8.23}
$$

It can be seen that for u, the contributions from u' and u'' are very close, while the contributions from v' and v'' are roughly the opposite. This agrees with the results using 3 basis vectors. For v, the contribution from v' is dominant, while the contributions from u', u'' and v'' are much smaller. This means that v has very high correlation with v'. Actually as in the case of 3 basis vectors, v has very high correlation with v''. This is because that all the 3 views differ only by rotations around the near-vertical axes. If the coefficients were determined using the eigen vectors associated with the 4th smallest eigen values, then v would have a very high correlation with v''.

Using (8.23), an edge image is synthesized and superimposed with the input view (Fig.8.15). The synthesized edges are shown in black, and the edges in the original input view in gray. It is evident that the synthesized data is very close to the data in the input view.

8.4 SUMMARY

In this chapter we have discussed how to apply epipolar geometry to 3D object recognition and localtion. By posing the problem as matching between model

Figure 8.15 Synthesized edge image superimposed with the original input view

view(s) and input view, we can recover the epipolar geometry underlying the images, and at the same time match the images. If the images can be matched, then the object is said to be recognized and localized.

Specifically, we discussed the problem of using one model view for recognition, which is generally underconstrained, and proposed an approach to the image matching which is similar to the one used for matching uncalibrated stereo images.

When more than one model views are available, there is a unique solution to the problem. In this case, we can also apply the above technique to match each pair of model views and each pair of model view and input view. By finding the epipolar geometries underlying the images by matching the feature points, we can generate edge images to be superimposed with the input view if the images are really the projections of the same object.

Real image examples were given and they were successfully matched. Future research is to be done to deal with a larger object database.

9

CONCLUDING REMARKS

What we have tried to do in this book is to present a detailed description of the epipolar geometry which underlies every pair of images of the same scene, to formulate the problems of stereo, motion and view-based object recognition under a unified framework from the epipolar geometry viewpoint, and to solve these problems in a unified manner.

A few future research directions are listed as follows.

Although we are unable to obtain any metric information (measurements of lengths and angles) from two uncalibrated images, it is still possible to extract rich information such as coplanarity, collinearity, and ratios, which is sometimes sufficient for artificial systems, such as robots, to perform tasks such as navigation and object recognition. Although some interesting work has already appeared [181, 11, 125, 127, 12, 31], further investigation is required to understand precisely what type of information is really necessary in performing a given task.

Since the epipolar geometry exists between any two views, it can be applied to longer image sequences. There has been research using Kalman filter to combine information of epipolar equations recovered from successive views under the weak perspective projection [138]. It would be useful if this approach is also applied to recover structure and shape from long sequences of motion images under the full perspective projection.

Under certain reasonable conditions, for example, using a moving camera with fixed but unknown intrinsic parameters or a moving uncalibrated stereo rig, self-calibration is possible [99, 92, 191, 30], that is, camera parameters and metric structure can be computed from images alone without using classical calibration apparatus. Determination of the epipolar geometry is the basis of all these techniques, and better results can be expected by making full use of the information provided by the robust techniques described in this book. Indeed, Zeller [180] shows that the self-calibration based on Kruppa equations yields much better results if the uncertainty of the fundamental matrix is taken into account.

We have considered in this book discrete displacements between images, i.e. the motion between two images can be large or small. If we know that the motion is small, the problem can be simplified using for example the first order expansion. The reader interested in the latter situation is referred to [166].

There should be many applications of the techniques described here, besides the well-known robotic applications like navigation and hand-eye coordination. For example, in medical imaging, microcameras are inserted into the human body to get 3D information so that the physician can execute operation. The camera configuration cannot be fixed inside the body, and thus automatic calibration is required, to which our approach can be applied.

In certain applications like building CAD models from image sequences, 3D Euclidean information is usually necessary to produce reality. Using the epipolar geometry approach, the users are given more freedom [37, 14]. Images can be taken by a camera or by different cameras without calibration (as long as they can be reasonably modeled as a pinhole, or the lens distortion can be corrected), and metric constraint such as distances and angles can be imposed posteriorly to generate a Euclidean model.

Exact 3D structure and 3D position can be recovered using epipolar geometry, and virtual images can even be synthesized from 2D images by using recovered epipolar equations, without recovering the explicit 3D structure [36, 84, 133]. Synthesis of new images from recovered epipolar equations can also be applied to computer graphics, to amusement industry, and other disciplines that require image synthesis.

Building a large high-resolution image from multiple images (called image mosaicing) is important in many domains such as computer animation, aerial photomosaics, video conferencing, and image compression. The reader is referred to [149, 76, 95, 81, 150] for several pieces of related work. The important step is to automatically register/align images, which is an old yet very difficult problem (see Chap. 6).

REFERENCES

[1] G. Adiv. Determining three-dimensional motion and structure from optical flow generated by several moving objects. *IEEE Trans. on Pattern Analysis and Machine Intelligence*, 7:384–401, 1985.

[2] J. Aloimonos. Perspective approximations. *Image and Vision Computing*, 8(3):179–192, Aug. 1990.

[3] J. Aloimonos and D. Shulman. *Integration of Visual Modules: An Extension of the Marr Paradigm.* Academic Press, 1989.

[4] H. Asada and M. Brady. The curvature primal sketch. *IEEE Trans. on Pattern Analysis and Machine Intelligence*, 8:2–14, 1986.

[5] N. Ayache. *Artificial Vision for Mobile Robots.* MIT Press, 1991.

[6] N. Ayache and F. Lustman. Fast and reliable passive trinocular stereovision. In *Proc. First Int'l Conf. Comput. Vision*, pages 422–427, London, UK, June 1987. IEEE.

[7] S. Ayer, P. Schroeter, and J. Bigün. Segmentation of moving objects by robust motion parameter estimation over multiple frames. In J.-O. Eklundh, editor, *Proc. Third European Conf. Comput. Vision*, pages 316–327, Vol.II, Stockholm, Sweden, May 1994.

[8] A. Azarbayejani and A. Pentland. Recursive estimation of motion, structure and focal length. *IEEE Trans. on Pattern Analysis and Machine Intelligence*, 17(6):562–575, 1995.

[9] D. H. Ballard and C. M. Brown. *Computer Vision.* Prentice-Hall, Englewood Cliffs, NJ, 1982.

[10] S. Barnard and W. Thompson. Disparity analysis of images. *IEEE Trans. on Pattern Analysis and Machine Intelligence*, 2(4):333–340, July 1980.

[11] P. Beardsley, A. Zisserman, and D. Murray. Navigation using affine structure from motion. In J.-O. Eklundh, editor, *Proc. 3rd European Conference on Computer Vision*, volume 2, pages 85–96, Stockholm, Sweden, May 1994. Springer-Verlag.

[12] P. A. Beardsley, I. D. Reid, A. Zisserman, and D. W. Murray. Active visual navigation using non-metric structure. In *Proc. 5th International Conference on Computer Vision*, pages 58–64, Boston, MA, June 1995.

[13] B. Boufama and R. Mohr. Epipole and fundamental matrix estimation using the virtual parallax property. In *Proc. 5th International Conference on Computer Vision*, pages 1030–1036, Boston, MA, June 1995.

[14] B. Boufama, R. Mohr, and F. Veillon. Euclidean Constraints for Uncalibrated Reconstruction. Technical Report RT96 IMAG 17 LIFIA, LIFIA, INSTITUT IMAG, Mar. 1993.

[15] T. Boult and L. Brown. Factorization-based segmentation of motions. In *Proc. IEEE Conf. Comput. Vision Pattern Recog.*, pages 179–186, 1991.

[16] R. Brooks. Symbolic reasoning among 3d models and 2d images. *Artif. Intell.*, 20:285–348, 1981.

[17] D. C. Brown. Close-range camera calibration. *Photogrammetric Engineering*, 37(8):855–866, 1971.

[18] C. Chen and A. Kak. A robot vision system for recognizing 3d objects in low-order polynomial time. *IEEE Trans. on Systems, Man, and Cybernetics*, 19(6):1535–1563, 1989.

[19] J. Cheng and T. Huang. Image registration by matching relational structures. *Pattern Recog.*, 17(1):149–159, 1984.

[20] C.-H. Chou and Y.-C. Chen. Moment-preserving pattern matching. *Pattern Recog.*, 23(5):461–474, 1990.

[21] G. Chou. Kinetic occlusion and figure-ground computation. Tech Rep 94-7, Harvard Robotics Laboratory, Cambridge, MA, 1994.

[22] G. Csurka. *Modélisation projective des objets tridimensionnels en vision par ordinateur.* PhD thesis, University of Nice, Sophia-Antipolis, France, 1996.

[23] G. Csurka, C. Zeller, Z. Zhang, and O. Faugeras. Characterizing the uncertainty of the fundamental matrix. *CVGIP: Image Understanding*, 1996. to appear. Updated version of INRIA Research Report 2560, 1995.

[24] S. Demey, A. Zisserman, and P. Beardsley. Affine and projective structure from motion. In *British Machine Vision Conference*, pages 49–58, Leeds, UK, Sept. 1992.

[25] R. Deriche and T. Blaszka. Recovering and characterizing image features using an efficient model based approach. In *Proc. IEEE Conf. Comput. Vision Pattern Recog.*, New-York, June 14-17 1993.

[26] R. Deriche and O. Faugeras. 2D-curves matching using high curvatures points : Applications to stereovision. In *Proc. 10th Int'l Conf. Pattern Recog.*, pages 240–242, Vol.1, Atlantic City, June 16-21 1990.

[27] R. Deriche and O. Faugeras. Tracking line segments. In O. Faugeras, editor, *Proc. First European Conf. Comput. Vision*, pages 259–268, Antibes, France, Apr. 1990. Springer, Berlin, Heidelberg.

[28] R. Deriche and G. Giraudon. Accurate corner detection: An analytical study. In *Proc. Third Int'l Conf. Comput. Vision*, pages 66–70, Osaka, 1990.

[29] R. Deriche and G. Giraudon. A computational approach for corner and vertex detection. *Int'l J. Comput. Vision*, 10(2):101–124, 1993.

[30] R. Enciso. *Auto-Calibration des Capteurs Visuels Actifs. Reconstruction 3D Active.* PhD thesis, Université Paris XI Orsay, Dec. 1995.

[31] R. Enciso, Z. Zhang, and T. Viéville. Dense reconstruction using fixation and stereo cues. In *World Automation Congress, ISIAC Symposia*, Montpellier, May 1996.

[32] M. Etoh and Y. Shirai. Segmentation and 2d motion estimation by region fragments. In *Proc. Fourth Int'l Conf. Comput. Vision*, pages 192–199, Germany, 1993. IEEE.

[33] W. Faig. Calibration of Close-Range Photogrammetry Systems: Mathematical Formulation. *Photogrammetric Engineering and remote Sensing*, 41(12):1479–1486, 1975.

[34] O. Faugeras. What can be seen in three dimensions with an uncalibrated stereo rig. In G. Sandini, editor, *Proc. 2nd European Conference on Computer Vision*, volume 588, pages 563–578, Santa Margherita Ligure, Italy, May 1992. Springer-Verlag.

[35] O. Faugeras. Stratification of 3-D vision: projective, affine, and metric representations. *Journal of the Optical Society of America A*, 12(3):465–484, Mar. 1995.

[36] O. Faugeras and S. Laveau. Representing three-dimensional data as a collection of images and fundamental matrices for image synthesis. In *Proc. International Conference on Pattern Recognition*, pages 689–691, Jerusalem, Israel, Oct. 1994.

[37] O. Faugeras, S. Laveau, L. Robert, C. Zeller, and G. Csurka. 3-d reconstruction of urban scenes from sequences of images. In A. Gruen, O. Kuebler, and P. Agouris, editors, *Automatic Extraction of Man-Made Objects from Aerial and Space Images*, pages 145–168, Ascona, Switzerland, Apr. 1995. ETH, Birkhauser Verlag. also INRIA Technical Report 2572.

[38] O. Faugeras, T. Luong, and S. Maybank. Camera self-calibration: theory and experiments. In G. Sandini, editor, *Proc 2nd ECCV*, volume 588 of *Lecture Notes in Computer Science*, pages 321–334, Santa Margherita Ligure, Italy, May 1992. Springer-Verlag.

[39] O. Faugeras and S. Maybank. Motion from point matches: multiplicity of solutions. *The International Journal of Computer Vision*, 4(3):225–246, 1990.

[40] O. Faugeras and G. Toscani. The calibration problem for stereo. In *Proc. International Conference on Computer Vision and Pattern Recognition*, IEEE Publication 86CH2290-5, pages 15–20, Miami Beach, FL, June 1986.

[41] O. D. Faugeras. *Three-Dimensional Computer Vision: A Geometric Viewpoint.* MIT Press, Cambridge, MA, 1993.

[42] M. Fischler and O. Firschein. *Intelligence: The Eye, the Brain, and the Computer.* Addison-Wesley Publishing Company, 1987.

[43] P. Fua. A parallel stereo algorithm that produces dense depth maps and preserves image features. *Machine Vision and Applications*, 6(1), Winter 1993.

[44] S. Geman and D. Geman. Stochastic relaxation, gibbs distribution and bayesian restoration of images. *IEEE Trans. on Pattern Analysis and Machine Intelligence*, 6:721–741, 1984.

[45] G. Giraudon. Chaînage efficace contour. Rapport de Recherche 605, INRIA, Sophia-Antipolis, France, Feb. 1987.

[46] G. Golub and C. van Loan. *Matrix Computations.* The John Hopkins University Press, 1989.

[47] A. Goshtasby, S. H. Gage, and J. F. Bartholic. A two-stage cross correlation approach to template matching. *IEEE Trans. on Pattern Analysis and Machine Intelligence*, 6(3):374–378, May 1984.

[48] W. Grimson. A computer implementation of a theory of human stereo vision. *Philosophical Trans. of the Royal Society of London, B.*, 292(1058):217–253, 1981.

[49] W. Grimson and T. Lozano-Perez. Model-based recognition and localization from sparse range or tactile data. *Int'l J. Robotics Res.*, 5(3):3–34, Fall 1984.

[50] W. E. L. Grimson. *From Images to Surfaces.* The MIT Press, Cambridge, MA, 1981.

[51] F. Hampel. Robust estimation: A condensed partial survey. *Z. Wahrscheinlichkeitstheorie Verw. Gebiete*, 27:87–104, 1973.

[52] M. Hannah. A system for digital stereo image matching. *Photogrammetric Engeneering and Remote Sensing*, 55(12):1765–1770, Dec. 1989.

[53] R. Haralick. Computer vision theory: The lack thereof. *Comput. Vision, Graphics Image Process.*, 36:372–386, 1986.

[54] R. Haralick et al. Pose estimation from corresponding point data. *IEEE Trans. on Systems, Man, and Cybernetics*, 19(6):1426–1446, Nov. 1989.

[55] C. Harris. Determination of ego-motion from matched points. In *Proceedings Alvey Conference*, 1987.

[56] C. Harris and M. Stephens. A combined corner and edge detector. In *Proceedings Alvey Conference*, pages 189–192, 1988.

[57] R. Hartley. Lines and points in three views-an integrated approach. In *Proc. ARPA Image Understanding Workshop*. Defense Advanced Research Projects Agency, Morgan Kaufmann Publishers, Inc., 1994.

[58] R. Hartley. Projective reconstruction and invariants from multiple images. *IEEE Trans. on Pattern Analysis and Machine Intelligence*, 16(10):1036–1040, 1994.

[59] R. Hartley. In defence of the 8-point algorithm. In *Proc. 5th International Conference on Computer Vision*, pages 1064–1070, Boston, MA, June 1995.

[60] R. Hartley, R. Gupta, and T. Chang. Stereo from uncalibrated cameras. In *Proc. International Conference on Computer Vision and Pattern Recognition*, pages 761–764, Urbana Champaign, IL, June 1992.

[61] R. Hartley and P. Sturm. Triangulation. In *Proc. ARPA Image Understanding Workshop*, pages 957–966. Defense Advanced Research Projects Agency, Morgan Kaufmann Publishers, Inc., 1994.

[62] R. I. Hartley. Cheirality invariants. In *Proc. ARPA Image Understanding Workshop*, pages 745–753, Washington, DC, Apr. 1993. Defense Advanced Research Projects Agency, Morgan Kaufmann Publishers, Inc.

[63] D. Heeger. A model for the extraction of image flow. *Journal of the Optical Society of America A*, A4:1455–1471, 1987.

[64] O. Hesse. Die cubische Gleichung, von welcher die Lösung des Problems der Homographie von M. Chasles abhängt. *J. reine angew. Math.*, 62:188–192, 1863.

[65] E. Hildreth. Computations underlying the measurement of visual motion. *Artif. Intell.*, 23:309–354, 1984.

[66] R. Horaud, S. Christy, and F. Dornaika. Object pose: The link between weak perspective, para perspective, and full perpective. Technical Report 2356, INRIA, Sept. 1994.

[67] R. Horaud and F.Veillon. Finding geometric and relational structures in an image. In *Proc. First European Conference On Computer Vision*, pages 374–384, 1990.

[68] R. Horaud and T. Skordas. Stereo correspondence through feature grouping and maximal cliques. *IEEE Trans. on Pattern Analysis and Machine Intelligence*, 11(11):1168–1180, 1989.

[69] B. Horn and B. Schunk. Determining optical flow. *Artif. Intell.*, 20:199–228, 1981.

[70] B. K. P. Horn. *Robot Vision.* MIT Press, 1986.

[71] T. Huang and C. Lee. Motion and structure from orthographic projections. *IEEE Trans. on Pattern Analysis and Machine Intelligence*, 11:536–540, 1989.

[72] T. Huang and A. Netravali. Motion and structure from feature correspondences: A review. *Proceedings of IEEE*, 82(2):252–268, Feb. 1994.

[73] D. Hubel and T. Wiesel. Receptive fields and functional architecture of monkey striate cortex. *J. Physiol. London*, 195:215–244, 1968.

[74] P. Huber. *Robust Statistics.* John Wiley & Sons, New York, 1981.

[75] K. Ikeuchi and T. Kanade. Applying sensor models to automatic generation of object recognition programs. In *Proc. Second Int'l Conf. Comput. Vision*, pages 228–237, Tampa, FL, Dec. 1988.

[76] L. Kanal, B. Lambird, D. Levine, and G. Stockman. Digital registration of images from similar and dissimilar sensors. In *Proc. Int'l Conf. Cybern. Society*, pages 347–351, 1981.

[77] M. Kass, A. Witkin, and D. Terzopoulos. Snakes: Active contour models. In *Proc. First Int'l Conf. Comput. Vision*, pages 321–330, London, England, 1987. IEEE.

[78] K. Kobayashi and G. Xu. An improved det operator. Tech rep, Department of Systems Enineering, Osaka University, Japan, 1993.

[79] J. Koenderink and A. van Doorn. Affine structure from motion. *Journal of the Optical Society of America A*, 8:337–385, 1991.

[80] J. J. Koenderink and A. J. van Doorn. Affine Structure from Motion. *Journal of the Optical Society of America*, A8:377–385, 1991.

[81] R. Kumar, P. Anandan, M. Irani, J. Bergen, and K. Hanna. Representation of scenes from collections of images. In *IEEE Workshop on Representation of Visual Scenes*. IEEE, June 1995.

[82] R. Kumar and A. Hanson. Analysis of different robust methods for pose refinement. In *Proc. Int'l Workshop Robust Comput. Vision*, pages 167–182, Seattle, WA, Oct. 1990.

[83] Y. Lamdan and H. J. Wolfson. Geometric hashing: A general and efficient model-based recognition scheme. In *Proc. Second Int'l Conf. Comput. Vision*, pages 238–249, Tampa, FL, Dec. 1988. IEEE.

[84] S. Laveau. *Géométrie d'un système de N caméras. Théorie. Estimation. Applications.* PhD thesis, École Polytechnique, May 96.

[85] J. Lavest. *Stéréovision axiale par zoom pour la robotique.* PhD thesis, Université Blaise Pascal de Clermont-Ferrand, France, 1992.

[86] C. Lee and T. Huang. Finding point correspondences and determining motion of a rigid object from two weak perspective views. *Comput. Vision, Graphics Image Process.*, (52):309–327, 1990.

[87] R. Lenz and R. Tsai. Techniques for calibrating of the scale factor and image center for high accuracy 3D machine vision metrology. In *International Conference on Robotics and Automation*, pages 68–75, Raleigh, NC, 1987.

[88] S. Li. Inexact matching of 3D surfaces. Technical Report VSSP-TR-3/90, Vision Speech & Signal Processing, Dept. Electronic and Electrical Engineering, University of Surrey, UK, Feb. 1990.

[89] Y. Liu and T. Huang. A linear algorithm for determining motion and structure from line correspondences. *Comput. Vision, Graphics Image Process.*, 44(1):35–57, 1988.

[90] H. Longuet-Higgins. A computer algorithm for reconstructing a scene from two projections. *Nature*, 293:133–135, 1981.

[91] D. Lowe. *Perceptual Organization and Visual Recognition.* Kluwer Academic, Boston, MA, 1985.

[92] Q.-T. Luong. *Matrice Fondamentale et Calibration Visuelle sur l'Environnement-Vers une plus grande autonomie des systèmes robotiques.* PhD thesis, Université de Paris-Sud, Centre d'Orsay, Dec. 1992.

[93] Q.-T. Luong and T. Viéville. Canonic representations for the geometries of multiple projective views. In J.-O. Eklundh, editor, *Proc. 3rd European Conference on Computer Vision*, pages 589–599, Vol. 1, Stockholm, Sweden, May 1994. Springer-Verlag.

[94] H. Maître and Y. Wu. Improving dydamic programming to solve image registration. *Pattern Recog.*, 20(4):443–462, 1987.

[95] S. Mann and R. W. Picard. Virtual bellows: Constructing high quality stills from video. In *International Conference on Image Processing*, pages 363–367, Nov. 1994.

[96] D. Marr. *Vision.* W.H. Freeman, San Francisco, CA, 1982.

[97] D. Marr and T. Poggio. Cooperative computation of stereo disparity. *Science*, 194:283–287, 1976.

[98] S. Maybank. *Theory of reconstruction From Image Motion.* Springer-Verlag, 1992.

[99] S. J. Maybank and O. D. Faugeras. A theory of self-calibration of a moving camera. *The International Journal of Computer Vision*, 8(2):123–152, Aug. 1992.

[100] P. McLauchlan, I. Reid, and D. Murray. Recursive affine structure and motion from image sequences. In J.-O. Eklundh, editor, *Proc. 3rd European Conference on Computer Vision*, volume I, pages 217–224, Stockholm, Sweden, May 1994. Springer-Verlag.

[101] G. Medioni and Y. Yasumuto. Corner detection and curve representation using cubic b-spline. In *Proc. Int'l Conf. Robotics Automation*, pages 764–769, 1986.

[102] P. Meer, D. Mintz, A. Rosenfeld, and D. Kim. Robust regression methods for computer vision: A review. *Int'l J. Comput. Vision*, 6(1):59–70, 1991.

[103] R. Mohr, B. Boufama, and P. Brand. Accurate projective reconstruction. In J. Mundy and A. Zisserman, editors, *Applications of Invariance in Computer Vision*, volume 825 of *Lecture Notes in Computer Science*, pages 257–276, Berlin, 1993. Springer-Verlag.

[104] R. Mohr, F. Veillon, and L. Quan. Relative 3d reconstruction using multiple uncalibrated images. In *Proc. International Conference on Computer Vision and Pattern Recognition*, pages 543–548. 1993.

[105] J. L. Mundy and A. Zisserman, editors. *Geometric invariance in computer vision.* MIT Press, 1992.

[106] D. W. Murray and B. F. Buxton. Scene segmentation from visual motion using global optimization. *IEEE Trans. on Pattern Analysis and Machine Intelligence*, 9(2):220–228, Mar. 1987.

[107] J. Nelder and R. Mead. A simplex method for function minimization. *Computer Journal*, (7):308–313, 1965.

[108] R. Nelson and J. Aloimonos. Using flow field divergence for obstacle avoidance in visual navigation. In *Proc. Second Int'l Conf. Comput. Vision*, pages 548–559, Tampa, FL, Dec. 1988. IEEE.

[109] E. Nishimura, G. Xu, and S. Tsuji. Motion segmentation and correspondence using epipolar constraint. In *Proc. 1st Asian Conf. Computer Vision*, pages 199–204, Osaka, Japan, 1993.

[110] J. Noble. Finding corners. *Image and Vision Computing*, 6:121–128, May 1988.

[111] J.-M. Odobez and P. Bouthemy. Robust multiresolution estimation of parametric motion models applied to complex scenes. Publication interne 788, IRISA-INRIA Rennes, France, Jan. 1994.

[112] S. Olsen. Epipolar line estimation. In *Proc. Second European Conf. Comput. Vision*, pages 307–311, Santa Margherita Ligure, Italy, May 1992.

[113] K. Pahlavan, T. Uhlin, and J.-O. Ekhlund. Dynamic fixation. In *Proc. 4th International Conference on Computer Vision*, pages 412–419, Berlin, Germany, May 1993.

[114] R. P. Paul, editor. *Robot Manipulators: Mathematics, Programming, and Control.* The MIT Press, 1981.

[115] T. Poggio and S. Edelman. A network that learns to recognize 3d objects. *Nature*, 343:263–266, 1990.

[116] T. Poggio and C. Koch. Ill-posed problems: From computational theory to analog networks. *Proc Royal Soc. London*, B, 1985.

[117] T. Poggio, V. Torre, and C. Koch. Computational vision and regularization theory. *Nature*, 317:214–319, 1985.

[118] S. Pollard, J. Mayhew, and J. Frisby. PMF : a stereo correspondence algorithm using a disparity gradient constraint. *Perception*, 14:449–470, 1985.

[119] W. H. Press, B. P. Flannery, S. A. Teukolsky, and W. T. Vetterling. *Numerical Recipes in C.* Cambridge University Press, 1988.

[120] L. Quan. Affine stereo calibration for relative affine shape reconstruction. In *Proc. fourth British Machine Vision Conference*, pages 659–668, Surrey, England, 1993.

[121] L. Quan and R. Mohr. Towards structure from motion for linear features through reference points. In *IEEE Workshop on Visual Motion*, New Jersey, 1991.

[122] B. Radig. Image sequence analysis using relational structures. *Pattern Recog.*, 17(1):161–167, 1984.

[123] I. Reid and D. Murray. Active tracking of foveated feature clusters using affine structure. *Int'l J. Comput. Vision*, 1994.

[124] W. J. Rey. *Introduction to Robust and Quasi-Robust Statistical Methods.* Springer, Berlin, Heidelberg, 1983.

[125] L. Robert, M. Buffa, and M. Hebert. Weakly-calibrated stereo perception for rover navigation. In *Proc. 5th International Conference on Computer Vision*, pages 46–51, Boston, MA, June 1995.

[126] L. Robert and O. Faugeras. Relative 3d positioning and 3d convex hull computation from a weakly calibrated stereo pair. In *Proc. 4th International Conference on Computer Vision*, pages 540–544, Berlin, Germany, May 1993. also INRIA Technical Report 2349.

[127] L. Robert, C. Zeller, O. Faugeras, and M. Hébert. Applications of non-metric vision to some visually-guided robotics tasks. In Y. Aloimonos, editor, *Visual Navigation: From Biological Systems to Unmanned Ground Vehicles*, chapter ? Lawrence Erlbaum Associates, 1996. to appear, also INRIA Technical Report 2584.

[128] A. Rosenfeld, R. Hummel, and S. Zucker. Scene labeling by relaxation operations. *IEEE Trans. on Systems, Man, and Cybernetics*, 6(4):420–433, 1976.

[129] C. Rothwell, G. Csurka, and O. Faugeras. A comparison of projective reconstruction methods for pairs of views. In *Proc. 5th International Conference on Computer Vision*, pages 932–937, Boston, MA, June 1995.

[130] C. Rothwell, D. Forsyth, A. Zisserman, and J. Mundy. Extracting projective structure from single perspective views of 3d point sets. In *4th International Conference on Computer Vision*, pages 573–582, Berlin, 1993.

[131] P. Rousseeuw and A. Leroy. *Robust Regression and Outlier Detection.* John Wiley & Sons, New York, 1987.

[132] R. Schalkoff. *Pattern Recognition: Statistical, Structural and New Approaches.* Wiley, 1992.

[133] S. Seitz and C. Dyer. Physically-valid view synthesis by image interpolation. In *Proc. IEEE Workshop on Representation of Visual Scenes*, pages 18–25, Cambridge, Massachusetts, USA, June 1995.

[134] L. Shapiro. *Affine Analysis of Image Sequences.* PhD thesis, University of Oxford, Department of Engineering Science, Oxford, UK, Nov. 1993.

[135] L. Shapiro and M. Brady. Rejecting outliers and estimating errors in an orthogonal-regression framework. *Phil. Trans. on Royal Soc. of Lon. A*, 350:407–439, 1995.

[136] L. Shapiro and R. Haralick. Structural description and inexact matching. *IEEE Trans. on Pattern Analysis and Machine Intelligence*, 3:504–519, Sept. 1981.

[137] L. Shapiro, A. Zisserman, and M. Brady. Motion from point matches using affine epipolar geometry. In *Proc. Third European Conf. Comput. Vision*, 1994.

[138] L. Shapiro, A. Zisserman, and M. Brady. 3d motion recovery via affine epipolar geometry. *Int'l J. Comput. Vision*, 16:147–182, 1995.

[139] A. Shashua. Projective structure from uncalibrated images: structure from motion and recognition. *IEEE Trans. on Pattern Analysis and Machine Intelligence*, 16(8):778–790, 1994.

[140] A. Shashua. Algebraic functions for recognition. *IEEE Trans. on Pattern Analysis and Machine Intelligence*, 17(8):779–789, Aug. 1995.

[141] M. Shizawa. Transparent 3d motions and structures from point correspondences in two frames: A quasi-optimal, closed-form, linear algorithm and degeneracy analysis. In *1st Asian Conference on Computer Vision*, pages 329–334, Osaka, 1993.

[142] D. Sinclair. Motion segmentation and local structure. In *Proc. Fourth Int'l Conf. Comput. Vision*, pages 366–373, 1993.

[143] C. C. Slama, editor. *Manual of Photogrammetry.* American Society of Photogrammetry, fourth edition, 1980.

[144] S. Soato and P. Perona. Three dimensional transparent strucuture segmentation and mutiple 3d motion estimation from monocular perspective image sequences. Tech Rep CIT-CNS 33/93, CalTech, 1993.

[145] M. Spetsakis. A linear algorithm for point and line-based structure from motion. *CVGIP: Image Understanding*, 56(2):230–241, Sept. 1992.

[146] M. Spetsakis and J. Aloimonos. A unified theory of structure from motion. Technical Report CAR-TR-482, Computer Vision Laboratory, University of Maryland, Dec. 1989.

[147] M. E. Spetsakis and J. Aloimonos. Structure from Motion Using Line Correspondences. *The International Journal of Computer Vision*, 4:171–183, 1990.

[148] R. Sturm. Das problem der projektivität und seine anwendung auf die flächen zweiten grades. *Math. Ann.*, 1:533–574, 1869.

[149] R. Szeliski. Image mosaicing for tele-reality applications. Technical Report CRL94/2, DEC-CRL, May 1994.

[150] R. Szeliski and S. Kang. Direct methods for visual scene reconstruction. In *IEEE Workshop on Representation of Visual Scenes*. IEEE, June 1995.

[151] D. Terzopoulos. The computation of visible-surface representation. *IEEE Trans. on Pattern Analysis and Machine Intelligence*, 10(4):417–438, 1988.

[152] D. Thompson and J. Mundy. Three-dimensional model matching from an unconstrained viewpoint. In *Proc. International Conference on Robotics and Automation*, Raleigh, NC, pages 208–220, 1987.

[153] W. Thompson, P. Lechleider, and E. Stuck. Detecting moving objects using the rigidity constraint. *IEEE Trans. on Pattern Analysis and Machine Intelligence*, 15(2):162–166, Feb. 1993.

[154] C. Tomasi and T. Kanade. Factoring image sequences into shape and motion. In *Proc. IEEE Workshop on Visual Motion*, pages 21–28, Princeton, NJ, Oct. 1991.

[155] C. Tomasi and T. Kanade. Shape and motion from image streams under orthography: a factorization method. *Int'l J. Comput. Vision*, 9(2):137–154, 1992.

[156] P. Torr. *Motion Segmentation and Outlier Detection*. PhD thesis, Department of Engineering Science, University of Oxford, 1995.

[157] P. Torr and D. Murray. Outlier detection and motion segmentation. In S, editor, *Sensor Fusion VI, SPIE Vol.2059*, pages 432–443, Boston, 1993.

[158] G. Toscani. *Système de Calibration optique et perception du mouvement en vision artificielle*. PhD thesis, Paris-Orsay, 1987.

[159] R. Tsai. Multiframe image point matching and 3d surface reconstruction. *IEEE Trans. on Pattern Analysis and Machine Intelligence*, 5:159–174, 1986.

[160] R. Tsai. Synopsis of recent progress on camera calibration for 3D machine vision. In O. Khatib, J. J. Craig, and T. Lozano-Pérez, editors, *The Robotics Review*, pages 147–159. MIT Press, 1989.

[161] R. Tsai and T. Huang. *Three-dimensional motion and structure from image sequences*. Springer-Verlag, 1983.

[162] R. Tsai and T. Huang. Uniqueness and estimation of three-dimensional motion parameters of rigid objects with curved surface. *IEEE Trans. on Pattern Analysis and Machine Intelligence*, 6(1):13–26, Jan. 1984.

[163] R. Y. Tsai. A versatile camera calibration technique for high-accuracy 3D machine vision metrology using off-the-shelf tv cameras and lenses. *IEEE Journal of Robotics and Automation*, 3(4):323–344, Aug. 1987.

[164] S. Ullman. *The Interpretation of Visual Motion.* MIT Press, Cambridge, MA, 1979.

[165] S. Ullman and R. Basri. Recognition by linear combinations of models. *IEEE Trans. on Pattern Analysis and Machine Intelligence*, 13(10):992–1106, 1991.

[166] T. Viéville and O. Faugeras. Motion analysis with a camera with unknown, and possibly varying intrinsic parameters. In *Proc. Int'l Conf. Computer Vision*, pages 750–756, Cambridge, Massachusetts, USA., June 1995.

[167] T. Viéville, Q. Luong, and O. Faugeras. Motion of points and lines in the uncalibrated case. *Int'l J. Comput. Vision*, 17(1):7–41, 1996.

[168] J. Weber and J. Malik. Rigid body segmentation and shape description from dense optical flow under weak perspective. In *Proc. Fifth Int'l Conf. Comput. Vision*, pages 251–256, 1995.

[169] G. Wei and S. Ma. Two plane camera calibration: A unified model. In *Proc. International Conference on Computer Vision and Pattern Recognition*, pages 133–138, Hawaii, June 1991.

[170] G. Wei and S. Ma. Implicit and explicit camera calibration: Theory and experiments. *IEEE Trans. on Pattern Analysis and Machine Intelligence*, 16(5):469–480, 1994.

[171] D. Weinshall and C. Tomasi. Linear and incremental acquisition of invariant shape models from image sequences. In *Proc. 4th International Conference on Computer Vision*, pages 675–682, Berlin, Germany, May 1993.

[172] J. Weng, N. Ahuja, and T. Huang. Optimal motion and structure estimation. *IEEE Trans. on Pattern Analysis and Machine Intelligence*, 15(9):864–884, Sept. 1993.

[173] J. Weng, P. Cohen, and N. Rebibo. Motion and structure estimation from stereo image sequences. *IEEE Trans. on Robotics and Automation*, 8(3):362–382, June 1992.

[174] J. Weng, T. Huang, and N. Ahuja. Motion and structure from two perspective views: Algorithms, error analysis and error estimation. *IEEE Trans. on Pattern Analysis and Machine Intelligence*, 11(5):451–476, May 1989.

[175] G. Xu. Unifying stereo, motion and object recognition via epipolar geometry. In *Proc. Second Asian Conf. Comput. Vision*, 1995. Invited paper.

[176] G. Xu, E. Segawa, and S. Tsuji. Robust active contours with insensitive parameters. *Pattern Recog.*, 27(7):879–884, 1994.

[177] G. Xu and S. Tsuji. Is a single view sufficient for 3d object recognition? Tech rep, Dept. Control Eng, Osaka Univ, 1992.

[178] Y. Yang, A. Yuille, and J. Liu. Local, global, multilevel stereo matching. In *Proc. IEEE Conf. Comput. Vision Pattern Recog.* IEEE, 1993.

[179] A. Yuille and N. Grzywacz. A methematical analysis of the motion coherence theory. *Int'l J. Comput. Vision*, 3:155–175, 1989.

[180] C. Zeller. *Calibration Projective Affine et Euclidienne en Vision par Ordinateur*. PhD thesis, École Polytechnique, Feb. 1996.

[181] C. Zeller and O. Faugeras. Applications of non-metric vision to some visual guided tasks. In *Proc. International Conference on Pattern Recognition*, pages 132–136, Jerusalem, Israel, Oct. 1994. A longer version in INRIA Tech Report RR2308.

[182] Z. Zhang. Estimating motion and structure from correspondences of line segments between two perspective images. Research Report 2340, INRIA Sophia, 1994.

[183] Z. Zhang. Token tracking in a cluttered scene. *Int'l J. of Image and Vision Computing*, 12(2):110–120, Mar. 1994. Also Research Report No.2072, INRIA Sophia-Antipolis, 1993.

[184] Z. Zhang. Estimating motion and structure from correspondences of line segments between two perspective images. *IEEE Trans. on Pattern Analysis and Machine Intelligence*, 17(12):1129–1139, Dec. 1995.

[185] Z. Zhang. Motion and structure of four points from one motion of a stereo rig with unknown extrinsic parameters. *IEEE Trans. on Pattern Analysis and Machine Intelligence*, 17(12):1222–1227, Dec. 1995.

[186] Z. Zhang. Motion of a stereo rig: Strong, weak and self calibration. In *Proc. Second Asian Conf. Comput. Vision*, pages I274–281, Dec. 1995.

[187] Z. Zhang. On the epipolar geometry between two images with lens distortion. In *International Conference on Pattern Recognition*, volume I, pages 407–411, Vienna, Austria, Aug. 1996.

[188] Z. Zhang, R. Deriche, O. Faugeras, and Q.-T. Luong. A robust technique for matching two uncalibrated images through the recovery of the unknown epipolar geometry. *Artificial Intelligence Journal*, 78:87–119, Oct. 1995.

[189] Z. Zhang, O. Faugeras, and R. Deriche. Calibrating a binocular stereo through projective reconstruction using both a calibration object and the environment. In R. Mohr and C. Wu, editors, *Proc. Europe-China Workshop on Geometrical modelling and Invariants for Computer Vision*, pages 253–260, Xi'an, China, Apr. 1995.

[190] Z. Zhang and O. D. Faugeras. *3D Dynamic Scene Analysis: A Stereo Based Approach*. Springer, Berlin, Heidelberg, 1992.

[191] Z. Zhang, Q.-T. Luong, and O. Faugeras. Motion of an uncalibrated stereo rig: self-calibration and metric reconstruction. *IEEE Trans. on Robotics and Automation*, 12(1):103–113, Feb. 1996.

[192] X. Zhuang, T. Wang, and P. Zhang. A highly robust estimator through partially likelihood function modeling and its application in computer vision. *IEEE Trans. on Pattern Analysis and Machine Intelligence*, 14(1):19–34, Jan. 1992.

[193] A. Zisserman. Notes on geometric invariants in vision. BMVC92 Tutorial, 1992.

[194] S. Zucker, Y. Leclerc, and J. Mohammed. Continuous relaxation and local maxima selection: Conditions for equivalence. *IEEE Trans. on Pattern Analysis and Machine Intelligence*, 3:117–127, Mar. 1981.

INDEX

Computational Imaging and Vision

1. B.M. ter Haar Romeny (ed.): *Geometry-Driven Diffusion in Computer Vision.* 1994 ISBN 0-7923-3087-0
2. J. Serra and P. Soille (eds.): *Mathematical Morphology and Its Applications to Image Processing.* 1994 ISBN 0-7923-3093-5
3. Y. Bizais, C. Barillot, and R. Di Paola (eds.): *Information Processing in Medical Imaging.* 1995 ISBN 0-7923-3593-7
4. P. Grangeat and J.-L. Amans (eds.): *Three-Dimensional Image Reconstruction in Radiology and Nuclear Medicine.* 1996 ISBN 0-7923-4129-5
5. P. Maragos, R.W. Schafer and M.A. Butt (eds.): *Mathematical Morphology and Its Applications to Image and Signal Processing.* 1996 ISBN 0-7923-9733-9
6. G. Xu and Z. Zhang: *Epipolar Geometry in Stereo, Motion and Object Recognition.* A Unified Approach. 1996 ISBN 0-7923-4199-6

'luwer Academic Publishers – Dordrecht / Boston / London